Computing with Multi-Value Logic in Quantum Dot Cellular Automata

IEEE Press
445 Hoes Lane
Piscataway, NJ 08854

Computing with Multi-Value Logic in Quantum Dot Cellular Automata

Reza Sabbaghi-Nadooshan
Islamic Azad University
Tehran, Iran

Reza Akbari-Hasanjani
Islamic Azad University
Tehran, Iran

Leila Dehbozorgi
Islamic Azad University
Tehran, Iran

Majid Haghparast
University of Jyväskylä
Jyväskylä, Finland

Hamid Reza Akbari-Hasanjani
Damghan University
Damghan, Iran

IEEE*Press*
WILEY

Published by John Wiley & Sons, Inc., Hoboken, New Jersey.
Published simultaneously in Canada.

Library of Congress Cataloging-in-Publication Data:

Names: Sabbaghi-Nadooshan, Reza, author.
Title: Computing with multi-value logic in quantum dot cellular automata / Reza Sabbaghi-Nadooshan, Reza Akbari-Hasanjani, Leila Dehbozorgi, Majid Haghparast, and Hamid Reza Akbari-Hasanjani.
Description: Hoboken, New Jersey : Wiley, [2024] | Includes index.
Identifiers: LCCN 2024023397 (print) | LCCN 2024023398 (ebook) | ISBN 9781394253944 (hardback) | ISBN 9781394253968 (adobe pdf) | ISBN 9781394253951 (epub)
Subjects: LCSH: Quantum dots. | Cellular automata. | Many-valued logic.
Classification: LCC TK7874.88 .S23 2017 (print) | LCC TK7874.88 (ebook) | DDC 621.39/5–dc23/eng/20240629
LC record available at https://lccn.loc.gov/2024023397
LC ebook record available at https://lccn.loc.gov/2024023398

Cover Design: Wiley
Cover Image: © Eugene Mymrin/Getty Images

Set in 9.5/12.5pt STIXTwoText by Straive, Chennai, India

Contents

List of Figures *xi*
List of Tables *xvii*
About the Authors *xxi*
Preface *xxiii*
Introduction *xxv*

1 Quantum Dots *1*
1.1 Introduction to Quantum Dots *1*
1.2 Physical Characteristics of Semiconductor Quantum Dots *2*
1.3 Quantum Dots Structure *3*
1.3.1 Core-Structure Quantum Dots *4*
1.3.2 Core–Shell Quantum Dots: Enhanced Properties and Diverse Applications *5*
1.3.3 Alloyed Quantum Dots *6*
1.4 Surface Structure of Quantum Dots *7*
1.4.1 Surface Passivation *7*
1.4.1.1 Organically Capped Quantum Dots *7*
1.4.1.2 Inorganically Passivated Quantum Dots *8*
1.4.1.3 Multishell Structure *8*
1.4.2 Characterization of Shell Structures *9*
1.5 Properties of Quantum Dots *10*
1.5.1 Optical and Electronic Properties *10*
1.5.2 Unique Properties and Applications of Quantum Dots *12*
1.5.2.1 Confinement Effect *12*
1.5.2.2 Excitation and Emission Spectra *12*
1.5.2.3 Stability Against Light Radiation *13*
1.5.2.4 Luminescence Effect *13*

2 Synthesis Methods of Quantum Dots and Applications *15*
2.1 Synthesis Processes *15*
2.2 Top-Down Synthesis Processes *15*
2.2.1 Etching Processes *15*
2.2.2 Focused Ion Beam Techniques *15*
2.2.3 Electron Beam Lithography *16*
2.3 Bottom-Up Synthesis Processes *17*
2.3.1 Wet Chemical Methods *17*

2.3.1.1 Sol–Gel 17
2.3.1.2 Microemulsion 17
2.3.1.3 Hot-Solution Decomposition 18
2.3.2 Vapor-Phase Method 18
2.3.2.1 Heteroepitaxial Growth 18
2.3.2.2 Molecular Beam Epitaxy 18
2.3.2.3 Physical Vapor Deposition 19
2.3.2.4 Chemical Vapor Deposition 19
2.4 Applications of Quantum Dots 19
2.4.1 White Light-Emitting Diode 19
2.4.2 Quantum Dot in Solar Cell 20
2.4.3 Quantum Computers 21
2.4.4 Quantum Dot Cellular Automata 21

3 Overview of QCA 23
3.1 Introduction to QCA Technology 23
3.2 Binary QCA Concept 24
3.3 Ternary QCA Cell Structure 24
3.4 Clock in QCA Technology 25
3.5 Manufacturing and Implementation of QCA 26
3.6 Overview of Quantum Gates 26
3.6.1 Quantum Wire 26
3.6.2 Binary NOT Gate 27
3.6.3 Binary Majority Gate 29

4 Polarization and Polarization Calculations in QCA (Quantum Calculations in QCA) 33
4.1 Introduction 33
4.2 Polarization and Calculation of Hamiltonian Matrix in QCA 33
4.2.1 Polarization and Hamiltonian Matrix in bQCA 33
4.2.1.1 Polarization in bQCA 33
4.2.1.2 Hamiltonian Matrix in Binary QCA 35
4.2.2 Polarization and Hamiltonian Matrix in Ternary QCA 35
4.2.2.1 Polarization in Ternary QCA 35
4.2.2.2 Hamiltonian Matrix in Ternary QCA 39
4.2.3 Polarization and Hamiltonian Matrix in Quaternary QCA 39
4.2.3.1 Polarization in Quaternary QCA 39
4.2.3.2 Hamiltonian Matrix in Quaternary QCA 44
4.2.4 Polarization and Hamiltonian Matrix in Quinary QCA 46
4.2.4.1 Polarization in Quinary QCA 46
4.2.4.2 Hamiltonian Matrix in Quinary QCA 56

5 Multi-Valued Cells Based on Polarization Calculations 59
5.1 Ternary QCA 59
5.1.1 Proposed Ternary QCA 59
5.1.2 Energy Calculations in Proposed TQCA 59
5.1.2.1 Internal Electrostatic Energy 59
5.1.2.2 Ground State Energy 60

5.1.2.3 Kinetic Energy *61*
5.1.2.4 External Electrostatic Energy *61*
5.2 Quaternary QCA *62*
5.2.1 Proposed QQCA *62*
5.2.2 Integration in Input Applied to Proposed QQCA Cell *64*
5.2.2.1 First Structure of Input and Output Drives in Proposed Cell *64*
5.2.2.2 Second Structure of Input and Output Drives *67*
5.2.3 Energy in Proposed QQCA Cell *69*
5.2.3.1 Internal Electrostatic Energy *69*
5.2.3.2 Ground State Energy *69*
5.2.3.3 Kinetic Energy *70*
5.2.3.4 External Electrostatic Energy *70*
5.3 Quinary QCA *72*
5.3.1 Proposed QuQCA Cells *72*
5.3.2 Integration in Input to QuQCA Cells *73*
5.3.2.1 First Structure of Input and Output Drive *74*
5.3.2.2 Second Structure of Input and Output Drive *78*
5.3.3 Energy in QuQCA Cell *79*
5.3.3.1 Internal Electrostatic Energy *79*
5.3.3.2 Ground State Energy *80*
5.3.3.3 Kinetic Energy *80*
5.3.3.4 External Electrostatic Energy *80*
5.4 Hypothesis Based on *n*-Value QCA Cell *82*
5.5 Fuzzy Logic Design Using MIN and MAX Functions *84*
5.6 Quantum Information *87*
5.6.1 Quantum Computation *87*
5.6.2 Quantum Communication *88*
5.6.3 Quantum Cryptography *88*

6 Effect of Polarization for Two Adjacent Cells and Power Consumption in Multi-Valued Cells *91*
6.1 Effect of Two Adjacent Cells in Terms of Polarization *91*
6.1.1 Effect of Two Adjacent Cells in TQCA *91*
6.1.2 Effect of Two Quaternary Cells *92*
6.1.3 Effect of Two QuQCA Cells *94*
6.2 Power Consumption in QCA Technology *95*
6.2.1 Power Consumption in a TQCA Cell with Changes in Time and Polarization *95*
6.2.2 Power Consumption in QQCA Cell with Changes in Time and Polarization *96*
6.2.2.1 Power Consumption in QQCA Using Quantum Computation *96*
6.2.2.2 Power Consumption of QQCA Cell *96*
6.2.3 Power Consumption in QuQCA Cell with Changes in Time and Polarization *97*
6.2.3.1 Power Consumption in QuQCA Using Quantum Computation *97*
6.2.3.2 Power Consumption in QuQCA Cell *99*

7 Structure of Basic Gates Using the Proposed Cells in Multi-Value QCA *101*
7.1 Structure of Basic Gates Using TQCA *101*
7.1.1 NOT Gate *101*

7.1.2 Simulation of Two Adjacent Cells *101*
7.1.3 Implementation of AND and OR Functions *103*
7.2 Structure of Basic QQCA Gates *105*
7.2.1 QQCA-Based NOT Gate *106*
7.2.2 Simulation of Two Adjacent Cells *106*
7.2.3 Implementation of AND and OR Functions *108*
7.3 Structure of Basic Gates Using QuQCA *111*
7.3.1 QuQCA-Based NOT Gate *111*
7.3.2 Simulation of Two Adjacent Cells in QuQCA *112*
7.3.3 Implementation of AND and OR Functions *112*

8 Implementation of Ternary and Quaternary Basic Gates Using QQCASim and TQCASim *119*
8.1 TQCA Simulator *120*
8.1.1 Majority Gate *120*
8.1.2 AND Gate *122*
8.1.3 OR Gate *122*
8.1.4 Ternary Wire *122*
8.2 QQCA Simulator for QQCA Circuit Simulation *125*
8.2.1 Majority Gate *127*
8.2.2 AND Gate *129*
8.2.3 OR Gate *130*
8.2.4 NOT Gate *130*
8.2.5 Quaternary Quantum Wire *131*

9 A Review of Literature on Memory Structures *135*
9.1 Introduction *135*
9.2 A Review of Basic Memory Structures *136*
9.2.1 D Flip-Flop in Binary QCA *136*
9.2.2 RAM Cell in Binary QCA *136*
9.2.3 PIM and Akers Logic Array *136*
9.3 Review of Literature on Memory Design *138*

10 Proposed PIM Cells and Their Fault Analysis in Binary QCA *141*
10.1 Introduction *141*
10.2 Binary QCA-Based Structures *143*
10.2.1 First Model of Binary PIM *143*
10.2.2 Second Model of Binary PIM *143*
10.2.3 Third Structure of PIM *143*
10.2.4 Fourth Model of PIM *147*
10.3 Fault Analysis *147*
10.3.1 Fault Tolerance Analysis Against Cell Omission and Extra-Cell Deposition Defects in Proposed Models *148*
10.3.1.1 Fault Evaluation of BPIM1 *149*
10.3.1.2 Fault Evaluation of BPIM2 *149*
10.3.1.3 Fault Evaluation of BPIM3 *153*
10.3.1.4 Fault Evaluation of BPIM4 *154*

11 Design Details of Binary Boolean Operators Using Basic PIM Cells: Proposing and Analyzing Basic-Extended Hypothesis *157*
11.1 Introduction *157*
11.2 Basic-Extended Hypothesis *157*
11.3 Designing Logic Gates with Binary QCA-Based PIM Capability *157*
11.3.1 AND Gate Design Using PIM Basic Cells in Binary QCA *157*
11.3.2 OR Gate Design Using Binary QCA-Based Basic PIM Cells *158*
11.3.3 NAND Gate Design Using Binary QCA-Based Basic PIM Cells *164*
11.3.4 NOR Gate Using Basic PIM Cells in Binary QCA *166*
11.3.5 XOR Gate Design Using Basic PIM Cells in Binary QCA *169*

12 Proposed PIM Cells in Ternary QCA and Their Fault Analysis *173*
12.1 Introduction *173*
12.2 Simulation of a Flip-Flop in Ternary QCA Using MATLAB *173*
12.3 Structures in Ternary QCA *174*
12.3.1 First Model of PIM Cell (TPIM1) *174*
12.3.2 Second Model of PIM Cell (TPIM2) *176*
12.3.3 Third Model of Ternary QCA-Based PIM Cell (TPIM3) *178*
12.4 Fault Analysis *181*
12.4.1 Fault Tolerance for Cell Omission or Extra-Cell Deposition Defects in Proposed Models *181*
12.4.1.1 Fault Analysis for First Model of PIM Cell (TPIM1) *182*
12.4.1.2 Fault Analysis for Second Model of PIM Cell (TPIM2) *184*
12.4.1.3 Fault Analysis for Third Model of PIM Cell (TPIM3) *185*

13 Design Details of Ternary Boolean Operators Using Basic PIM Cells and Analysis of Basic-Extended Hypothesis *189*
13.1 Introduction *189*
13.2 Designing Logic Gates With Ternary QCA-Based PIM Capability *189*
13.2.1 AND Gate Design Using PIM Basic Cells in Ternary QCA *189*
13.2.2 OR Gate Design Using Ternary QCA-Based Basic PIM Cells *192*
13.2.3 XOR Gate Design Using Ternary QCA-Based PIM Model *192*
13.3 Evaluation, Analysis, and Comparison of Results *196*

14 Conclusions and Suggestions *201*
14.1 Introduction *201*
14.2 Summary and Conclusion *201*
14.3 Suggestions *203*

References *205*

A Matrices *217*
A.1 Particle State Matrix in QQCA *217*
A.2 Particle State Matrix in QQCA *218*

B Electrostatic Energy Calculation *223*

Index *237*

List of Figures

Figure 1.1 The role of energy levels in the reduction of semiconductor gaps with the increase in nanocrystal size. *3*

Figure 1.2 Basic structure of an uncoated QD. *4*

Figure 1.3 The bandgap alignments. *6*

Figure 1.4 Schematic representation of the emission mechanism in QDs. *12*

Figure 1.5 Photoelectrochemical property of quantum dots under light irradiation: (a) generation of anodic current in the presence of electron donor compound and (b) generation of cathodic current in the presence of electron acceptor compound. *13*

Figure 2.1 Comparison of top-down and bottom-up methods. *16*

Figure 3.1 Polarizations of bQCA cells. *24*

Figure 3.2 TQCA cell. *24*

Figure 3.3 States of electron placement in a stable TQCA cell. *25*

Figure 3.4 Clock in QCA in four phases. *25*

Figure 3.5 90° wire. *26*

Figure 3.6 45° wire. *27*

Figure 3.7 Overlapping or plate wiring structure. *27*

Figure 3.8 Multilayer wiring structure. *28*

Figure 3.9 Transmission wire with two-phase clock difference. *28*

Figure 3.10 NOT gate using two bQCA cells diagonally. *28*

Figure 3.11 NOT gate using two diagonal NOT gates. *29*

Figure 3.12 Inverter gate with reduced area. *29*

Figure 3.13 Majority gate with three inputs. *29*

Figure 3.14 The majority gate with five inputs. *30*

Figure 3.15 AND gate structure. *30*

Figure 3.16 OR gate structure. *31*

Figure 4.1 TQCA polarization simulation diagram. *39*

Figure 4.2 Polarization diagram in QQCA based on calculations. *44*

Figure 4.3 Polarization diagram in QuQCA. *56*

Figure 5.1 Polarizations of the ternary model. *60*
Figure 5.2 TQCA cell proposed in Ref. [129]. *60*
Figure 5.3 Distance between items of two proposed TQCA cells. *61*
Figure 5.4 Distances between two ternary cells to calculate the external electrostatic energy. *62*
Figure 5.5 Shows the proposed QQCA model. *63*
Figure 5.6 Specifications of the used layers. *63*
Figure 5.7 QQCA cell symbol. *64*
Figure 5.8 Input drive block in QQCA. *66*
Figure 5.9 The first proposed output drive. *67*
Figure 5.10 Second structure of the input drive. *68*
Figure 5.11 Distances between two ternary cells in the first layer. *71*
Figure 5.12 Distances between two QQCA cells in the second layer. *71*
Figure 5.13 Symbolic representation of two QCA cells. *72*
Figure 5.14 Proposed QuQCA model. *72*
Figure 5.15 Specifications of the ternary layer. *73*
Figure 5.16 QuQCA cell symbol. *73*
Figure 5.17 Schematic of the input drive using decoder. *75*
Figure 5.18 Schematic of the first output drive. *77*
Figure 5.19 Simplified output drive model [131]. *77*
Figure 5.20 Second input drive in QuQCA. *78*
Figure 5.21 TQCA-based half-adder circuit. *79*
Figure 5.22 Distances between electrons in the first layer. *81*
Figure 5.23 Distances between electrons in the second layer. *82*
Figure 5.24 Symbolic representation of two adjacent cells in QuQCA. *82*
Figure 5.25 Block diagram of the studied fuzzy system. *85*
Figure 5.26 Fuzzy system rule base simulation: (a) the circuit representation and (b) the simulation results of the rule base. *86*
Figure 5.27 Quantum gates. *88*
Figure 6.1 Effect of two neighboring TQCA cells at different kinetic energies. *92*
Figure 6.2 Interactions of two cells in the ternary layer. *93*
Figure 6.3 Interactions of two cells in the binary layer. *93*
Figure 6.4 Interactions of two neighboring cells in the ternary layers of QuQCA cells. *94*
Figure 6.5 Power consumption in TQCA as per polarization set $\{-2, 0, 2\}$. *96*
Figure 6.6 Energy consumption in QQCA using quantum computation. *97*
Figure 6.7 Power consumption in the proposed QQCA cell. *98*
Figure 6.8 Power consumption using QuQCA quantum calculations. *98*
Figure 6.9 Power consumption in the proposed QuQCA cell. *99*

Figure 7.1 Ternary NOT gate. *102*
Figure 7.2 Two adjacent TQCA cells (wire). *102*
Figure 7.3 Majority ternary gate. *104*
Figure 7.4 (a) AND gate structure with a fixed input in state A and (b) OR gate structure with a fixed input in state B. *105*
Figure 7.5 QQCA-based NOT gate. *106*
Figure 7.6 Two QQCA cells (QQCA-based wire). *107*
Figure 7.7 QQCA majority gates: (a) standard majority gate and (b) diagonal majority gate. *109*
Figure 7.8 AND and OR gates with two inputs and one fixed input (a) AND gate structure with one fixed input in state A and (b) OR gate structure with one fixed input in state D. *110*
Figure 7.9 Representation of a NOT gate using the QuQCA-based symbolic model. *112*
Figure 7.10 Two neighboring cells using the QuQCA symbolic model. *113*
Figure 7.11 QuQCA majority gate. *115*
Figure 7.12 QQCA-based AND and OR gate structures: (a) AND gate structure with a fixed input in state A and (b) OR gate structure with a fixed input in state E. *117*
Figure 8.1 Environment of TQCASim software version 1.0.11.1. *120*
Figure 8.2 TQCA cell options. *120*
Figure 8.3 Simulation of a three-value majority gate: (a) majority gate designed with TQCASim, (b) majority gate waveform, and (c) the truth table obtained from the software in TQCA. *121*
Figure 8.4 Simulation of ternary AND gates; (a) AND gates designed with TQCASim software, (b) AND gate waveform obtained by TQCASim, and (c) AND gate truth table obtained by TQCASim. *123*
Figure 8.5 Simulation of the ternary OR gate: (a) OR gate designed by TQCASim, (b) OR gate waveform using TQCASim, and (c) OR gate truth table obtained from TQCASim. *124*
Figure 8.6 Simulation of two TQCA cells: (a) the two cells simulated in TQCASim, (b) two-cell interaction waveform using TQCASim, and (c) the truth table of two cells simulated in TQCASim. *125*
Figure 8.7 Simulation of a ternary wire with an odd number of cells: (a) odd number of cells in a wire simulated in TQCASim, (b) interaction waveform of the wire in TQCASim, and (c) truth table of the wire in TQCASim. *126*
Figure 8.8 Simulation of a ternary wire with an even number of cells: (a) ternary wire with an even number of TQCA cells, (b) interaction waveform of the wire in TQCASim, and (c) truth table of the wire simulated in TQCASim. *127*
Figure 8.9 Environment of QQCASim software version 1.0.0.1. *128*
Figure 8.10 Specifications of QQCA cells. *128*
Figure 8.11 Simulation of a quaternary majority gate: (a) majority gate designed with QQCASim, (b) the majority gate waveform, and (c) truth table obtained from the software. *129*

Figure 8.12 Simulation of the quaternary AND gate: (a) AND gate designed using QQCASim, (b) AND gate waveform in QQCASim, and (c) AND gate truth table obtained by QQCASim. *130*

Figure 8.13 Simulation of the quaternary OR gate: (a) OR gate designed in QQCASim, (b) OR gate waveform using QQCASim, and (c) OR gate truth table obtained using QQCASim. *131*

Figure 8.14 Quaternary NOT gate simulation: (a) NOT gate in QQCASim, (b) NOT gate output waveform, and (c) NOT gate truth table simulation using QQCASim. *132*

Figure 8.15 A quaternary quantum wire: (a) quantum wire, (b) quantum wire output waveform, and (c) quantum wire truth table, all obtained using QQCASim. *133*

Figure 9.1 (a) Comparison of von Neumann and PIM structures, (b) representation of Akers array and a logic cell with three inputs X, Y, and Z and two identical outputs, and (c) stripping process in Akers array to obtain output. *137*

Figure 9.2 Three different fault-tolerant three-input majority gates. *140*

Figure 10.1 BPIM1: (a) implementation in binary QCA and (b) simulation results. *144*

Figure 10.2 BPIM2: (a) implementation in binary QCA and (b) simulation results. *145*

Figure 10.3 BPIM3: (a) block diagram of the proposed model, (b) implementation in binary QCA, and (c) simulation results. *146*

Figure 10.4 BPIM4: (a) block diagram, (b) implementation in binary QCA, and (c) simulation results. *148*

Figure 10.5 Cell numbering in BPIM1. *150*

Figure 10.6 Cell numbering in BPIM2. *151*

Figure 10.7 Cell numbering in BPIM3. *153*

Figure 10.8 Cell numbering in BPIM4. *154*

Figure 11.1 Block diagram of two-input AND gate using Akers array. *158*

Figure 11.2 Two-input AND gate using PIM structure in binary QCA: (a) BAND1 and (b) BAND2. *159*

Figure 11.3 AND gate simulation results using binary QCA-based PIM structure for (a) BAND1 and (b) BAND2. *160*

Figure 11.4 Block diagram of two-input OR gate using Akers array. *161*

Figure 11.5 Two-input OR gate using binary QCA-based PIM structure: (a) BOR1 and (b) BOR2. *161*

Figure 11.6 OR gate simulation results using binary QCA-based PIM: (a) BOR1 and (b) BOR2. *163*

Figure 11.7 Block diagram of two-input NAND gate in Akers array. *164*

Figure 11.8 Two-input NAND gate based on PIM structure in binary QCA: (a) BNAND1 and (b) BNAND2. *165*

Figure 11.9 NAND simulation results based on PIM structure in binary QCA: (a) BNAND1 and (b) BNAND2. *166*

Figure 11.10 Block diagram of two-input NOR gate using Akers array. *167*

Figure 11.11 Two-input NOR gate using PIM structure in binary QCA: (a) BNOR1 and (b) BNOR2. *167*

Figure 11.12 NOR simulation results based on PIM structure in binary QCA: (a) BNOR1 and (b) BNOR2. *168*

Figure 11.13 (a) structure of basic cells and their layout in Akers array and (b) block diagram of two-input XOR gate in Akers array. *170*

Figure 11.14 Two-input BXOR gate using PIM in binary QCA. *171*

Figure 11.15 Simulation results of BXOR using PIM in binary QCA. *172*

Figure 12.1 DFF structure. *174*

Figure 12.2 Proposed DFF implemented and simulated using MATLAB. *174*

Figure 12.3 Output diagrams of the proposed ternary QCA-based DFF designed using MATLAB for (a) AND1, (b) NOT, (c) AND2, and (d) OR1 structures. *175*

Figure 12.4 First model of PIM cell, TPIM1: (a) implementation in ternary QCA and (b) simulation results. *176*

Figure 12.5 Second model of PIM cell, TPIM2: (a) implementation in ternary QCA and (b) simulation results. *177*

Figure 12.6 Third model of PIM cell (TPIM3): (a) implementation in ternary QCA and (b) simulation results. *179*

Figure 12.7 (a) Diagram and operating principle of switches in the QCA-based Akers basic cell, (b) operating principle of switches in the Akers cell, (c) operating principle of switches in the binary cell, and (d) operating principle of switches in the ternary cell. *180*

Figure 12.8 Numbering of cells in TPIM1. *182*

Figure 12.9 Numbering of cells in TPIM2. *184*

Figure 12.10 Numbering of cells in TPIM3. *186*

Figure 13.1 Two-input AND gate using PIM structure in ternary QCA: (a) TAND1 model and (b) TAND2 model. *190*

Figure 13.2 AND gate simulation results using ternary QCA-based PIM structure for (a) TAND1 and (b) TAND2. *191*

Figure 13.3 Two-input OR gate using PIM structure in ternary QCA: (a) TOR1 model and (b) TOR2 model. *193*

Figure 13.4 OR gate simulation results using ternary QCA-based PIM structure for (a) TOR1 and (b) TOR2. *194*

Figure 13.5 XOR gate using ternary QCA-based PIM: (a) TXOR model and (b) truth table and simulation results. *195*

Figure 13.6 Comparison of fault tolerance against extra-cell deposition and cell omission defects for PIM structures in binary and ternary QCA. *198*

List of Tables

Table 3.1 Binary NOT gate response for different polarizations. *29*
Table 3.2 Binary AND gate truth table in QCA technology. *30*
Table 3.3 Binary gate truth table in QCA technology. *31*
Table 4.1 Spin state of two electrons. *35*
Table 4.2 Spin states of a three-particle system (with electrons). *40*
Table 4.3 Vectors of quaternary input states. *43*
Table 4.4 Polarization of a three-particle system in QQCA. *44*
Table 4.5 Spin states of a four-electron system. *46*
Table 4.6 Input states of the quinary system. *54*
Table 4.7 Polarizations of the quinary system. *56*
Table 5.1 QQCA cell polarizations. *65*
Table 5.2 Truth table for the quaternary input drive. *66*
Table 5.3 Truth table of the first structure of the quaternary output drive. *67*
Table 5.4 The truth table of the second structure of the quaternary input drive. *68*
Table 5.5 The truth table for determination of quaternary output using bQCA adder. *69*
Table 5.6 Polarizations in QuQCA. *74*
Table 5.7 Truth table of five-value input drive. *76*
Table 5.8 Input states for the simplified output model in the quinary system. *78*
Table 5.9 Truth table of the second input drive in QuQCA. *79*
Table 5.10 QuQCA output truth table. *79*
Table 5.11 Specifications of n-valued cell structures. *83*
Table 5.12 Truth table of the proposed fuzzy pressure control system. *85*
Table 7.1 Output of ternary NOT gate. *102*
Table 7.2 Outputs of the ternary inverter gate for all polarization states. *102*
Table 7.3 Interactions of two adjacent ternary cells (TQCA quantum wire). *103*
Table 7.4 Outputs of the interactions of two adjacent ternary cells for all output states. *103*

Table 7.5 Majority ternary gate output. *104*
Table 7.6 Results of AND and OR gates with three values. *105*
Table 7.7 Output of the QQCA-based NOT gate. *106*
Table 7.8 Results of the QQCA-based NOT gate in different states and corresponding final values. *107*
Table 7.9 Effect of two adjacent four-valued cells (QQCA-based wire). *107*
Table 7.10 Output states of the interactions of two quaternary cells and the final results. *108*
Table 7.11 QQCA majority gate output. *109*
Table 7.12 Validity of QQCA-based AND and OR gates. *110*
Table 7.13 Quaternary AND and OR gate truth table. *111*
Table 7.14 NOT gate states in QuQCA. *112*
Table 7.15 QuQCA NOT gate for all states and determining the final value (× 1e-20). *113*
Table 7.16 QuQCA-based wire truth table. *113*
Table 7.17 Output states of the interactions of two QuQCA cells and final results (× 1e-20). *114*
Table 7.18 Majority gate truth table in QuQCA. *116*
Table 7.19 Truth tables of the quinary AND and OR gates. *118*
Table 10.1 Parameters used in QCADesigner. *142*
Table 10.2 Outputs of different inputs applied to BPIM3188. *147*
Table 10.3 Cell omission fault evaluation in BPIM1. *150*
Table 10.4 Extra-cell deposition fault evaluation in BPIM1187. *151*
Table 10.5 Cell omission fault evaluation in BPIM2. *152*
Table 10.6 Extra-cell deposition fault evaluation in BPIM2. *152*
Table 10.7 Cell omission fault evaluation in BPIM3. *154*
Table 10.8 Extra-cell deposition fault evaluation in BPIM3. *154*
Table 10.9 Cell omission fault evaluation in BPIM4. *155*
Table 10.10 Extra-cell deposition fault evaluation in BPIM4. *155*
Table 12.1 Cell omission defect analysis for TPIM1. *183*
Table 12.2 Extra-cell deposition defect analysis for TPIM1. *183*
Table 12.3 Cell omission defect analysis for TPIM2. *185*
Table 12.4 Extra-cell deposition defect analysis for TPIM2. *185*
Table 12.5 Cell omission defect analysis for TPIM3. *186*
Table 12.6 Extra-cell deposition defect analysis for TPIM3. *187*
Table 13.1 Parameters of proposed binary and ternary RAM and PIM cells. *197*
Table 13.2 Design parameters of binary and ternary gates. *198*
Table 13.3 Comparison of binary and ternary XOR structures. *199*
Table B.1 Output results of the ternary AND gate in all cases. *224*
Table B.2 Output of ternary OR gates in all cases. *225*

Table B.3 Outputs of the quaternary AND gates in all cases. *226*

Table B.4 Output results of quaternary OR gate. *228*

Table B.5 Output results of Coulomb energy values in the quaternary AND gate for all states (× 1e-20 (J)). *230*

Table B.6 Output results of Coulomb energy values for the quinary OR gate in all states (× 1e-20 (J)). *233*

About the Authors

Reza Sabbaghi-Nadooshan received his BS and MS degrees in electrical engineering from the Iran University of Science and Technology, Tehran, Iran, in 1991 and 1994, respectively, and his PhD degree in electrical engineering from the Science and Research Branch, Islamic Azad University, Tehran, in 2010. He is a Professor with the Department of Electrical Engineering in the Central Tehran Branch, Islamic Azad University. Having published over 100 research papers in various international journals and conferences, his research interests include multiple-valued logic, quantum dot cellular automata (QCA), and nanocomputing. He is on the panel of reviewers for various international journals.

Reza Akbari-Hasanjani was born in Iran in 1987. He received his BSc, MSc, and PhD degrees in electrical engineering from Islamic Azad University, Central Tehran Branch, Tehran, Iran, in 2012, 2015, and 2022, respectively. His interests are in various research areas, including nanocomputing, reversible computation, multiple-value logic, and QCA. These fields have captivated his attention and driven him to explore innovative solutions within the realm of electrical engineering.

Leila Dehbozorgi was born in Iran in 1981. She received her PhD degree in electrical engineering from Islamic Azad University, Central Tehran Branch, Tehran, Iran, in 2023. Her interests are in various research areas, including nanocomputing, QCA, reversible computation, multiple-value logic, and neural networks.

Majid Haghparast is a researcher at the University of Jyväskylä, Finland, and a senior member of IEEE. He has authored over 90 papers in peer-reviewed journals and conferences. In addition, he serves as Editorial Board Member for *Cluster Computing* (Springer), *Journal of Computational Electronics* (Springer), and *Optical and Quantum Electronics* (Springer). Throughout his career, Majid has supervised or advised on over 10 PhD and 40 master's theses. His primary focus is on quantum computing and quantum software engineering, where he has been involved in various projects.

Hamid Reza Akbari-Hasanjani was born in Iran in 1989. He received his PhD in analytical chemistry from Damghan University, Semnan, Iran, in 2019. He has been a member of the Iranian Chemical Society (ICS) since 2019. His interests are quantum dot synthesis, electrochemical biosensors, electroanalysis, and bioinformatics.

Preface

Examining new technologies for electrical and computer scientists is more interesting today. This appeal is likely to endure indefinitely as technologies are advancing rapidly and scientists are trying every day to enhance human comfort through technological advancements. Addressing the question of which technologies could replace complementary metal–oxide–semiconductor (CMOS) transistors has spurred researchers to explore alternatives such as quantum dot cellular automata (QCA). The evolution of technologies paves the way for scientists to develop new computing systems that seamlessly support multi-valued calculations. One such computational approach involves leveraging the spin states in QCA cells, leading us to compile and publish a book on multi-value QCA technology. This book provides a composite solution for multi-value logic in QCA circuits. It comprises 14 chapters in which multi-valued cells are investigated, from computation to the design of logical gates. Intended for senior undergraduate and graduate levels in nanoelectronics, computer arithmetic, memory, and processing in memory (PIM), the book is also valuable resource for researchers exploring emerging nanotechnologies and their architectural implications.

2024 IRAN-Tehran

Reza Sabbaghi-Nadooshan
Reza Akbari-Hasanjani
Leila Dehbozorgi
Majid Haghparast
Hamid Reza Akbari-Hasanjani

Introduction

In recent years, the downscaling of complementary metal–oxide–semiconductor (CMOS) technology has presented challenges, prompting the introduction of alternative technologies to address them. Among these, quantum dot cellular automata (QCA) is a suitable candidate, which has received significant attention in the design of digital circuits. The advantages of QCA technology encompass its tiny dimensions, high speed, low delay, and low power consumption.

So far, binary quantum dot cellular automata (bQCA) has been thoroughly investigated, while multi-value QCA remains unexplored. This technology is one of the most current topics that has received attention, particularly due to its proximity to fuzzy logic. By studying the previous research on the strengths and weaknesses of QCA cells, we found that no comprehensive model considers the calculations of Hamiltonian matrix polarization and power consumption in a cell in multi-valued cells.

This book introduces computational methods for designing multi-valued cells and their basic gates. We calculate the polarization values by adjusting the number of particles within the desired systems to accomplish this objective. In designing a quaternary cell, the combination of bQCA and ternary quantum dot cellular automata (TQCA) cells is used to achieve stability in the design of basic gates.

The external electrostatic model is employed in designing basic gates using the proposed cells. This model is based on the interaction between the electrons of each cell and those of other cells. In the subsequent step, the value of the Hamiltonian matrix is calculated to determine the power consumption of each cell.

In the next step, the basic gates are simulated using MATLAB software to validate the proposed cells, and the amount of electrostatic energy is calculated in all cases. Then, to verify all the calculations for simulating the gates and ternary circuits, two simulators with appropriate dimensions are presented for ternary and quaternary circuits.

In the following chapters, the structures of processing in memory (PIM) and more extended bQCA and TQCA circuits will be designed and analyzed. Multi-value PIM structures represent a new subject in the digital realm and have not been implemented in TQCA thus far. In this book, two different designs are proposed for PIM cells. One design explores the impact of memory relocation, while the other examines the effects of reducing cell count and optimizing components of PIM in bQCA and TQCA. Moreover, the effect of faults/defects on the designed structures is investigated. The results show that compared to binary structures, many of the proposed ternary structures, especially the PIM structures, exhibit greater fault tolerance against cell omission and extra-cell deposition defects, and they are more optimal in terms of cost function and occupied area. Consequently, PIMs have the potential to serve as a suitable alternative to random-access memories (RAMs) in the future.

The simulation results obtained with the ternary and quaternary simulators are entirely consistent with the MATLAB simulation results and previous studies conducted on other technologies, such as CMOS. These results demonstrate significant progress in designing multi-valued cells, indicating the feasibility of designing multi-valued circuits with functionality closely resembling fuzzy logic.

1

Quantum Dots

1.1 Introduction to Quantum Dots

Nanotechnology, specifically the field of nanostructured materials and quantum dots (QDs), has gained significant interest in recent years. It has applications in various fields, such as medicine, chemistry, physics, engineering, electronics, optoelectronics, and biology [1–3].

Nanostructured materials, which can be categorized as two-dimensional (e.g., thin films or quantum wells), one-dimensional (e.g., quantum wires), or zero-dimensional (dots), can bridge the gap between bulk materials and molecular structures. The optical and electronic properties of these materials vary with particle size, especially when the size is less than 100 nm. QDs, referred to as zero-dimensional structures, exhibit discrete quantized energy states due to the limited number of electrons they can accommodate [4].

The discovery of QDs dates back to 1981, when they were first observed in a glass matrix. Since then, significant advancements have been made in synthesizing and understanding their properties. QDs, composed of semiconductor nanocrystals, show distinct advantages over traditional organic dyes, such as tunable broad excitation and narrow emission spectra, high signal brightness, and photostability. They have found applications in fluorescence imaging, sensing, light-emitting diodes (LEDs), solar cells, and more [5].

Furthermore, QDs have been integrated into electrochemical biosensors to improve their analytical performance. These biosensors, which utilize nanomaterials, combine a biological recognition element with an electrochemical transduction element to provide specific quantitative analytical information. QDs enhance the electron-transfer speed, amplify electrochemical signals through increased surface area, and serve as signal labels, enabling the development of ultrasensitive electrochemical biosensors. These nanosensors have the potential to revolutionize personalized detection tools and "point-of-care" diagnostics [6, 7].

The study of nanostructured materials and QDs has seen significant research efforts in the past two decades. Compound semiconductor-based nanostructures have garnered particular attention due to their size-dependent optoelectronic and optical properties. By manipulating the size and composition of these materials, a wide range of fundamental properties can be achieved.

Historically, the synthesis of QDs dates back over 2000 years, with the Romans and Greeks using low-cost natural materials to create colored cosmetics. Over time, techniques for controlling the size and color of QDs in silicate glasses were developed. Their optical and optoelectronic properties were explored, leading to the advancement of quantum size effects and the understanding of how to control the color of glass by altering the size or stoichiometry of semiconductor particles [8].

Computing with Multi-Value Logic in Quantum Dot Cellular Automata, First Edition.
Reza Sabbaghi-Nadooshan, Reza Akbari-Hasanjani, Leila Dehbozorgi, Majid Haghparast, and Hamid Reza Akbari-Hasanjani.

Experimental and theoretical research on QDs has significantly increased in the past two decades, driven by the exploration of their fundamental properties and the potential for commercialization. Scientists and researchers continue to investigate the applications of QDs in various fields, propelled by the continuous advancement of nanotechnology.

In summary, the combination of nanostructured materials and QDs has opened up new avenues for scientific exploration and technological applications. Their unique properties and versatility make them promising candidates for advancements in medicine, chemistry, physics, engineering, electronics, the design of quantum dot cellular automata (QCA), and biosensing. The ongoing research in this field holds great potential for future breakthroughs and innovations.

1.2 Physical Characteristics of Semiconductor Quantum Dots

Semiconductors are materials with intermediate conductivity between conductive and insulating materials. As it is known, there is a forbidden band between the valence and conduction bands, which is actually the energy difference between the two bands or layers. In semiconductor materials, electrons in the valence band can be excited to the conduction band by absorbing energy. The vacancy of the transferred electron is theoretically assumed to be a hole with a hypothetical positive charge, and the combination of this electron hole is called an exciton [9]. From a scientific point of view, this type of exciton is called the "Wannier–Mott" exciton. The generated electron–hole pairs are kept together through electrostatic attraction.

In QDs that are composed of semiconductor materials, the concepts related to valence band, conduction, and bandgap are the same as those discussed for semiconductor materials in Section 1.1. However, there is a major difference between them. In bulk materials, electron transfer takes place easily due to the availability of sufficient space for electron movement. In contrast, in QDs, electron transfer occurs within a very small radial distance, and this small distance is called the "Bohr radius."

The main point is that if the dimensions of QDs or quantum crystals are as small as the Bohr radius, the electron cannot move easily in the material, and the laws of electron transfer are altered completely. This problem leads to unique optical properties, including the effect on light absorption and reflection in semiconducting crystals with dimensions within the range of the Bohr radius [10, 11].

Another physical characteristic of QDs is that a change in the number of atoms leads to variation in the bandgap. In other words, the energy difference between the conduction and valence bands is changed. In addition to the number of atoms, the way in which the atoms are arranged at the level of QDs also affects the amount of this energy difference.

In QDs, for an electron transition from the valence band to the conduction band, more energy is required compared to the bulk state. This energy difference is often referred to as the "bandgap energy" or "forbidden energy" of QDs, which is greater than that of the corresponding bulk material. Since energy has an inverse relationship with wavelength, higher energy corresponds to a shorter wavelength. Therefore, in QDs, compared to bulk materials, shorter wavelengths of light, often in the blue range, are emitted. This causes a shift toward the blue end of the spectrum. According to the presented information, the resulting colloids exhibit different colors and emit different wavelengths depending on the dimensions of the QDs [9, 12, 13].

Due to their small size, electrons in QDs are confined to a limited space, and when the radius of the semiconductor nanocrystal is smaller than the radius of the Bohr exciton, energy levels are determined by the Pauli exclusion principle.

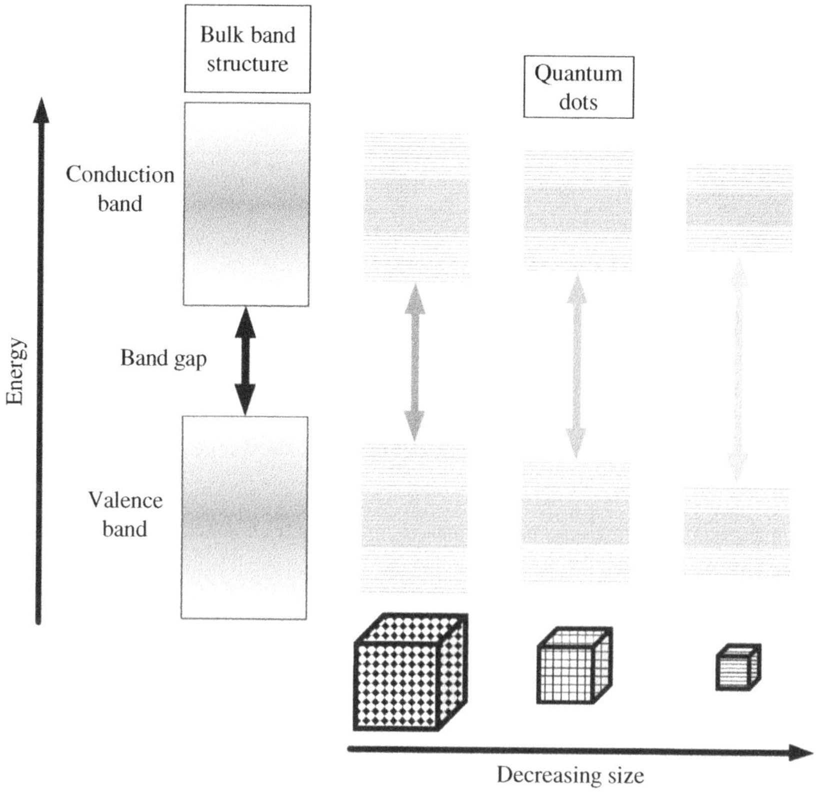

Figure 1.1 The role of energy levels in the reduction of semiconductor gaps with the increase in nanocrystal size.

The quantized energy levels in QDs are more reminiscent of individual atoms rather than bulk materials. Generally, as the size of the QD decreases, the energy levels decrease, resulting in an increased energy difference between the highest valence band and the lower conduction band. As a result, a change in color from red to blue is observed that show in Figure 1.1 as  and  respectively. As a phenomenon of QDs, any light color can be emitted by changing the energy difference. QDs can be classified based on their composition and structure. QDs can be single component with a uniform internal composition of chalcogens (sulfides or selenides) of metals such as cadmium (e.g. CdSe and CdS) and tri-*n*-octyl-phosphine (TOP). The electroluminescent images can change simply by changing the size of the crystals [14]. Figure 1.1 shows that the energy level in QDs increases due to the quantum confinement effect and semiconductor gaps, while the size of nanocrystals decreases.

1.3 Quantum Dots Structure

QDs are core–shell nanocrystals in which the core comprises elements of Groups II–VI, III–V, or IV–VI of the periodic table and is covered by a shell of semiconductor compounds. Due to the effects of QDs, quantum cells have unique properties such as light emission, more chemical stability, a broader absorption spectrum, and a symmetrical emission spectrum. One of the prominent

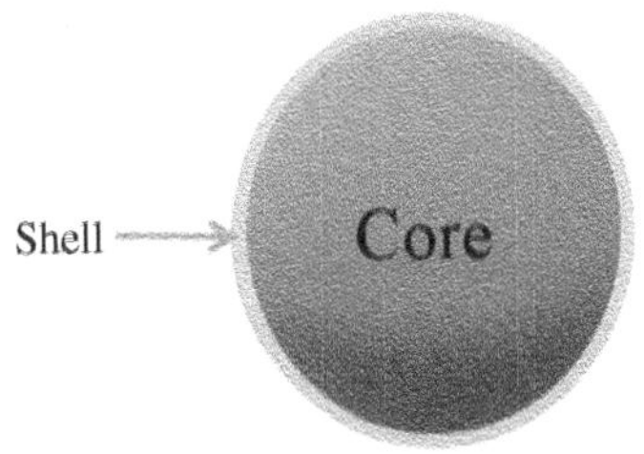

Figure 1.2 Basic structure of an uncoated QD.

features of QDs is that their emission wavelength can be adjusted by varying their dimensions and chemical composition [15, 16].

QDs can be classified in a number of different ways, one of which is their chemical composition, including core-type, core–shell, and alloyed QDs. In fact, the properties of a QD are not only determined by its size but also by its composition and structure. Figure 1.2 shows the basic structure of an uncoated QD.

1.3.1 Core-Structure Quantum Dots

Core-type QDs are a category of QDs composed of single-component materials with consistent internal compositions. These materials consist of selenides, sulfides, or tellurides of metals such as cadmium, lead, or zinc. Core-type QDs exhibit exceptional fluorescence properties, which have led to their widespread applications in biosensing and fluorescence imaging. Among core-type QDs, those based on cadmium compounds, specifically CdX (X = Se, S, or Te), have undergone extensive investigation due to their outstanding optical and electrochemical characteristics. However, the release of cadmium ions from these QDs has raised concerns about their cytotoxicity, limiting their practical use. Heavy metal-free QDs, including Group IV QDs such as carbon-based QDs (C/GQDs) and silicon or germanium QDs (Si/GeQDs), have been developed to address this issue.

Carbon and graphene QDs are highly promising carbon-based nanomaterials, particularly in the field of sensing research. They possess advantageous properties such as low toxicity, high solubility in various solvents, remarkable electronic characteristics, chemical stability, large specific surface areas, numerous edge sites for functionalization, and versatility. The accidental discovery of carbon quantum dots (CQDs) occurred in 2004 when Lee et al. employed preparative electrophoresis to purify arc-synthesized single-walled carbon nanotubes (SWCNTs) [17]. Subsequently, graphene quantum dots (GQDs) were fabricated by Ponomarenko and Geim in 2008, building upon previous work on CQDs. GQDs differ from quasi-spherical CQDs as they possess a graphene lattice structure within the dots. They are smaller than 100 nm in size and consist of less than 10 layers. GQDs possess unique properties arising from the combination of graphene's structure with the quantum confinement and edge effects of CQDs [18, 19].

Silicon (Si) has long been recognized as an indirect bandgap semiconductor with limited optical properties. However, in the 1990s, Canham reported the emission of visible light from mesoporous Si layers, which led to the proposal of quantum confinement as an explanation for the observed features in the absorption spectra of porous silicon by Lehman and Gösele [20, 21]. This stimulated increased interest in silicon quantum dots (SiQDs). SiQDs are semiconductor Si nanoparticles ranging in size from 1 to 10 nm and hold significant potential for optoelectronic devices. Their biocompatibility, in contrast to conventional toxic QDs based on Group II–VI and Group III–V metals, makes them even more attractive for practical applications to prevent environmental pollution and harm to living organisms. The properties of SiQDs offer promising opportunities for biological applications, including biosensors and bioimaging [22]. Similarly, germanium quantum dots (GeQDs) have attracted attention due to their exceptional electrical and optical properties, which are correlated with the quantum confinement effect. GeQDs can be widely utilized in thermoelectric devices to enhance electrical conductivity while reducing thermal conductivity. Furthermore, viability tests have demonstrated that GeQDs show lower toxicity compared to traditional metal-based QDs [23]. Previous studies have shown that the toxicity of cadmium sulfide quantum dots (CdSQDs) decreases after doping with germanium, significantly improving their biocompatibility. Consequently, these QDs can be utilized in the design of electrochemical sensors for the detection of biological compounds [24].

1.3.2 Core–Shell Quantum Dots: Enhanced Properties and Diverse Applications

Core–shell QDs are a result of advanced engineering in the structures of QDs. These engineered QDs exhibit superior optical properties compared to simple QDs, thanks to the addition of a shell around the QD core. The shell's growth enhances the stability and photoluminescence efficiency of the QDs. Photoluminescence in QDs occurs when electron–hole pairs (excitons) recombine through radiative pathways. However, nonradiative decay pathways can also reduce the fluorescence quantum yield (QY). To address this limitation, a method used to improve the efficiency and brightness of semiconductor nanocrystals involves growing shells of higher bandgap semiconducting materials around them. These structures, known as core–shell QDs, consist of small regions of one material embedded in another with a wider band gap. For example, when a CdSe core is coated with a ZnS shell, the resulting core–shell QDs exhibit a QY greater than 50%. The shell coating not only enhances the QY by minimizing nonradiative recombination sites but also provides robustness to the QDs under various processing conditions, making them suitable for a wide range of applications. Researchers have extensively explored this approach to tailor the photophysical properties of QDs. Common combinations for the core and shell materials include type II–VI, IV–VI, and III–V semiconductors, such as (CdS)ZnS, (CdSe)ZnS, (CdSe)CdS, and (InAs)CdSe [25].

Core–shell QDs can be categorized based on the relative positions of the valence and conduction bands and the energy gap between them in the semiconductors. In type I configurations, the bandgap of the core is smaller than that of the shell, or the bandgap of the core falls within the bandgap of the shell. In inverse type I configurations, the shell's bandgap is smaller than that of the core. In type II configurations, either the valence band edge of the core falls within the shell's bandgap or the conduction band edge of the shell falls within the bandgap of the core [26]. Figure 1.3 illustrates the bandgap alignments for each of the four types.

Beyond their enhanced properties, core–shell QDs offer the ability to control their cytotoxicity through further engineering of their structures. Researchers have recently focused on doped QDs, where transition metal ions are incorporated into the QDs. These doped QDs not only retain the advantages of undoped QDs but also possess additional benefits, such as reduced toxicity. By minimizing the presence of toxic cadmium in undoped QDs, doped QDs become more suitable for biological studies. Moreover, high-quality core–shell QDs can mitigate the toxicity associated with heavy metal-based QDs. The presence of a shell around the core QDs acts as a protective barrier, preventing oxidation and the potential formation of free heavy metal ions, especially when the core contains transition metals like cadmium. Consequently, core–shell QDs are highly desirable for various biological applications, as the shell passivates the core (typically using a wider bandgap material), leading to improved fluorescent properties and preventing leaching [27]. Core–shell

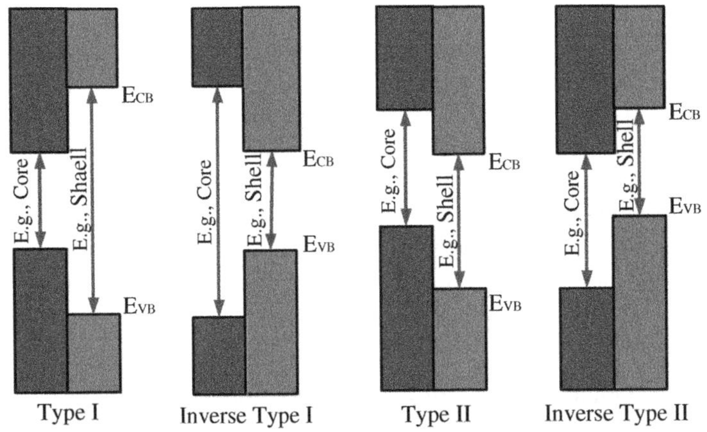

Figure 1.3 The bandgap alignments.

QDs, whether composed of semiconductor or metal cores or shells, hold great potential for applications in the field of electronics.

While core–shell QDs show immense potential for sensing applications, their primary application has been in the field of biocompatible nanoprobes for fluorescence bioimaging. Given the concerns surrounding toxicity, core–shell structures have been widely adopted in imaging studies to address this issue. As research on QDs progresses rapidly, new structures are being introduced to fabricate safer and more efficient QDs for biological applications [28]. These advancements aim to enhance the biocompatibility, stability, and overall performance of core–shell QDs, opening up new possibilities for their utilization in various fields, such as biomedical research, diagnostics, and optoelectronics.

1.3.3 Alloyed Quantum Dots

Alloyed QDs have emerged as a prominent and dynamic area of research within semiconductor nanotechnology. These QDs, such as cadmium selenium sulfide (CdSeS), exhibit remarkable properties due to their uniform or gradient internal structures, resulting in enhanced luminescence. Unlike conventional core–shell structures, gradient QDs possess a smoothly transitioning composition from the core to the surface, enabling continuous adjustment of their optical and electronic characteristics. This capability facilitates precise tuning of their properties to meet specific requirements.

The composition and internal structure of alloyed QDs play vital roles in determining their optical and electronic behavior. By manipulating these factors, researchers can achieve the desired characteristics crucial for various applications. Homogeneous QDs, characterized by a consistent composition throughout, provide a uniform behavior across the QD. On the other hand, gradient QDs exhibit varying compositions along their radial axis, resulting in unique optical and electrical properties that differ from their homogeneous counterparts [29, 30].

The distinctive properties of gradient alloy nanocrystals make them highly valuable for a wide range of applications. Their tunable optical and electronic properties enable their utilization in fields such as optoelectronics, lighting, displays, sensors, and biomedical imaging. The ability to continuously adjust their properties without altering the particle size opens up new avenues for innovative technologies and further advancements in the field of semiconductor nanotechnology [31].

1.4 Surface Structure of Quantum Dots

The optical properties of QDs are greatly influenced by their surface structure, which is attributed to the high surface-to-volume ratio. Surface states, electronic quantum states associated with the surface, play a significant role in these optical properties. For instance, in a 5 nm CdS QD, approximately 15% of the atoms reside at the surface [32]. This high surface-to-volume ratio can impact the transfer rate of photogenerated charge carriers due to the abundance of surface sites.

The surface states of QDs have implications on various aspects, including optical absorption (photoluminescence excitation [PLE]), quantum efficiency, luminescent intensity, spectrum, and aging effects. These surface states arise from unsatisfied bonds on the reconstructed surface and can be influenced by factors like nonstoichiometry and voids [33]. Their energy levels typically lie within the bandgap of the QDs, enabling them to trap charge carriers (electrons or holes) and act as reducing or oxidizing agents [34].

The electrochemical reactions and behavior occurring at the QD surface have a significant impact on the overall conductivity and optical properties. Hence, surface states play a crucial role in shaping the optical and optoelectronic characteristics of QDs. One approach to mitigate the effects of surface states is through surface passivation, which involves coating the QDs with a passivation layer. This passivation layer helps confine the charge carriers within the QD core and enhances the optical properties. However, it is important to note that the passivation layer can act as an insulator or a barrier for charge conduction.

1.4.1 Surface Passivation

Defects on the surface of QDs can hinder their performance by trapping electrons, holes, or excitons, leading to reduced QYs and diminished radiative recombination. Therefore, it is crucial to cap or passivate the surface of QDs to enhance their photostability. Ideally, a perfectly passivated QD surface would have all dangling bonds saturated, resulting in the absence of surface states and the confinement of near band-edge states. However, if the anion dangling bonds on a compound semiconductor surface are not passivated, a band of surface states is expected to form just above the valence band edge. On the other hand, passivating the anions with surface cations would create a broad band of surface states just below the conduction band edge. Consequently, modifying the surface of QDs is a challenging task, often achieved by depositing an organic or inorganic capping layer [35]. Surface engineering techniques aim to modify the surface states by controlling factors such as ligand chemistry, surface functionalization, and surface treatment methods. These approaches can help minimize surface defects, enhance charge carrier mobility, and improve the overall performance of QDs.

1.4.1.1 Organically Capped Quantum Dots

Typically, monodispersed QDs are created by introducing organic molecules that adsorb onto the QD surface and act as capping agents. Organic capping layers offer several advantages, including colloidal suspension and the ability to bioconjugate the QDs [36]. However, selecting organic ligands that bond effectively with the QD surface atoms is a delicate matter. Phosphenes (e.g., tri-*n*-octyl phosphene oxide or TOPO) and mercaptans (–SH) are commonly used ligands. Most organic capping molecules have distorted shapes and larger sizes compared to a surface site, which can hinder the coverage of surface atoms. Another challenge is simultaneously passivating both anionic and cationic surface sites using these capping agents, which is achievable but complex. Furthermore, organic-capped QDs tend to be photounstable. The bonding between the capping molecules and surface atoms is generally weak, leading to inadequate passivation and the creation of new surface states under ultraviolet (UV) irradiation. These surface states are known to cause photodegradation and luminescence quenching in QDs. Chemically reduced bovine serum albumin (BSA) was used to simultaneously passivate and functionalize the surface of CdTe QDs, making them water soluble [37]. The stability and photoluminescence quantum yield (PL-QY) of the dBSA-coated CdTe QDs were optimal in a pH range of 6–9, with significant decreases observed at higher and lower pH values. Additionally, excessively high or low concentrations of dBSA in the QDs solution resulted in decreased PL-QY. Similarly, DNA-passivated CdS QDs have been reported as stable and nontoxic for biological systems [38].

1.4.1.2 Inorganically Passivated Quantum Dots

An alternative approach to passivating QD surfaces involves using inorganic layers, particularly materials with larger band gaps. The passivating shell can be grown epitaxially or as a nonepitaxial crystalline or amorphous layer on the QD core. Applying a defect-free and uniform shell coating to QDs improves their QY. When the shell material matches the lattice parameters of the core during epitaxial growth, coherency strains may arise, influencing the properties of core–shell QD systems. For instance, strain can cause redshifts in the absorption and emission spectra of core–shell QDs [39]. The maximum PL efficiency or QY of core–shell QDs also depends on the thickness of the shell layer. Optimum properties are achieved when the shell layer is less than two monolayers for CdSe/CdS core–shell nanoparticles. Thicker capping layers result in the formation of nonradiative recombination sites called misfit dislocations, which decrease the PL-QY.

1.4.1.3 Multishell Structure

Researchers are investigating double-shell QDs to improve their optical properties. As mentioned earlier, the lattice mismatch and bandgap differences play significant roles in the properties of core–shell QDs. The bandgaps and band offsets of the core and shell materials are critical for suppressing the tunneling of charge carriers from the core to the surface states of the shell. In the case of CdSe/CdS QDs, the lattice mismatch is small, but so are the band offsets. Conversely, CdSe/ZnS QDs have a large lattice mismatch along with large band offsets. The advantages of both shell materials are combined in core–shell–shell CdSe/CdS/ZnS QDs [40]. These double-shell nanostructures reduce lattice strain at the interface while maintaining large band offsets.

QDs often exhibit low QY due to Auger recombination, a process strongly influenced by confinement. To minimize nonradiative decay caused by Auger recombination, it is important to minimize the wave function overlap of charge carriers. A recent development in this area is the introduction of quantum well quantum dots (QWQDs) [41].

In addition to the existing information, it is worth noting that surface passivation is a critical aspect of QD research. The presence of surface defects can have a detrimental effect on the performance of QDs by reducing their QY and radiative recombination. Surface passivation aims to eliminate or minimize these defects by capping the surface with organic or inorganic layers.

Organically capped QDs offer advantages such as colloidal suspension and the ability to bio-conjugate with other molecules, making them useful for various applications. However, the selection of organic ligands is crucial, and achieving simultaneous passivation of both anionic and cationic surface sites can be challenging. Moreover, the photostability of organic-capped QDs is a concern due to weak bonding at the interface, which can lead to the creation of new surface states under UV irradiation [42].

Inorganically passivated QDs, on the other hand, utilize inorganic materials with larger bandgaps to create a protective shell around the core QD. This shell coating improves the QY by reducing defects and nonradiative recombination sites. The thickness of the shell layer is an important factor affecting the properties of core–shell QDs, and a balance must be struck to avoid the formation of misfit dislocations.

Multishell structures such as double-shell QDs are being explored to further enhance the optical properties. These structures combine different materials with varying bandgaps and band offsets to optimize the suppression of charge carrier tunneling and reduce lattice strain [21, 43].

1.4.2 Characterization of Shell Structures

Recent research conducted by Yu et al. [44] explored the distribution of shell materials on CdSe/ZnS core–shell QDs using scanning transmission electron microscopy (STEM) combined with electron energy loss spectroscopy (EELS). Their analysis revealed that the ZnS shell was well defined around the CdSe core. However, the distribution of the shell material was found to be highly anisotropic, potentially due to variations in the chemical activity of different crystal faces of the CdSe core [45]. It is important to note that the accuracy of the measurement may be questionable, as the localized EELS spectra were obtained from a very small area on a single QD, which can be influenced by nearby areas of the same QD or neighboring QDs due to their movement under the high-energy electron beam. Z-contrast STEM analysis offers advantages over conventional transmission electron microscopy (TEM) [46].

Z-contrast STEM relies on the collection of incoherently scattered electrons using a high-angle annular dark field (HAADF) detector. The intensity of scattered electrons, following the Rutherford scattering equation, is directly related to the square of the atomic number within a specific angular range. This enables the direct observation of mass contrast in the images. While previous studies demonstrated the imaging of both bare CdSe QDs and core–shell QDs using aberration-corrected Z-contrast STEM, the interpretation of bonding information from TEM analysis remains challenging. Additionally, obtaining Z-contrast STEM images of QDs with an amorphous shell, such as silica, is extremely difficult [46].

X-ray photoelectron spectroscopy (XPS) is a crucial tool for surface analysis of QDs [47]. XPS studies have extensively investigated the nature of CdSe QDs, including their core-level positions for Cd and Se. These positions were found to be consistent with those of bulk CdSe when the QDs were covalently bound on a substrate [47]. However, when the CdSe QDs were washed and exposed to air, signs of oxidation were observed. XPS has also been used to confirm the presence of shells in core–shell QDs, providing information about surface ligands, shell coverage, and surface oxidation [48]. However, conventional XPS has limitations in analyzing the QD surface, shell structure, and core–shell interface bonding due to the comparable escape length in XPS to the size of QDs and the relatively larger spot size in XPS analysis.

Characterizing the shell structures of QDs is crucial for understanding their properties and optimizing their performance. Techniques such as STEM and XPS have been employed to investigate the distribution of shell materials, analyze surface composition, and verify the presence of shells in core–shell QDs. These characterization techniques provide valuable insights into the structural and chemical properties of QDs, enabling researchers to customize their design and enhance functionality. However, challenges remain in accurately interpreting bonding information from TEM analysis and obtaining high-resolution images of QDs with amorphous shells using STEM. Further advancements in characterization techniques are necessary to gain a comprehensive understanding of the surface, shell structure, and core–shell interface bonding of QDs at the nanoscale.

1.5 Properties of Quantum Dots

QDs possess unique properties that are characterized by their optical, electronic, dimensional, and quantum mechanical features. These properties make them comparable to real atoms and have earned them the nickname "artificial atoms" in the field of quantum mechanics [49].

One notable feature of QDs is that electrons within them occupy discrete energy levels, similar to the states of electrons in an atom. This discrete nature of energy levels gives rise to a distinct light emission spectrum that corresponds to these energy levels, akin to the emission spectra observed in real atoms.

In contrast to real atoms, semiconductor QDs offer the advantage of being able to manipulate the number of free electrons they contain using external tools. This allows for the creation of artificial atoms with varying numbers of electrons, such as 1, 2, 3, or more. Consequently, by adding or removing electrons from QDs, a wide range of artificial materials can be produced. It is important to note, however, that these artificially created materials do not possess all the characteristics of their original counterparts. In essence, the resulting materials may be viewed as nanorobots that can be controlled by humans in terms of their appearance and functionality.

Semiconductors, including QDs, consist of energy bands that contain discrete energy levels. Electrons in these materials can occupy different energy levels within the bands. The energy levels are typically very close together, forming a nearly continuous interval, with adjacent levels having a negligible energy difference. Between the energy bands, an energy gap exists where electrons cannot reside. This gap, known as the energy gap or bandgap, separates the valence band, which contains the lower energy levels occupied by valence electrons, from the conduction band, which contains higher energy levels that can accommodate conduction electrons. In bulk semiconductor materials, the majority of electrons occupy the valence band, rendering the material nonconductive in its normal state. Additional energy must be supplied to overcome the energy gap to move electrons into the conduction band and achieve conductivity. This can be achieved through excitation by light, electric fields, or heat. During excitation, a temporary "hole" is created in the valence band as an electron moves to the conduction band.

In the case of QDs, the allowed energy levels are discrete, introducing a fundamental difference compared to bulk semiconductor materials. Consequently, the properties of QDs deviate from those of bulk materials, particularly in terms of absorption and emission behavior. The discrete energy levels of QDs can be modified by adding or removing atoms, as well as by altering the arrangement of atoms on their surface. These modifications affect the energy gap, which is larger in QDs than in bulk materials. As a result, electrons in QDs require more energy to escape the gap, leading to the emission of light with shorter wavelengths and a blueshifted spectrum. This property enables the precise tuning of the radiation wavelength emitted by QDs, allowing for the selection of desired colors.

The optical behavior of QDs is further influenced by their size. Smaller QDs have larger energy gaps, while larger QDs have smaller energy gaps. When exposed to UV light, smaller QDs absorb photons and undergo excitation to higher energy levels. As the excited electrons return to their stable states, they release the excess energy by traversing a larger energy gap. The emitted light, therefore, has higher energy, resulting in a tendency toward blue light emission [50].

1.5.1 Optical and Electronic Properties

Nanoparticles have unique optical and electrical properties due to their small size, and electrons play a key role in the emergence of these properties. In general, electrons in semiconductor materials have a specific range of energy. When electrons are excited by light, electric field, or heat, some of them move from a lower energy level (valence band) to a higher energy level. In this case, the valence level that becomes empty is called a hole because, during the mentioned event, a temporary hole is created in the valence band.

Electrons that are excited to a higher level return to the valence band after a while. This return of electrons is accompanied by the radiation of energy in the form of light. Due to this characteristic of semiconductor materials, there is a gap in the continuous range of their energy (that is, electrons are not allowed to have energy in this gap), and light is emitted only at a certain wavelength [14, 50–52].

In QDs, electrons have a specific range of energy. However, due to the very small size of these nanoparticles, their energy levels are not continuous, and removing or adding several atoms from/to the QDs causes a change in the energy gap. Changing the way atoms are arranged on the surface of QDs also changes the energy gap, which is again due to the very small size of the QDs. The size of the energy gap in QDs is always larger than that in the bulk material. This feature makes it possible to adjust the radiation wavelength, although the wavelength of the radiated photons does not depend on the type of material; it depends on the size of the particles. Therefore, by controlling the size of nanocrystals, it is possible to specify the radiation wavelengths so that large QDs emit red light and small QDs emit blue light. This feature of QDs is called the size quantization effect [53].

Each color can be chosen for QDs. QDs are nanocrystals that are composed of only a few hundred atoms. Because of their small size, these nanocrystals have some nonclassical and interesting properties, such that they are like a fluorophore (a chromophore that can absorb light, reach the state of excitation, and emit light) and exhibit fluorescence properties. Depending on the type of material from which QDs are synthesized, and especially according to their size, they can be excited under the influence of laser light, and all of them have fluorescence emission in the range of visible light [54, 55].

Since QDs are very similar to atoms, sometimes they are also called artificial atoms. Now, if we consider an artificial atom to be a structural building unit, we can form arrays of QDs. Electrons inside QD arrays can move around and interact with each other with interesting effects on magnetic, optical, and transmission properties. By pairing some QDs, an artificial molecule structure can be achieved. The important point about artificial molecules is that to pair dots of different types, the input voltage or the distances between dots can be changed. The feature of adjustability allows the occurrence of a variety of phenomena. As a result of arranging QDs in an alternating network and pairing them continuously with each other, a structural bond is obtained. This structural network was first introduced by "Sakai" under the name of quantum dots superlattice (QDSL) or hyperlattice [56].

In natural light-material semiconductors, electrons (larger than 10 nm) exist in a wide range of energy levels. Many energy levels can be described as a continuum. This means that there is a negligible energy difference between them. Also, the region where the electron is located or migrated from is known as the band gap. The band gap is specific to the type of semiconductor material. The electrons in the energy levels below this band gap are in the valence band, while the electrons that occupy the upper level are in the conduction band. A stimulus, such as photon, voltage, or heat, can move the electron from the valence band to the other side of the gap. By leaving the valence band, an empty space is created, known as a hole, which has a positive charge. This electron and hole form an exciton as a pair. Conductor electrons and valence holes have an average physical distance known as the exciton Bohr radius (EBR) [57]. Absorption causes the emission properties of semiconductor materials to change. When the conduction band is excited, the electron moves toward the valence band, causing electromagnetic radiation with an average wavelength proportional to

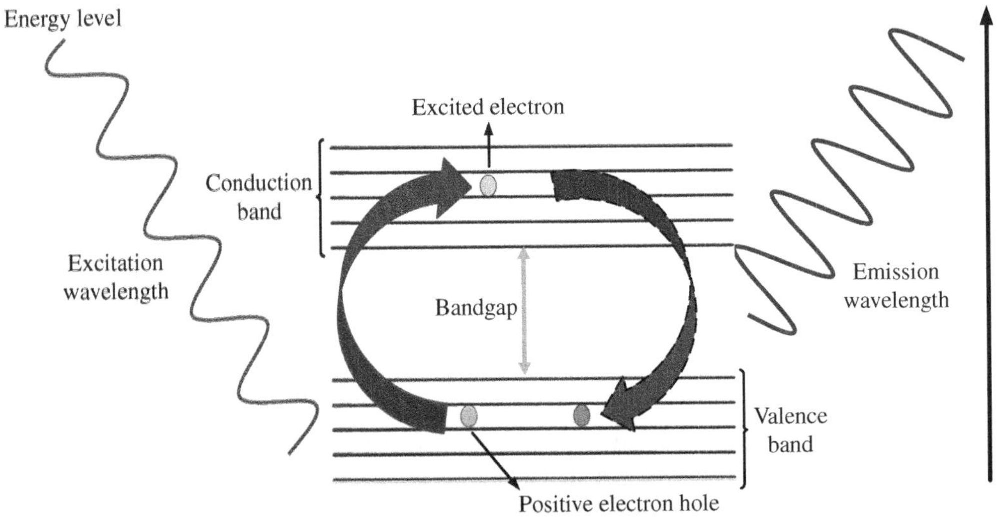

Figure 1.4 Schematic representation of the emission mechanism in QDs.

its energy in diffusion transition. Figure 1.4 shows the schematic representation of the emission mechanism in QDs.

1.5.2 Unique Properties and Applications of Quantum Dots

QDs possess distinctive electrical and optical properties due to the quantum confinement effect, which arises when their size is smaller than the EBR. This effect sets QDs apart from bulk materials [58].

1.5.2.1 Confinement Effect

The quantum confinement effect is a crucial feature of QDs that contributes to their unique properties. It occurs when the size of a QD is below a critical limit. As a result, the optical and electrical characteristics of QDs differ from those of bulk materials.

1.5.2.2 Excitation and Emission Spectra

QDs offer several advantages over traditional fluorophores, such as organic dyes and fluorescent proteins. A key factor in determining the usefulness of a fluorophore is the width of its excitation and emission spectra, as well as stability against light radiation and fluorescence lifetime.

- *Excitation Spectrum Width*: Organic dyes often have a narrow excitation spectrum, limiting them to specific wavelengths. In contrast, QDs exhibit a broad excitation spectrum, allowing them to be excited by a wide range of wavelengths. This versatility enables using a single excitation wavelength for QDs of different colors [59].
- *Emission Spectrum Width*: Organic fluorophores typically have wide emission spectra, resulting in spectral overlap between different colors. This limits the simultaneous detection of multiple fluorophores. QDs, on the other hand, have a narrow emission spectrum that can be precisely controlled by adjusting the size, composition, and surface coating of the nanoparticles. This tunability allows QDs to emit light across a wide range of wavelengths, from UV to infrared.

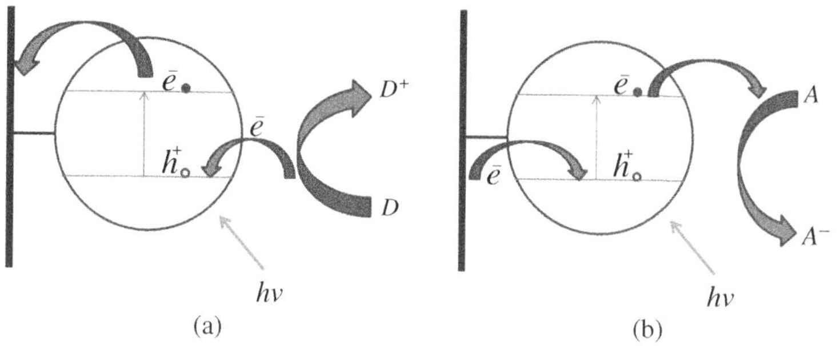

Figure 1.5 Photoelectrochemical property of quantum dots under light irradiation: (a) generation of anodic current in the presence of electron donor compound and (b) generation of cathodic current in the presence of electron acceptor compound.

1.5.2.3 Stability Against Light Radiation

QDs exhibit exceptional stability when exposed to light radiation, making them suitable for fluorescence-based applications. Unlike organic fluorophores that quickly lose their color upon light exposure, QDs can withstand repeated irradiation and high light intensities. Their stability enables long-term tracking of fluorescence signals used for labeling, making them valuable for applications such as whole-body fluorescence imaging.

1.5.2.4 Luminescence Effect

When QDs are in contact with an electrode surface and exposed to light, they exhibit electrochemical properties. Photons stimulate electron transfer from the valence band to the conduction band, creating an electron–hole pair (exciton). When the electron returns to its ground state, luminescence occurs. Under appropriate conditions, QDs can generate electric current through two mechanisms [60]:

Electron Donor Mechanism: In the presence of an electron-donating compound, electron transfer from the compound to the QDs valence band reduces the probability of excited electrons in the conduction band returning to the ground state. This leads to the generation of anodic current in the electrode.

Electron Acceptor Mechanism: In the presence of an electron-accepting compound, excited electrons transfer to the conduction band of the compound, while the hole in the valence band is neutralized by an electron transfer from the electrode. This generates a cathodic current.

Figure 1.5 shows the photoelectrochemical property of QDs under light irradiation during the generation of cathodic current in the presence of an electron acceptor compound.

2

Synthesis Methods of Quantum Dots and Applications

2.1 Synthesis Processes

Quantum dots (QDs) can be synthesized using either top-down or bottom-up approaches. The general schematic for comparing top-down and bottom-up methods is shown in Figure 2.1. In this section, the top-down and bottom-up synthesis processes will be discussed.

2.2 Top-Down Synthesis Processes

Top-down approaches involve the fabrication of QDs by thinning down a bulk semiconductor material. Several techniques are employed to achieve this, including electron beam lithography, reactive-ion etching (RIE), and wet chemical etching. These methods allow the production of QDs with a diameter of approximately 30 nm. Top-down synthesis approaches offer precise control over the size, shape, and arrangement of QDs, making them suitable for various applications. However, it is important to note that these techniques may introduce impurities into the QDs and can result in structural imperfections during the patterning process. Ongoing research focuses on refining these top-down methods and developing new techniques to enhance the efficiency, scalability, and purity of QD synthesis.

2.2.1 Etching Processes

Etching, a well-established technique for over two decades, plays a vital role in the nanofabrication of QDs. In dry etching, a reactive gas species is introduced into an etching chamber, where a radio frequency voltage creates a plasma. The energetic species formed by this process strike the surface, resulting in the etching of the sample. RIE is a specific type of dry etching where ions are used for etching. By employing a masking pattern, selective etching of the substrate can be achieved. For example, RIE with a boron trichloride and argon mixture has been utilized to fabricate GaAs/AlGaAs quantum structures as small as 40 nm [61]. RIE has also been employed to produce close-packed arrays of ZnTe QDs with varying interdot distances using CH_4 and H_2 gases.

2.2.2 Focused Ion Beam Techniques

Focused ion beam (FIB) techniques provide the possibility of fabricating QDs with extremely high lateral precision. These techniques involve the use of highly FIBs from a molten metal source,

Computing with Multi-Value Logic in Quantum Dot Cellular Automata, First Edition.
Reza Sabbaghi-Nadooshan, Reza Akbari-Hasanjani, Leila Dehbozorgi, Majid Haghparast, and Hamid Reza Akbari-Hasanjani.

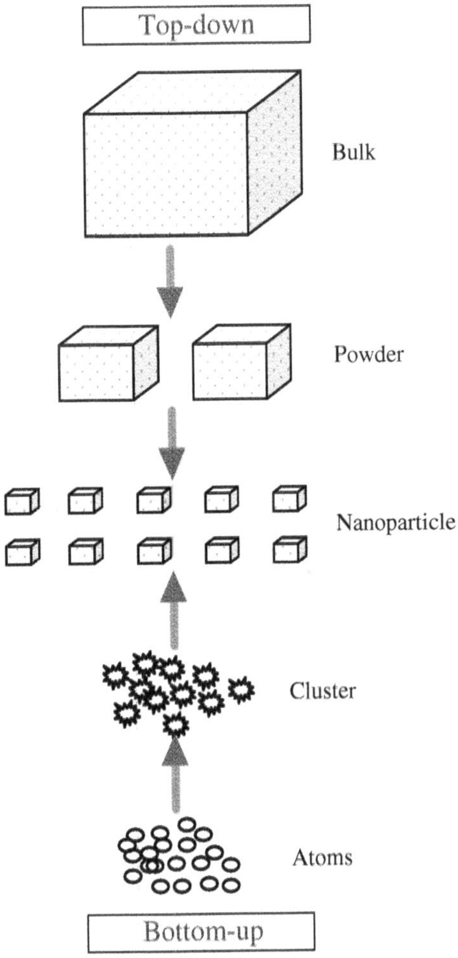

Figure 2.1 Comparison of top-down and bottom-up methods.

such as Ga, Au/Si, Au/Si/Be, or Pd/As/B, to sputter the surface of the semiconductor substrate. The shape, size, and interparticle distance of the QDs depend on the size of the ion beam. FIB systems with a minimum beam diameter of 8–20 nm have been reported, enabling the etching of QDs to dimensions below 100 nm. Additionally, FIB can be used for selective deposition of materials from precursor gases with a resolution of approximately 100 nm. However, FIB processes are slow, have low throughput, and can cause residual surface damage [62].

2.2.3 Electron Beam Lithography

Another method for achieving QD patterns is electron beam lithography followed by etching or lift-off processes. This approach offers a high degree of flexibility in designing nanostructured systems. It allows for the realization of QDs, wires, or rings with precise separation and periodicity. Electron beam lithography has been successfully employed for the synthesis of III–V and II–VI QDs with particle sizes as small as 30 nm [63].

2.3 Bottom-Up Synthesis Processes

QDs can be synthesized using different self-assembly techniques, which can be broadly divided into wet chemical and vapor-phase methods. Wet chemical methods include microemulsion, sol–gel, competitive reaction chemistry, hot-solution decomposition, and electrochemistry. These methods involve controlling various parameters to achieve desired QD properties. On the other hand, vapor-phase methods involve the self-assembly of nanostructures in materials grown through techniques like molecular beam epitaxy (MBE), sputtering, liquid metal ion sources, or aggregation of gaseous monomers.

2.3.1 Wet Chemical Methods

Wet chemical methods typically utilize precipitation techniques with precise control over parameters for a single solution or a mixture of solutions. The precipitation process involves nucleation and limited growth of nanoparticles. Nucleation can be homogeneous, heterogeneous, or secondary in nature [64]. Homogeneous nucleation occurs when solute atoms or molecules combine and reach a critical size without the assistance of a preexisting solid interface. By adjusting factors such as temperature, electrostatic double layer thickness, stabilizers or micelle formation, precursor concentrations, ratios of anionic-to-cationic species, and solvent, QDs of desired size, shape, and composition can be achieved.

Wet chemical methods offer various techniques for synthesizing QDs, each with its advantages and limitations. The sol–gel process is a versatile method but suffers from broad size distribution and defects. Microemulsion processes, particularly the reverse micelle method, provide better control over QD size but have lower yields and potential impurities. The hot-solution decomposition process is a well-established route for QD synthesis, relying on the pyrolysis of organometallic compounds at high temperatures. Some common synthesis processes are discussed in Sections 2.3.1.1 to 2.3.1.3.

2.3.1.1 Sol–Gel

The sol–gel process is a widely used wet chemical method for synthesizing QDs. In this method, a sol, which is a dispersion of nanoparticles in a solvent, is prepared by using a metal precursor (such as alkoxides, acetates, or nitrates) in an acidic or basic medium. The process involves three main steps: hydrolysis, condensation (sol formation), and growth (gel formation) [65]. First, the metal precursor undergoes hydrolysis in the medium, which involves breaking down the precursor into smaller fragments due to the reaction with water. This is followed by condensation, where the hydrolyzed fragments react with each other to form a sol, which consists of small nanoparticles suspended in the solvent. Finally, the sol undergoes a polymerization process, leading to the formation of a gel, which is a three-dimensional network of interconnected nanoparticles.

The sol–gel process allows for precise control over the size, shape, and composition of the QDs by manipulating various factors such as temperature, precursor concentrations, and solvent properties. However, it has some limitations, including a broad size distribution and a high concentration of defects, which can affect the optical and electronic properties of the QDs [65].

2.3.1.2 Microemulsion

The microemulsion process is another wet chemical method commonly used for synthesizing QDs, particularly at room temperature. This process involves the formation of emulsions by mixing polar and nonpolar liquids using surfactants. There are two types of microemulsions: normal microemulsion (oil-in-water) and reverse microemulsion (water-in-oil), although other polar solvents like alcohol can also be used instead of water.

In the reverse micelle process, which is the most popular for QD synthesis, two immiscible liquids (polar water and nonpolar long-chain alkane) are mixed and stirred to form an emulsion. Surfactants with hydrophilic and hydrophobic groups on opposite ends are added to stabilize the emulsion. These surfactants form tiny droplets called micelles in the continuous oil medium. The micelles act as nanoreactors, where the QD precursors are confined and react to form QDs. By adjusting the molar ratio of water-to-surfactant, the size of the resulting QDs can be controlled [66].

The microemulsion process offers advantages such as precise control over QD size, a narrow size distribution compared to other methods like sol-gel, and ease of QD dispersion. However, it has limitations, including relatively low yield and the possibility of incorporating impurities and defects during the synthesis process.

2.3.1.3 Hot-Solution Decomposition

The hot-solution decomposition process is a vapor-phase method that is commonly used for synthesizing QDs. This method involves the high-temperature pyrolysis (around 300 °C) of organometallic compounds. The process starts with the degassing and drying of a coordinating solvent, such as trioctylphosphine oxide (TOPO), at a high temperature under vacuum. In a separate step, a mixture of group II and group VI precursors is prepared and injected into the flask containing the dried coordinating solvent with vigorous stirring [67].

During the pyrolysis process, the precursors decompose and react to form QDs. The control of reaction parameters, such as temperature, precursor ratios, and reaction time, allows for the controlled synthesis of QDs with desired properties. The hot-solution decomposition process is known for its ability to produce high-quality QDs with well-defined sizes and compositions.

The hot-solution decomposition process offers control over QD synthesis by manipulating the reaction conditions. However, it requires high temperatures and can be technically demanding compared to wet chemical methods.

2.3.2 Vapor-Phase Method

Initially, nanostructured materials were created through the growth of highly strained materials using heteroepitaxial methods. These methods involve the vapor-phase deposition of atom-by-atom layers, leading to the self-assembly of QDs on a substrate without any specific patterning. The growth modes observed are the Frank–van der Merwe (FvdM) mode [68], Volmer–Weber (VW) mode [69], and Stranski–Krastonov (SK) mode [70], determined by factors such as interfacial/surface energies and lattice mismatch. The methods are heteroepitaxial growth, MBE [71], physical vapor deposition (PVD) [64], and chemical vapor deposition (CVD) [72].

2.3.2.1 Heteroepitaxial Growth

The first method discussed is heteroepitaxial growth. It involves the vapor-phase deposition of atom-by-atom layers, leading to the self-assembly of QDs on a substrate. The growth modes observed are the FvdM, VW, and SK modes, depending on factors such as interfacial/surface energies and lattice mismatch [68–70].

2.3.2.2 Molecular Beam Epitaxy

MBE is another method mentioned. It utilizes ultra-high vacuum conditions to deposit overlayers and grow semiconductor nanostructured materials on a heated substrate. MBE involves evaporation from a source to form a beam of atoms or molecules. It can use solid sources or a combination of solids and gases. III–V semiconductors and II–VI semiconductors are commonly used for QD self-assembly via MBE [71].

2.3.2.3 Physical Vapor Deposition

PVD is a method where layer growth occurs through the condensation of a solid from vapors produced by thermal evaporation or sputtering. Techniques such as electron beam heating, resistive or Joule heating, arc discharge, and pulsed laser ablation can be used for evaporation. The formation of QDs from deposited thin films is controlled by factors such as strain and surface energies [64].

2.3.2.4 Chemical Vapor Deposition

CVD is another method discussed. It involves introducing precursor molecules into a chamber at a specific pressure and temperature. The precursors diffuse to the heated substrate to form a film, and gas-phase byproducts are desorbed and removed from the chamber. QDs made of InGaAs and AlInAs have been synthesized using surface energy or strained-induced SK growth processes. However, inhomogeneous optoelectronic properties can occur due to fluctuations in QD sizes when using vapor-phase methods [73].

2.4 Applications of Quantum Dots

QDs have gained significant attention and found numerous applications in various fields. These nanoscale semiconductor particles, typically ranging from 1 to 10 nm in size, possess unique properties that make them highly desirable for a wide range of applications.

One prominent application of QDs is in optoelectronics and display technologies. Due to their size-dependent electronic and optical properties, QDs exhibit tunable emission spectra. Another significant application of QDs lies in the field of biological imaging and sensing. QDs possess exceptional fluorescence properties, including high brightness and photostability. These attributes make them excellent candidates for fluorescent labeling and tracking of biological molecules, cells, and tissues.

QDs also find applications in the field of energy, particularly in photovoltaics. By integrating QDs into solar cell designs, it is possible to enhance light absorption and improve the efficiency of solar energy conversion. QD-based solar cells have demonstrated the ability to harvest a broader range of the solar spectrum due to their tunable absorption properties. In the field of quantum computing and information processing, QDs hold promise as potential qubits, the fundamental units of quantum information. Qubits based on QDs offer long coherence times and controllable quantum states, which are essential for the development of robust and scalable quantum computing systems.

Furthermore, QDs have applications in areas such as chemical catalysis, environmental sensing, and even in the development of next-generation lighting technologies. In the continued section, the application of QDs in white light-emitting diode (LED), solar energy, quantum computer, and quantum dot cellular automata (QCA), which is the original subject of this book, will be studied.

2.4.1 White Light-Emitting Diode

White LEDs have revolutionized the lighting industry with their energy efficiency and versatility. QDs offer a compelling solution to further enhance the capabilities of white LEDs by enabling a greater range of colors and improved efficiency.

The size of the energy gap in QDs can be precisely controlled, allowing them to function as efficient light emitters. By adjusting the size of the QDs, it becomes possible to emit light at specific wavelengths across the visible spectrum. This tunability offers a significant advantage over conventional LEDs, which typically emit light at fixed wavelengths. With QDs, a broader range of colors can be achieved, enabling more vibrant and visually appealing lighting solutions [64].

One of the remarkable features of QDs is their ability to be combined with different dimensions. By using QDs of various sizes, it becomes possible to create highly efficient white light sources. When excited, these QDs emit light at different wavelengths corresponding to their respective sizes. By carefully selecting and combining QDs of different sizes, it is possible to cover a wide spectrum of colors. This technique, known as spectral mixing, allows for the creation of white light by combining three or more light rays with different wavelengths. This reverse process of combining colors mimics the natural phenomenon of splitting white light into its constituent colors, as seen in a rainbow.

In the context of QDs, all the different-sized QDs can be excited in the same way by using a light beam with a wavelength in the ultraviolet (UV) range [16]. Although QDs of various dimensions emit different wavelengths, they possess a common characteristic: they can all be stimulated by UV light. This characteristic simplifies the design and operation of white LEDs utilizing QDs, as a single UV light source can excite all the QDs simultaneously [67].

The use of QDs in white LEDs offers several advantages. First, it expands the achievable color gamut, providing a wider range of colors that can be accurately reproduced. This is particularly beneficial in applications such as displays and signage, where color accuracy and vibrancy are crucial. Second, QDs offer improved color rendering, allowing for a more accurate representation of colors compared to traditional LED technologies. Additionally, QD-based white LEDs exhibit enhanced energy efficiency, lowering power consumption and reducing environmental impact [16].

2.4.2 Quantum Dot in Solar Cell

The demand for clean and abundant energy has prompted the exploration of solar energy as a viable solution. Each day, the Sun provides a significant amount of energy to the Earth, approximately 9×10^{22} J, while human consumption amounts to only about 9×10^{18} J. Conventional silicon solar cells have an efficiency limit of around 31%. However, higher theoretical conversion efficiencies can be achieved by utilizing multijunction cells, which divide the solar spectrum into smaller segments using multiple subcell bandgaps. For example, configurations with three subcells can reach approximately 40% efficiency, while configurations with four or more subcells can exceed 58% efficiency [72].

The incorporation of QDs into the fabrication of solar cells offers several advantages. One advantage is the ability to tune the bandgaps of QDs based on their size, enabling the absorption of the entire spectrum of sunlight, ranging from UV to visible to infrared (IR) [72]. Additionally, QDs can mitigate hot carrier relaxation dynamics and enable multiple exciton generation (MEG), where a single photon can generate more than one exciton. Furthermore, QDs exhibit favorable heterojunction properties with hole conductors and demonstrate greater stability and resistance to oxygen, moisture, and UV radiation compared to polymers [72].

Hybrid solar cells, which combine both inorganic and organic materials, have been developed using various types of QDs, such as CdSe, TiO_2, ZnO, PbS, PbSe, $CuInS_2$, and $CuInSe_2$ [74]. These hybrid cells leverage the strengths of both materials to improve conversion efficiency. For instance, the incorporation of carbon nanotubes linked to CdTe QDs has shown enhanced charge transport and improved conversion efficiency. The integration of organic and inorganic components allows for better diffusion of excitons and overall higher efficiency [74].

Another approach to utilizing QDs in solar cells is through quantum dot-sensitized solar cells (QDSSCs). Traditionally, dyes have been used as sensitizers in energy-harvesting cells, but QDs offer advantages such as tunable bandgaps and favorable light-harvesting properties. By incorporating QDs into wide bandgap semiconductors like TiO_2, ZnO, NbO_2, or Ta_2O_5, efficient charge separation and electron transfer can be achieved. QDs with absorption ranging from UV to near-infrared (NIR) can be utilized, and surface passivation techniques can enhance their stability under light exposure. The sensitization of highly porous TiO_2 electrodes with QDs has demonstrated high photocurrent quantum yields and improved voltage outputs. Researchers have also explored the use of organic hole conductor layers and heterojunction-based solar cells to further optimize QDSSCs.

2.4.3 Quantum Computers

In fact, a quantum computer is a device that makes a physical phenomenon unique based on the laws of quantum mechanics to fundamentally recognize a new mode of information processing [75].

In the context of quantum computing, QDs can be used as qubits, the fundamental building blocks of quantum computers. Here are a few ways in which QDs are being explored for applications in quantum computers:

A. *Qubit Implementation:* QDs can be engineered to trap individual electrons or holes, which can serve as qubits. Information can be stored and processed by manipulating the charge or spin state of the electrons or holes confined in the QDs. QDs offer several advantages, such as scalability and compatibility with existing semiconductor fabrication techniques, making them a promising candidate for qubit implementation.
B. *Spin Qubits:* QDs can be used to create spin qubits, where the spin of an electron or hole trapped in the dot is used to encode and process quantum information. Spin qubits offer long coherence times and are less susceptible to certain types of noise compared to other qubit implementations. QDs allow for precise control of the spin state, making them a viable platform for spin qubits.
C. *Quantum Dot Arrays:* QDs can be arranged in arrays to create networks of qubits. This configuration allows for interactions between neighboring qubits, enabling the implementation of quantum gates and entanglement operations. QD arrays have the potential for scalability, as individual QDs can be controlled and addressed independently.
D. *Hybrid Systems:* QDs can be coupled with other types of qubits or quantum systems to create hybrid quantum systems. For example, coupling QDs with superconducting circuits or other solid-state qubits can leverage the advantages of both systems and enable new functionalities. Hybrid systems can enhance the coherence and control of qubits and facilitate communication between different types of qubits.
E. *Quantum Dot Sensors:* QDs can also be used as highly sensitive sensors in quantum computers. They can detect and measure physical quantities such as electric fields, magnetic fields, or charge fluctuations, which are essential for performing quantum operations and error correction.

2.4.4 Quantum Dot Cellular Automata

In QCA, QDs play a crucial role in implementing qubits, which are the fundamental units of information in quantum computing. QCA is a computational approach that utilizes quantum mechanics principles for performing computations.

QDs are nanoscale semiconductor structures that exhibit quantum confinement effects. They are tiny regions within larger semiconductor materials where electron movement is restricted in all three dimensions. This confinement leads to discrete energy levels within the QDs, similar to the energy levels found in atoms.

In the context of QCA, QDs are utilized as physical entities for encoding and manipulating qubits. Qubits in QCA are typically represented by the presence or absence of electrons within the QDs. The properties of the QDs, such as their size, shape, and composition, determine the energy levels and behavior of the qubits.

The interaction between neighboring QDs in a QCA device is crucial for information processing. The qubits encoded in the QDs can interact through phenomena like quantum tunneling or Coulombic repulsion. These interactions enable qubits to influence each other's states, facilitating quantum operations like logic gates and information storage.

Arranging QDs in a regular lattice or pattern makes it possible to create a computational grid where information can be processed and propagated. The state of a qubit in a QD can impact the state of neighboring qubits, forming the foundation for performing computations in QCA.

The advantage of employing QDs in QCA lies in their capacity to provide stable and controllable quantum states. QDs can maintain quantum properties, such as superposition and entanglement, even at low temperatures, making them well-suited for quantum computing applications.

3

Overview of QCA

3.1 Introduction to QCA Technology

The complementary metal–oxide–semiconductor (CMOS) technology has been at the forefront of the semiconductor industry for decades, but according to Moore's law, the number of transistors per unit area doubles almost every 2 years. As the number of transistors increases, new methods must be sought to reduce the dimensions of transistors and designed circuits in general. Reducing the dimensions of the transistors causes problems such as short-channel effects, increased leakage current, and increased gate resistance. These issues justify searching for an alternative to metal oxide semiconductor (MOS) transistor technology [76, 77]. Suggested technologies that can be used to replace CMOS technology include quantum dot cellular automata (QCA) [78–80], single electron transistors [81, 82], carbon nanotubes [83, 84], and molecular electronics [85].

Among the technologies mentioned, QCA has received the most attention. The reason is the possibility to design digital circuits with nanodimensions. Quantum dot cellular automation is one of the best solutions regarding the occupied area, latency, and energy consumption [86–90]. This technology was first introduced by Lent et al. in 1993 [78]. Notable advantages of this technology are high computing speed in the frequency range of THz, low power consumption of about 100 W/cm^2, and high circuit density, which reduces area consumption [86].

Research and design in the field of digital circuits using QCA indicate many advantages of this technology over CMOS technology. The operation of QCA is based on quantum effects and Coulomb interaction, and this technology requires no current flow. The first introduced QCA cell is called bQCA or binary QCA cell and is used in binary digital computing. However, today, with the advancement of multi-valued logic, multi-value QCA structures, such as ternary and quaternary QCA cells, are the subjects of interest in the literature.

In multi-valued logic, each bit can take more than one value. Accordingly, unlike binary logic, which includes a value of zero or one, multi-valued logic can take a value between zero and one. Consequently, multi-valued logic can process more information stored in multi-valued memories. Other advantages that can be mentioned for using multi-valued logic are as follows: a smaller number of inputs and outputs, lower costs, and numerical and algebraic analysis at higher speeds [91, 92].

Research studies on multi-valued logic have shown its advantages over binary logic, including easy presentation and simplified implementation of this logic, storing more information, higher speed, better performance, higher efficiency, and cost reduction [93].

Computing with Multi-Value Logic in Quantum Dot Cellular Automata, First Edition.
Reza Sabbaghi-Nadooshan, Reza Akbari-Hasanjani, Leila Dehbozorgi, Majid Haghparast, and Hamid Reza Akbari-Hasanjani.

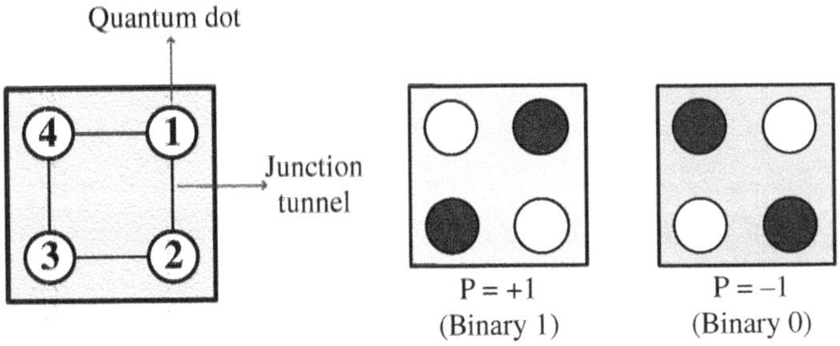

Figure 3.1 Polarizations of bQCA cells. Source: Das and De [100].

3.2 Binary QCA Concept

The structure of each bQCA quantum cell consists of four potential wells at the vertices of a square and two electrons capable of moving freely within the cell [86–96]. According to the motion of the electrons inside each cell, six different states are created, from which only two states are stable because, according to Coulomb's law, the two electrons are located at the farthest distance from each other (diagonally) [86, 90]. Two stable states of polarization are defined in the bQCA structure as P = +1 and P = −1 [87, 88, 94, 97–99]. In the equation of polarization of bQCA, the polarizations +1 and −1 are regarded as logic one and zero, respectively [86, 87]. Figure 3.1 shows the polarizations of bQCA cells [100], which can be calculated from quantum measurements of the particles.

In QCA technology, the electrons are allowed to move only inside the same cell, and no electron can tunnel from a cell to an adjacent cell. The Coulomb repulsion force transmits the state of electrons from one cell to another, so information flows from one cell to another. The state of cells adjacent to a cell is such that the Coulomb repulsion force is minimum [86].

3.3 Ternary QCA Cell Structure

The ternary quantum dot cellular automata (TQCA) cell consists of eight potential wells and two electrons [91, 101–109]. The two electrons in each cell can move by tunneling to create different polarization states. The polarization configuration of electrons depends on the electrostatic attraction and repulsion in each cell, so four stable states are obtained according to the number of electrons and the number of wells [93]. Figure 3.2 shows the TQCA cell structure [101].

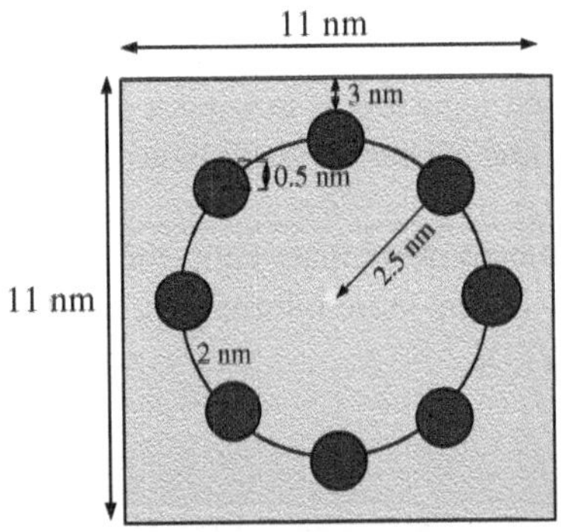

Figure 3.2 TQCA cell. Source: Mohaghegh et al. [101]/with permission of Elsevier.

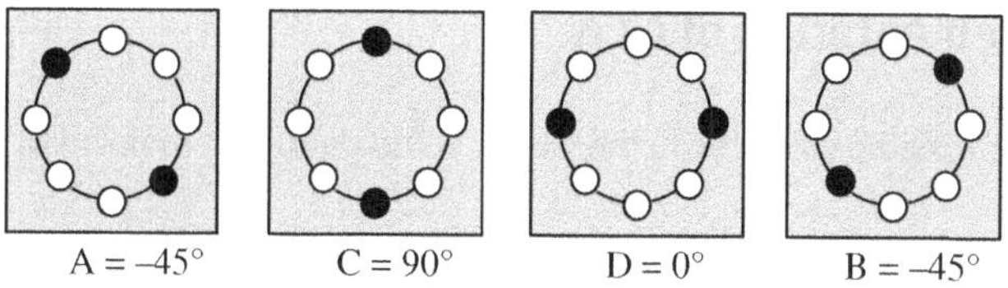

Figure 3.3 States of electron placement in a stable TQCA cell. Source: Pecar et al. [102]/with permission of The IOP Publishing.

In Refs. [101, 103], the polarization in TQCA is {−1, 0, 0, 1}, which can also be represented by {A, D, C, B}. Equivalent polarization values in ternary logic are as follows: A is equal to logic 0, C is equal to 1/2, D is equal to 1/2, and B is equal to logic 1 [102]. Figure 3.3 shows the electron position states in a TQCA cell.

3.4 Clock in QCA Technology

The clock in a QCA cell is like an electronic structure that controls the movement of electrons inside each cell [87, 110]. Another reason for using clocks in QCA cells is to control the transmission of information in a circuit. The mechanism of controlling information transfer requires that if there are several inputs in a circuit structure and some of these inputs reach the input of the circuit later, information is prevented from propagating in that area until other inputs arrive, ensuring synchronization in the circuit [87].

QCA-based cells are based on two characteristics: cellular automaton and quantum mechanics, since the information in the array is located inside the quantum cells and does not require any external power other than the clock. The clock in the QCA structure has four phases (1–4), which occur respectively [87, 110–114]. Figure 3.4 shows the clock in QCA technology.

Switch: In this phase, the potential barriers rise over time, and eventually, the electrons cannot freely move, and their movement is slow.
Hold: In this phase, the potential barrier inside each cell reaches its maximum, and the electrons are fixed entirely.
Release: In this phase, the potential barriers are lowered, and the electrons are slowly released.
Relax: In this phase, the potential barrier remains at its lowest level, and the electrons inside each cell are free.

Figure 3.4 Clock in QCA in four phases. Source: Adapted from Zahmatkesh et al. [87].

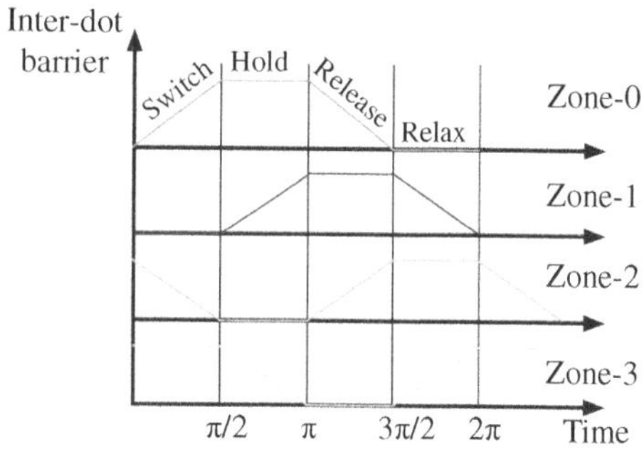

3.5 Manufacturing and Implementation of QCA

Implementation of QCA-based circuits is feasible using the metal island, semiconductor, molecular, and magnetic methods.

In the implementation of the metal island method, the first manufacturing technology for QCA circuits, quantum wells are made of aluminum islands, but the dimensions obtained in this method are in the range of micrometers; these dimensions are large because quantum properties appear at these scales. Moreover, as a disadvantage of this method, low operating temperatures are needed [115].

In semiconductor implementation, potential wells are semiconductors. Quantum cell fabrication in this method encounters some problems, such as slow lithography, high cost, and unsuitability for mass production [116].

The third implementation method is the molecular method, in which each QCA cell is made of a single molecule. The advantages of this method are incredibly high symmetry, high switching speed, and dimensionality, as well as the possibility of room-temperature implementation and mass production [117, 118].

The latest manufacturing technology uses magnetic structures in which the cells are made of nanomagnetic materials. This fabrication method is based on magnetic connections between nanoparticles. One of the advantages of this method is its efficiency at room temperature and low power loss [119].

3.6 Overview of Quantum Gates

QCA-based quantum gates include wire, majority gate (AND and OR gate designs), and inverter gate.

3.6.1 Quantum Wire

The adjacency of several binary cells in a row (vertically or horizontally) forms a binary wire [120–127]. When an input signal is applied to the first cell in a wire, the polarization of the cell changes. According to Coulomb repulsion, the input signal state is propagated by the cell electrons along the wire. Finally, the input signal state is transferred to the last cell as the output of the wire. In general, two structures have been introduced for bQCA wires. The first structure includes a 90° alignment of the cells, as shown in Figure 3.5.

The second wire structure includes the 45° alignment of the cells, as shown in Figure 3.6. In a 45° wire structure, polarizations +1 and −1 are placed next to each other alternatively. If the number of cells is odd, the input signal is directly transmitted to the output, but if the number of cells is even, the inverted input state is transferred to the output [110].

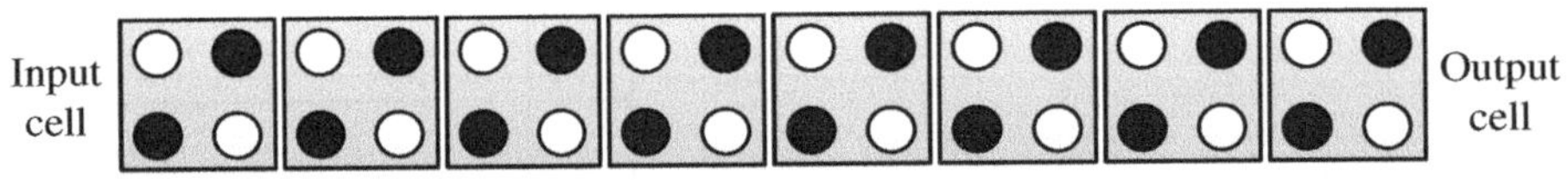

Figure 3.5 90° wire. Source: Adapted from Maharaj and Muthurathinam [125].

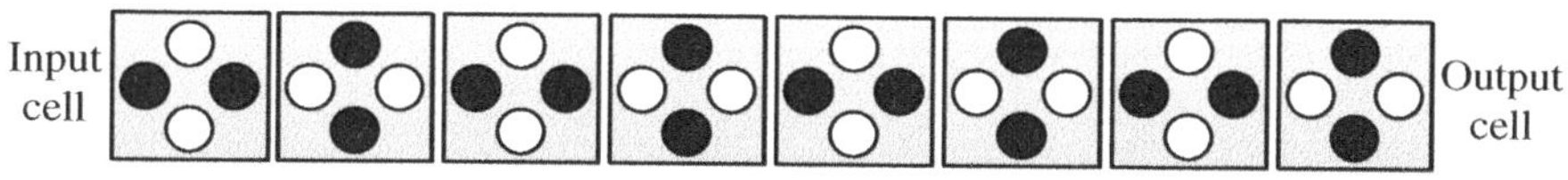

Figure 3.6 45° wire. Source: Adapted from Roohi et al. [110].

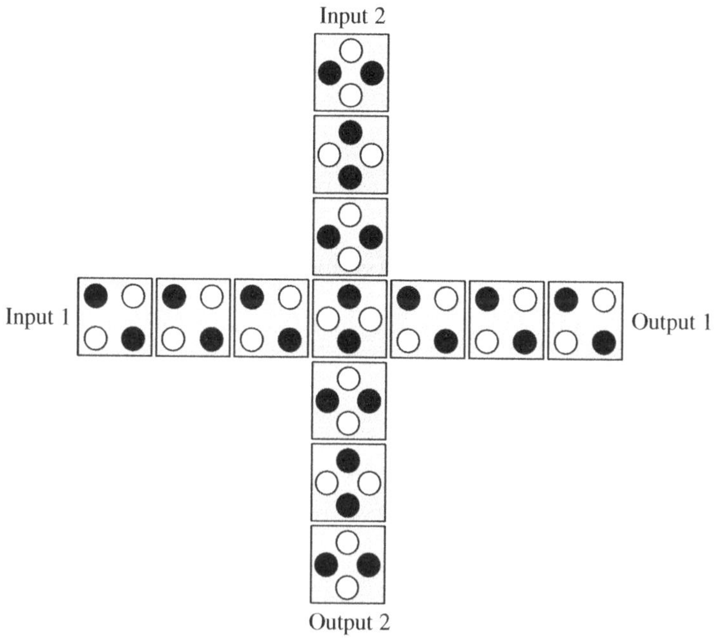

Figure 3.7 Overlapping or plate wiring structure. Source: References [110, 128].

In complex QCA-based circuits, it is sometimes necessary to pass wires over each other. This is one of the important issues in designing QCA-based circuits. Various methods have been proposed to pass wires over each other. These methods include overlapping, the use of multiple layers (multilayer), and utilizing clocks in the design of the passing wires [110, 128].

Figure 3.7 shows the overlapping or plate wiring structure. As shown in the figure, in this method, one wire with standard (90°) cells and another one with 45° cells can cross with 45° cells placed on the first wire. Figure 3.8 shows the multilayer wiring structure. This method uses several layers, which transfer the signal to another layer, then to the desired location, and back to the original state. In the third method, the clock is used to pass the wires over each other such that the wire that wants to pass over the other one must have a two-phase clock difference with the lower wire; thus, the signal of the two wires does not interfere [111]. In Figure 3.9, the passing wire with a two-phase clock difference is shown.

3.6.2 Binary NOT Gate

The binary NOT gate is one of the most widely used gates with two structures, which are shown in Figures 3.10 and 3.11, respectively [80, 90, 95, 126–128]. In the first structure, two bQCA cells are used diagonally. The second structure comprises two NOT gates presented in Figure 3.10. Table 3.1 shows the response of the NOT gate for different polarizations.

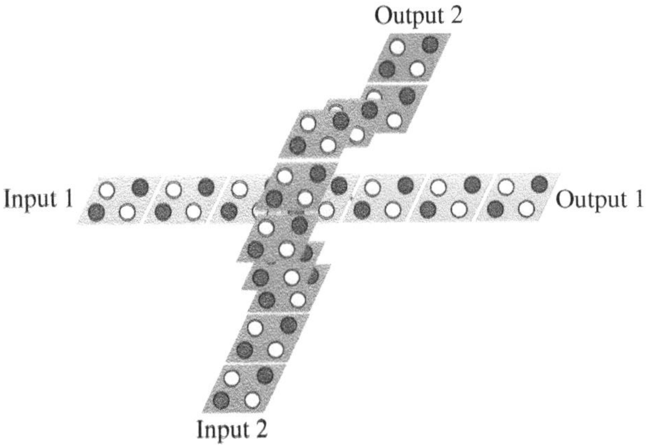

Figure 3.8 Multilayer wiring structure. Source: Adapted from Roohi et al. [110].

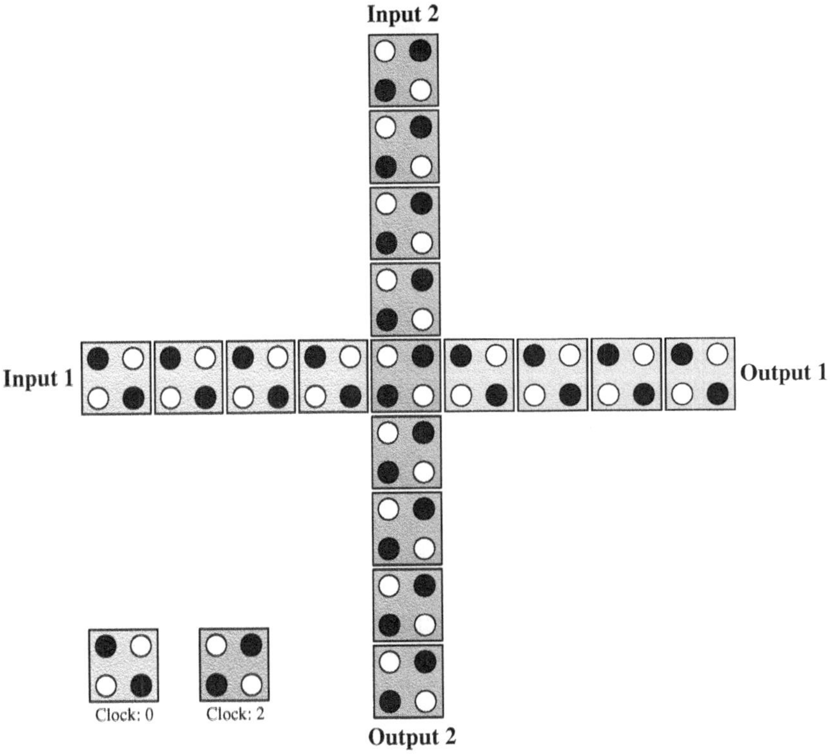

Figure 3.9 Transmission wire with two-phase clock difference. Source: Hashemi and Navi [111]/John Wiley & Sons/CC-BY 4.0.

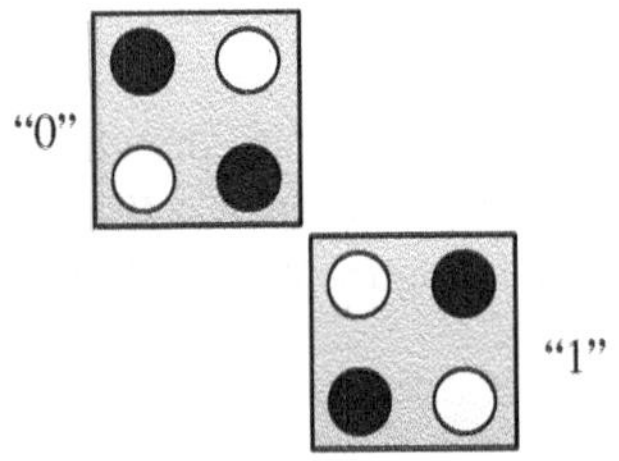

Figure 3.10 NOT gate using two bQCA cells diagonally. Source: Adapted from Refs. [87, 90, 95, 126–128].

Figure 3.11 NOT gate using two diagonal NOT gates. Source: Adapted from Refs. [80, 90, 95, 126–128].

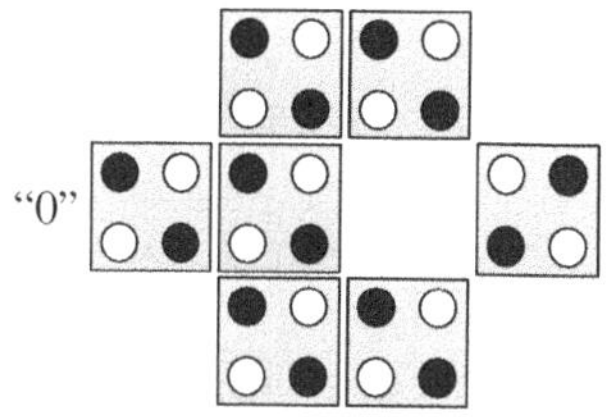

Table 3.1 Binary NOT gate response for different polarizations.

Input		Output	
Symbol	Polarization	Symbol	Polarization
0	−1	1	1
1	1	0	−1

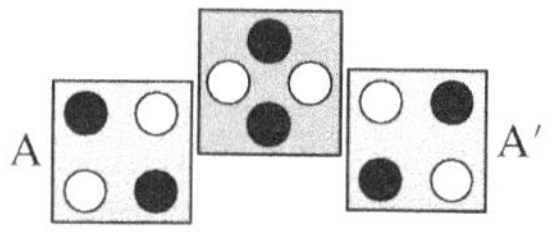

Figure 3.12 Inverter gate with reduced area. Source: Adapted from Zahmatkesh et al. [87].

In a new structure of the NOT gate, three bQCA cells are used as in Figure 3.12. The area occupation in this structure is less than the inverter gate with two cells [87].

3.6.3 Binary Majority Gate

Another important gate in QCA is the majority gate, which contains three or more inputs and only one output [95, 125–128]. The majority gates with three and five inputs are shown in Figures 3.13 and 3.14, respectively.

In the majority gate with three inputs, the AND and OR gates can be designed by considering one of the inputs to be fixed. If one of the inputs is in the steady state in polarization −1, the gate functions as an AND gate, as given in Eq. (3.1). Figure 3.15 and Table 3.2 present the AND gate circuit and its truth table based on bQCA cells [127], respectively.

$$Y = \min(X_1, X_2) = X_1 \text{ AND } X_2 \tag{3.1}$$

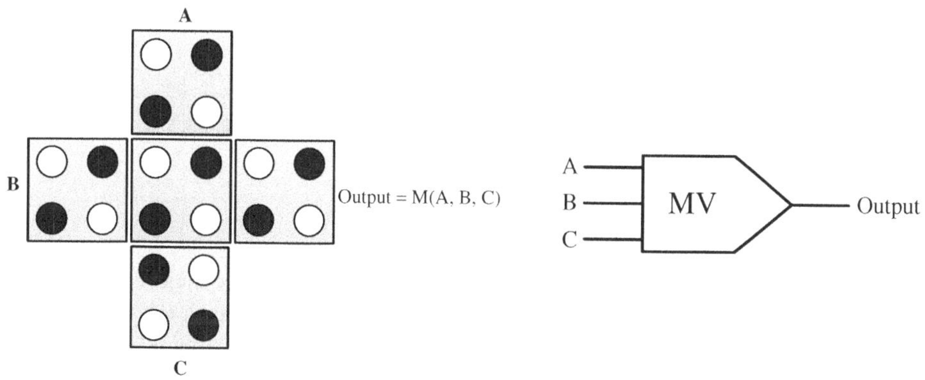

Figure 3.13 Majority gate with three inputs. Source: Adapted from Refs. [125–128].

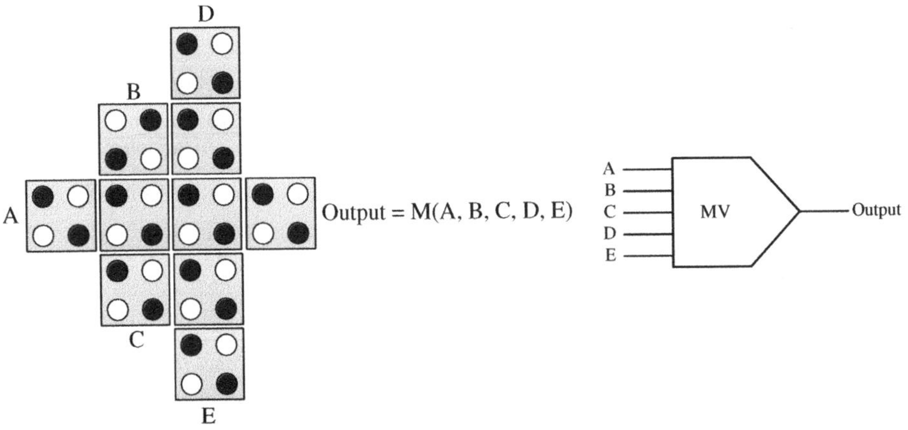

Figure 3.14 The majority gate with five inputs. Source: Adapted from Kianpour and Sabbaghi-Nadooshan [95].

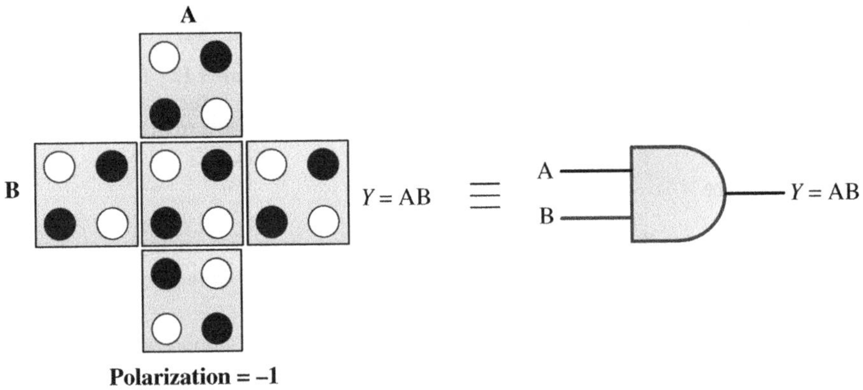

Figure 3.15 AND gate structure. Source: Adapted from Laajimi and Niu [127].

Table 3.2 Binary AND gate truth table in QCA technology.

	Input		
Fix polarization	**A**	**B**	**Output**
−1	−1	−1	−1
−1	−1	1	−1
−1	1	−1	−1
−1	1	1	1

Now, if one of the majority gate inputs with bQCA is in polarization +1, the majority gate functions as an OR gate. The OR gate structure is shown in Figure 3.16, and its truth table is given in Table 3.3.

$$Y = \max\left(X_1, X_2\right) = X_1 \text{ OR } X_2 \tag{3.2}$$

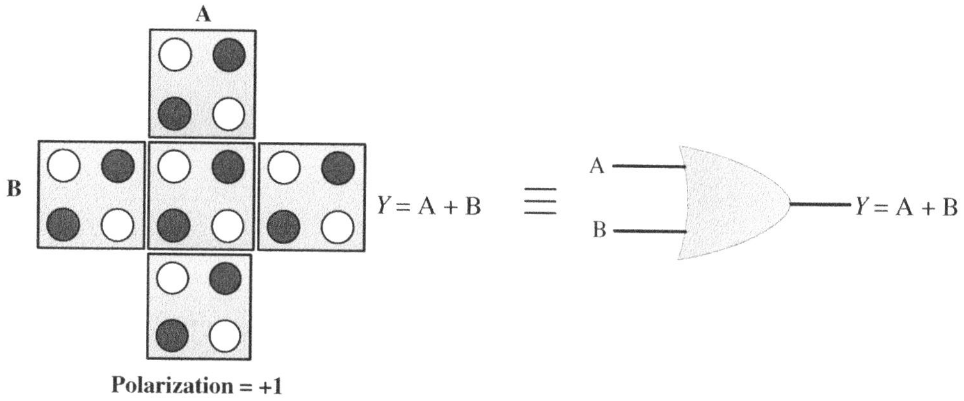

Figure 3.16 OR gate structure. Source: Adapted from Laajimi and Niu [127].

Table 3.3 Binary gate truth table in QCA technology.

Fix polarization	Input		Output
	A	B	
1	−1	−1	−1
1	−1	1	1
1	1	−1	1
1	1	1	1

4

Polarization and Polarization Calculations in QCA (Quantum Calculations in QCA)

4.1 Introduction

This chapter investigates the quantum effects of multi-value quantum dot cellular automata (QCA) cells, including the Hamiltonian matrix and polarization calculations. Polarization calculations determine the number of levels a structure can create, which is considered the logical level and used in the design of quantum gates. The number of particles per cell determines the levels per QCA cell. The particles in QCA are electrons with a spin of 1/2, which can be represented by multilevel structures [129–132].

The Hamiltonian matrix is another expression of the Schrödinger equation that can be used to calculate the amount of power consumed per cell. In this chapter, polarizations up to five levels will be evaluated.

4.2 Polarization and Calculation of Hamiltonian Matrix in QCA

As mentioned in the introduction, the number of levels is considered the number of particles. The particles in the QCA structure are electrons with spin 1/2 and quantum numbers −1/2 and +1/2. In calculating the Hamiltonian matrix in QCA, the Hartree–Fock method is used [129–134]. This approximation is used for fermion particles.

4.2.1 Polarization and Hamiltonian Matrix in bQCA

4.2.1.1 Polarization in bQCA

As mentioned, there are two electrons in binary quantum dot cellular automata (bQCA) with two quantum numbers and one spin, so based on the QCA structure, two acceptable polarizations are obtained. The two quantum numbers allow for designing a binary system with an electron in a system. However, since there is no force, the state does not change, so another electron is introduced into the cell to control the system state. The quantum matrix of the bQCA is given in Eq. (4.1) [129, 132].

$$S = \begin{bmatrix} \left\langle \frac{1}{2}, \frac{1}{2} \middle| S \middle| \frac{1}{2}, \frac{1}{2} \right\rangle & \left\langle \frac{1}{2}, \frac{1}{2} \middle| S \middle| \frac{1}{2}, -\frac{1}{2} \right\rangle \\ \left\langle \frac{1}{2}, -\frac{1}{2} \middle| S \middle| \frac{1}{2}, \frac{1}{2} \right\rangle & \left\langle \frac{1}{2}, -\frac{1}{2} \middle| S \middle| \frac{1}{2}, -\frac{1}{2} \right\rangle \end{bmatrix} \tag{4.1}$$

Computing with Multi-Value Logic in Quantum Dot Cellular Automata, First Edition.
Reza Sabbaghi-Nadooshan, Reza Akbari-Hasanjani, Leila Dehbozorgi, Majid Haghparast, and Hamid Reza Akbari-Hasanjani.

The operator S is the special function of the electron spin matrix in which each element can be obtained by Eq. (4.2) [129–132].

$$\vec{S^2} = \langle j, m' | j^2 | j, m \rangle = \hbar^2 j(j+1)\delta_{j',j}\delta_{m',m} \tag{4.2}$$

where $\delta_{j',j}$ and $\delta_{m',m}$ are the Kronecker delta with the following properties:

$$\delta_{ij} = \begin{cases} 1 & \text{if } i = j \\ 0 & \text{if } i \neq j \end{cases} \tag{4.3}$$

Now, after obtaining matrix elements $\vec{S^2}$ based on Eqs. (4.2) and (4.3), we can obtain a relation for S^2 as follows:

$$S^2 = \frac{3}{4}\hbar^2 \begin{bmatrix} 1 & 0 \\ 0 & 1 \end{bmatrix}$$

where $\hbar$ is the reduced Planck constant. After calculating S^2, we need to calculate the spin angular momentum in three directions, x, y, and z, to perform quantum measurements using the obtained values. The value of S_z can be obtained using Eq. (4.4) [129–132].

$$S_z = \langle j, m' | j_z | j, m \rangle = \hbar m \delta_{j',j}\delta_{m',m} \tag{4.4}$$

According to Eq. (4.4), the value of S_z is equal to:

$$S_z = \begin{bmatrix} \hbar/2 & 0 \\ 0 & -\hbar/2 \end{bmatrix} = \frac{\hbar}{2}\begin{bmatrix} 1 & 0 \\ 0 & -1 \end{bmatrix} = \frac{\hbar}{2}\delta_z$$

To calculate the momentum in the x and y directions, the raising operator (S_+) and lowering operator (S_-), which are defined respectively in Eqs. (4.5) and (4.6), are needed [129–154].

$$S_+ = \hbar\sqrt{j(j+1) - m(m+1)}\,\delta_{j',j}\delta_{m',m+1} \tag{4.5}$$

$$S_- = \hbar\sqrt{j(j+1) - m(m-1)}\,\delta_{j',j}\delta_{m',m-1} \tag{4.6}$$

Using Eqs. (4.1), (4.5), and (4.6), the values of S_+ and S_- are equal to:

$$S_+ = \hbar\begin{bmatrix} 0 & 1 \\ 0 & 0 \end{bmatrix} \quad S_- = \hbar\begin{bmatrix} 0 & 0 \\ 1 & 0 \end{bmatrix}$$

Now, using the S_+ and S_- operators, we can calculate the momentum in the x and y directions from Eqs. (4.7) and (4.8), respectively [129, 132].

$$S_x = \frac{1}{2}(S_+ + S_-) \tag{4.7}$$

$$S_y = \frac{i}{2}(S_- - S_+) \tag{4.8}$$

According to Eqs. (4.7) and (4.8), the momentum matrices in the x and y directions are as follows [132]:

$$S_y = \frac{\hbar}{2}\begin{bmatrix} 0 & -i \\ i & 0 \end{bmatrix} \quad S_x = \frac{\hbar}{2}\begin{bmatrix} 0 & 1 \\ 1 & 0 \end{bmatrix}$$

The matrices with the coefficient $\frac{\hbar}{2}$ in S_x, S_y, and S_z are called Pauli matrices and are represented by δ_x, δ_y, and δ_z as follows:

$$\delta_x = \begin{bmatrix} 0 & 1 \\ 1 & 0 \end{bmatrix} \quad \delta_y = \begin{bmatrix} 0 & -i \\ i & 0 \end{bmatrix} \quad \delta_z = \begin{bmatrix} 1 & 0 \\ 0 & -1 \end{bmatrix}$$

To calculate the polarization of the kth cell, a quantum measurement must be made in the z-direction as given in Eq. (4.9) [129–132].

$$P_k = -\langle \delta_{zk} \rangle \tag{4.9}$$

An input is needed to calculate the polarization of a cell. In bQCA, there are two input states, |0> and |1>, which represent $\begin{bmatrix}0\\1\end{bmatrix}$ and $\begin{bmatrix}1\\0\end{bmatrix}$, respectively. The polarization values are obtained as follows [129, 132]:

Polarization for input | 0>:

$$P_{|0\rangle} = -[1\ \ 0]\begin{bmatrix}1 & 0\\0 & -1\end{bmatrix}\begin{bmatrix}1\\0\end{bmatrix} = -1$$

Polarization for input | 1>:

$$P_{|1\rangle} = -\begin{bmatrix}0 & 1\end{bmatrix}\begin{bmatrix}1 & 0\\0 & -1\end{bmatrix}\begin{bmatrix}0\\1\end{bmatrix} = 1$$

According to calculations in QCA, polarization −1 is equivalent to logic 0, and polarization 1 is equivalent to logic 1. The analysis of a QCA cell with four potential wells indicates that the presence of the second electron stabilizes the cell [132].

4.2.1.2 Hamiltonian Matrix in Binary QCA

In calculating the Hamiltonian matrix in QCA, the Hartree–Fock approximation is used as given in Eq. (4.10) [129–132]. In this equation, γ_j is the tunneling potential, E_k is the amount of the kink energy, and P_j is the polarization, which is equal to −1 and +1 in bQCA [129–132].

$$\hat{H}_j = -\gamma_j\delta_x(j) - \frac{E_k}{2}\delta_z(j)P_j \tag{4.10}$$

Then, by inserting δ_x and δ_z into Eq. (4.10), the Hamiltonian matrix in bQCA is obtained as follows [129, 132]:

$$H(j) = \begin{bmatrix}-\frac{E_k}{2}P_j & -\gamma j\\ -\gamma j & \frac{E_k}{2}P_j\end{bmatrix}$$

4.2.2 Polarization and Hamiltonian Matrix in Ternary QCA

4.2.2.1 Polarization in Ternary QCA

In this section, the TQCA modification based on polarization calculations is evaluated to check the correctness of the modified model. The behavior of AND, OR, and NOT logic gates and wires must be consistent with the ternary system. The simulations performed in ternary quantum dot cellular automata (TQCA) are reviewed in the later chapters.

In the following sections, we will examine polarization in TQCA. In the performed calculations, we consider the spin states of each electron separately so that all states can be examined. Table 4.1 shows the spin states of the two electrons.

According to Table 4.1, each electron can take either the +1/2 or −1/2 state, so the state matrix, S, is a 4×4 matrix [129]. In Eq. (4.11), S is related to a structure with two electrons [129].

Table 4.1 Spin state of two electrons.

First electron	Second electron
1/2	1/2
1/2	−1/2
−1/2	1/2
−1/2	−1/2

$$S=\begin{bmatrix}\left\langle \frac{1}{2},\frac{1}{2};\frac{1}{2},\frac{1}{2}\right|S\left|\frac{1}{2},\frac{1}{2};\frac{1}{2},\frac{1}{2}\right\rangle & \left\langle \frac{1}{2},\frac{1}{2};\frac{1}{2},\frac{1}{2}\right|S\left|\frac{1}{2},\frac{1}{2};\frac{1}{2},-\frac{1}{2}\right\rangle & \left\langle \frac{1}{2},\frac{1}{2};\frac{1}{2},\frac{1}{2}\right|S\left|\frac{1}{2},-\frac{1}{2};\frac{1}{2},\frac{1}{2}\right\rangle & \left\langle \frac{1}{2},\frac{1}{2};\frac{1}{2},\frac{1}{2}\right|S\left|\frac{1}{2},-\frac{1}{2};\frac{1}{2},-\frac{1}{2}\right\rangle \\ \left\langle \frac{1}{2},\frac{1}{2};\frac{1}{2},-\frac{1}{2}\right|S\left|\frac{1}{2},\frac{1}{2};\frac{1}{2},\frac{1}{2}\right\rangle & \left\langle \frac{1}{2},\frac{1}{2};\frac{1}{2},-\frac{1}{2}\right|S\left|\frac{1}{2},\frac{1}{2};\frac{1}{2},-\frac{1}{2}\right\rangle & \left\langle \frac{1}{2},\frac{1}{2};\frac{1}{2},-\frac{1}{2}\right|S\left|\frac{1}{2},-\frac{1}{2};\frac{1}{2},\frac{1}{2}\right\rangle & \left\langle \frac{1}{2},\frac{1}{2};\frac{1}{2},-\frac{1}{2}\right|S\left|\frac{1}{2},-\frac{1}{2};\frac{1}{2},-\frac{1}{2}\right\rangle \\ \left\langle \frac{1}{2},-\frac{1}{2};\frac{1}{2},\frac{1}{2}\right|S\left|\frac{1}{2},\frac{1}{2};\frac{1}{2},\frac{1}{2}\right\rangle & \left\langle \frac{1}{2},-\frac{1}{2};\frac{1}{2},\frac{1}{2}\right|S\left|\frac{1}{2},\frac{1}{2};\frac{1}{2},-\frac{1}{2}\right\rangle & \left\langle \frac{1}{2},-\frac{1}{2};\frac{1}{2},\frac{1}{2}\right|S\left|\frac{1}{2},-\frac{1}{2};\frac{1}{2},\frac{1}{2}\right\rangle & \left\langle \frac{1}{2},-\frac{1}{2};\frac{1}{2},\frac{1}{2}\right|S\left|\frac{1}{2},-\frac{1}{2};\frac{1}{2},-\frac{1}{2}\right\rangle \\ \left\langle \frac{1}{2},-\frac{1}{2};\frac{1}{2},-\frac{1}{2}\right|S\left|\frac{1}{2},\frac{1}{2};\frac{1}{2},\frac{1}{2}\right\rangle & \left\langle \frac{1}{2},-\frac{1}{2};\frac{1}{2},-\frac{1}{2}\right|S\left|\frac{1}{2},\frac{1}{2};\frac{1}{2},-\frac{1}{2}\right\rangle & \left\langle \frac{1}{2},-\frac{1}{2};\frac{1}{2},-\frac{1}{2}\right|S\left|\frac{1}{2},-\frac{1}{2};\frac{1}{2},\frac{1}{2}\right\rangle & \left\langle \frac{1}{2},-\frac{1}{2};\frac{1}{2},-\frac{1}{2}\right|S\left|\frac{1}{2},-\frac{1}{2};\frac{1}{2},-\frac{1}{2}\right\rangle \end{bmatrix} \tag{4.11}$$

Equations (4.11) can be separated into matrices given in Eqs. (4.12) and (4.13) corresponding to the first and second electrons [129].

$$S_1 = \begin{bmatrix} \left\langle \frac{1}{2},\frac{1}{2}\middle| S \middle| \frac{1}{2},\frac{1}{2}\right\rangle & \left\langle \frac{1}{2},\frac{1}{2}\middle| S \middle| \frac{1}{2},\frac{1}{2}\right\rangle & \left\langle \frac{1}{2},\frac{1}{2}\middle| S \middle| \frac{1}{2},-\frac{1}{2}\right\rangle & \left\langle \frac{1}{2},\frac{1}{2}\middle| S \middle| \frac{1}{2},-\frac{1}{2}\right\rangle \\ \left\langle \frac{1}{2},\frac{1}{2}\middle| S \middle| \frac{1}{2},\frac{1}{2}\right\rangle & \left\langle \frac{1}{2},\frac{1}{2}\middle| S \middle| \frac{1}{2},\frac{1}{2}\right\rangle & \left\langle \frac{1}{2},\frac{1}{2}\middle| S \middle| \frac{1}{2},-\frac{1}{2}\right\rangle & \left\langle \frac{1}{2},\frac{1}{2};\middle| S \middle| \frac{1}{2},-\frac{1}{2}\right\rangle \\ \left\langle \frac{1}{2},-\frac{1}{2}\middle| S \middle| \frac{1}{2},\frac{1}{2}\right\rangle & \left\langle \frac{1}{2},-\frac{1}{2}\middle| S \middle| \frac{1}{2},\frac{1}{2}\right\rangle & \left\langle \frac{1}{2},-\frac{1}{2}\middle| S \middle| \frac{1}{2},-\frac{1}{2}\right\rangle & \left\langle \frac{1}{2},-\frac{1}{2}\middle| S \middle| \frac{1}{2},-\frac{1}{2}\right\rangle \\ \left\langle \frac{1}{2},-\frac{1}{2}\middle| S \middle| \frac{1}{2},\frac{1}{2}\right\rangle & \left\langle \frac{1}{2},-\frac{1}{2}\middle| S \middle| \frac{1}{2},\frac{1}{2}\right\rangle & \left\langle \frac{1}{2},-\frac{1}{2}\middle| S \middle| \frac{1}{2},-\frac{1}{2}\right\rangle & \left\langle \frac{1}{2},-\frac{1}{2}\middle| S \middle| \frac{1}{2},-\frac{1}{2}\right\rangle \end{bmatrix} \tag{4.12}$$

$$S_2 = \begin{bmatrix} \left\langle \frac{1}{2},\frac{1}{2}\middle| S \middle| \frac{1}{2},\frac{1}{2}\right\rangle & \left\langle \frac{1}{2},\frac{1}{2}\middle| S \middle| \frac{1}{2},-\frac{1}{2}\right\rangle & \left\langle \frac{1}{2},\frac{1}{2}\middle| S \middle| ;\frac{1}{2},\frac{1}{2}\right\rangle & \left\langle \frac{1}{2},\frac{1}{2}\middle| S \middle| \frac{1}{2},-\frac{1}{2}\right\rangle \\ \left\langle \frac{1}{2},-\frac{1}{2}\middle| S \middle| \frac{1}{2},\frac{1}{2}\right\rangle & \left\langle \frac{1}{2},-\frac{1}{2}\middle| S \middle| \frac{1}{2},-\frac{1}{2}\right\rangle & \left\langle \frac{1}{2},-\frac{1}{2}\middle| S \middle| \frac{1}{2},\frac{1}{2}\right\rangle & \left\langle \frac{1}{2},-\frac{1}{2}\middle| S \middle| \frac{1}{2},-\frac{1}{2}\right\rangle \\ \left\langle \frac{1}{2},\frac{1}{2}\middle| S \middle| \frac{1}{2},\frac{1}{2}\right\rangle & \left\langle \frac{1}{2},\frac{1}{2}\middle| S \middle| \frac{1}{2},-\frac{1}{2}\right\rangle & \left\langle \frac{1}{2},\frac{1}{2}\middle| S \middle| \frac{1}{2},\frac{1}{2}\right\rangle & \left\langle \frac{1}{2},\frac{1}{2}\middle| S \middle| \frac{1}{2},-\frac{1}{2}\right\rangle \\ \left\langle \frac{1}{2},-\frac{1}{2}\middle| S \middle| \frac{1}{2},\frac{1}{2}\right\rangle & \left\langle \frac{1}{2},-\frac{1}{2}\middle| S \middle| \frac{1}{2},-\frac{1}{2}\right\rangle & \left\langle \frac{1}{2},-\frac{1}{2}\middle| S \middle| \frac{1}{2},\frac{1}{2}\right\rangle & \left\langle \frac{1}{2},-\frac{1}{2}\middle| S \middle| \frac{1}{2},-\frac{1}{2}\right\rangle \end{bmatrix} \tag{4.13}$$

After calculating the matrix of each electron, the x, y, and z components of the spin angular momentum must be calculated for performing quantum measurements in TQCA. Using Eq. (4.4), S_z of the electrons can be written based on Eqs. (4.12) and (4.13) [129].

$$S_{1z} = \frac{\hbar}{2}\begin{bmatrix} 1 & 1 & 0 & 0 \\ 1 & 1 & 0 & 0 \\ 0 & 0 & -1 & -1 \\ 0 & 0 & -1 & -1 \end{bmatrix} \qquad S_{2z} = \frac{\hbar}{2}\begin{bmatrix} 1 & 0 & 1 & 0 \\ 0 & -1 & 0 & -1 \\ 1 & 0 & 1 & 0 \\ 0 & -1 & 0 & -1 \end{bmatrix}$$

For the x and y components of momentum of each electron, the raising (S_+) and lowering (S_-) operators can be obtained using Eqs. (4.5) and (4.6) [129]:

$$S_{1+} = \hbar\begin{bmatrix} 0 & 0 & 1 & 1 \\ 0 & 0 & 1 & 1 \\ 0 & 0 & 0 & 0 \\ 0 & 0 & 0 & 0 \end{bmatrix} \qquad S_{1-} = \hbar\begin{bmatrix} 0 & 0 & 0 & 0 \\ 0 & 0 & 0 & 0 \\ 1 & 1 & 0 & 0 \\ 1 & 1 & 0 & 0 \end{bmatrix}$$

$$S_{2+} = \hbar\begin{bmatrix} 0 & 1 & 0 & 1 \\ 0 & 0 & 0 & 0 \\ 0 & 1 & 0 & 1 \\ 0 & 0 & 0 & 0 \end{bmatrix} \qquad S_{2-} = \hbar\begin{bmatrix} 0 & 0 & 0 & 0 \\ 1 & 0 & 1 & 0 \\ 0 & 0 & 0 & 0 \\ 1 & 0 & 1 & 0 \end{bmatrix}$$

These matrices can be used to obtain S_x and S_y matrices for each particle using Eqs. (4.7) and (4.8) as follows [129]:

$$S_{1x} = \frac{\hbar}{2}\begin{bmatrix} 0 & 0 & 1 & 1 \\ 0 & 0 & 1 & 1 \\ 1 & 1 & 0 & 0 \\ 1 & 1 & 0 & 0 \end{bmatrix} \qquad S_{1y} = \frac{\hbar}{2}\begin{bmatrix} 0 & 0 & -i & -i \\ 0 & 0 & -i & -i \\ i & i & 0 & 0 \\ i & i & 0 & 0 \end{bmatrix}$$

$$S_{2x} = \frac{\hbar}{2}\begin{bmatrix} 0 & 1 & 0 & 1 \\ 1 & 0 & 1 & 0 \\ 0 & 1 & 0 & 1 \\ 1 & 0 & 1 & 0 \end{bmatrix} \qquad S_{2y} = \frac{\hbar}{2}\begin{bmatrix} 0 & -i & 0 & -i \\ i & 0 & i & 0 \\ 0 & -i & 0 & -i \\ i & 0 & i & 0 \end{bmatrix}$$

Next, the momenta of the whole system (cell) must be calculated using the linear combination of S_x, S_y, and S_z for each electron in the cell. Equations (4.14) to (4.16) use the linear combination (algebraic sum) of S_x, S_y, and S_z of each electron to obtain the momenta of the whole system [129].

$$S_x = S_{1x} + S_{2x} \tag{4.14}$$

$$S_y = S_{1y} + S_{2y} \tag{4.15}$$

$$S_z = S_{1z} + S_{2z} \tag{4.16}$$

According to Eqs. (4.14) to (4.16), the matrices S_x, S_y, and S_z can be formed as follows [129]:

$$S_x = \frac{\hbar}{2}\begin{bmatrix} 0 & 1 & 1 & 2 \\ 1 & 0 & 2 & 1 \\ 1 & 2 & 0 & 1 \\ 2 & 1 & 1 & 0 \end{bmatrix} \qquad S_y = \frac{\hbar}{2}\begin{bmatrix} 0 & -i & -i & -2i \\ i & 0 & 0 & -i \\ i & 0 & 0 & -i \\ 2i & i & i & 0 \end{bmatrix}$$

$$S_z = \frac{h}{2}\begin{bmatrix} 2 & 1 & 1 & 0 \\ 1 & 0 & 0 & -1 \\ 1 & 0 & 0 & -1 \\ 0 & -1 & -1 & -2 \end{bmatrix}$$

By eliminating $\hbar/2$ in S_x, S_y, and S_z, parameters δ_x, δ_y, and δ_z, which represent quasi-Pauli matrices (called so because these matrices do not have all properties of the Pauli matrices) are obtained [129].

In this section, polarizations in TQCA are represented by electrons as quantum particles. In the model proposed for TQCA in Refs. [101, 102, 109], four states are used. Two of these states are equal to zero (0(90°) and 0(0°)); one state is at the highest level, and the other state is at the lowest level [129].

After calculating quasi-Pauli matrices, quantum measurements are made on the matrices in the z-direction using Eq. (4.9). Then, the measurements are inserted into a system to obtain polarizations [129]. Cell states are expressed as follows:

$$\begin{bmatrix} 1 \\ 0 \\ 0 \\ 0 \end{bmatrix} = \text{Lowest level} \qquad \begin{bmatrix} 0 \\ 1 \\ 0 \\ 0 \end{bmatrix} = 0(0^\circ) \qquad \begin{bmatrix} 0 \\ 0 \\ 1 \\ 0 \end{bmatrix} = 0(90^\circ) \qquad \begin{bmatrix} 0 \\ 0 \\ 0 \\ 1 \end{bmatrix} = \text{Highest level}$$

Therefore, for each of the abovementioned cases, we find the cell polarizations as follows [129]:

$$P_{|00\rangle} = -2 \qquad P_{|01\rangle} = 0 \qquad P_{|10\rangle} = 0 \qquad P_{|11\rangle} = 2$$

According to calculations, the polarization set in TQCA with electrons as particles is {−2 0 0 2}, which can be represented by {A C D B}. In ternary logic, A or its corresponding polarization, −2, is equivalent to logic zero, B or its corresponding polarization, 2, is equal to logic 1, C or its corresponding polarization, 0, is equivalent to logic $\frac{1}{2}$, and D is introduced as an intracellular

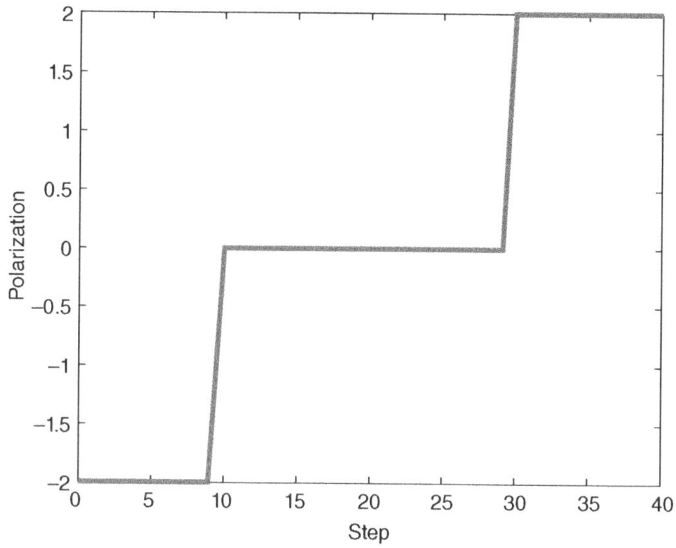

Figure 4.1 TQCA polarization simulation diagram. Source: Akbari-Hasanjani and Sabbaghi-Nadooshan [131]/with permission of John Wiley & Sons.

state, which is explained in Section 4.2.2.2. Figure 4.1 shows the polarization diagram of a TQCA [131].

According to Figure 4.1, the calculated polarization values are the same as the values simulated in MATLAB, so the accuracy of the calculations is confirmed. In this figure, the horizontal axis is the number of time steps required for one polarization.

4.2.2.2 Hamiltonian Matrix in Ternary QCA

To calculate the Hamiltonian matrix in TQCA, the Hartree–Fock approximation given in Eq. (4.10) is used, with the difference that in this equation, the polarization related to TQCA ({− 2, 0, 0, 2}) and the quasi-Pauli matrices calculated for TQCA are used [129]. The Hamiltonian matrix in TQCA looks like this:

$$\hat{H}(j) = \begin{bmatrix} -E_{\mathrm{k}}P_j & -\frac{E_{\mathrm{k}}}{2}P_j - \gamma_j & -\frac{E_{\mathrm{k}}}{2}P_j - \gamma_j & -2\gamma_j \\ -\frac{E_{\mathrm{k}}}{2}P_j - \gamma_j & 0 & -2\gamma_j & \frac{E_{\mathrm{k}}}{2}P_j - \gamma_j \\ -\frac{E_{\mathrm{k}}}{2}P_j - \gamma_j & -2\gamma_j & 0 & \frac{E_{\mathrm{k}}}{2}P_j - \gamma_j \\ -2\gamma_j & \frac{E_{\mathrm{k}}}{2}P_j - \gamma_j & \frac{E_{\mathrm{k}}}{2}P_j - \gamma_j & E_{\mathrm{k}}P_j \end{bmatrix}$$

4.2.3 Polarization and Hamiltonian Matrix in Quaternary QCA

4.2.3.1 Polarization in Quaternary QCA

This section will analyze quantum polarization in quaternary systems. A quaternary system as a three-particle system (with electrons) means three spins in the system [130]. A system with three spins has eight spin states, as shown in Table 4.2.

Table 4.2 Spin states of a three-particle system (with electrons).

First particle (electron I)	Second particle (electron II)	Third particle (electron III)
1/2	1/2	1/2
1/2	1/2	−1/2
1/2	−1/2	1/2
1/2	−1/2	−1/2
−1/2	1/2	1/2
−1/2	1/2	−1/2
−1/2	−1/2	1/2
−1/2	−1/2	−1/2

After determining the particle states in the system, the polarization values of the system must be calculated to find the value of the cellular model for the quaternary system [129–132].

As mentioned, a cell or quaternary system is considered a three-particle system. Here, the three particles are of electron type, with a spin value of 1/2 and a quantum number for each electron. The spin values of electrons are equal to +1/2 and −1/2 [130]. The spin state matrix for a three-particle system or cell (quaternary) is expressed as the S matrix in Eq. (4.17). Only a few elements of the S matrix are shown for the sake of brevity.

$$S = \begin{bmatrix} \left\langle \frac{1}{2},\frac{1}{2};\frac{1}{2},\frac{1}{2};\frac{1}{2},\frac{1}{2} \middle| S \middle| \frac{1}{2},\frac{1}{2};\frac{1}{2},\frac{1}{2};\frac{1}{2},\frac{1}{2} \right\rangle & \cdots\cdots & \left\langle \frac{1}{2},\frac{1}{2};\frac{1}{2},\frac{1}{2};\frac{1}{2},\frac{1}{2} \middle| S \middle| \frac{1}{2},-\frac{1}{2};\frac{1}{2},-\frac{1}{2};\frac{1}{2},-\frac{1}{2} \right\rangle \\ \left\langle \frac{1}{2},\frac{1}{2};\frac{1}{2},\frac{1}{2};\frac{1}{2},-\frac{1}{2} \middle| S \middle| \frac{1}{2},\frac{1}{2};\frac{1}{2},\frac{1}{2};\frac{1}{2},\frac{1}{2} \right\rangle & \cdots\cdots & \left\langle \frac{1}{2},\frac{1}{2};\frac{1}{2},\frac{1}{2};\frac{1}{2},-\frac{1}{2} \middle| S \middle| \frac{1}{2},-\frac{1}{2};\frac{1}{2},-\frac{1}{2};\frac{1}{2},-\frac{1}{2} \right\rangle \\ \vdots & & \vdots \\ & \ddots & \\ \vdots & \ddots & \vdots \\ \left\langle \frac{1}{2},-\frac{1}{2};\frac{1}{2},-\frac{1}{2};\frac{1}{2},\frac{1}{2} \middle| S \middle| \frac{1}{2},\frac{1}{2};\frac{1}{2},\frac{1}{2};\frac{1}{2},-\frac{1}{2} \right\rangle & & \left\langle \frac{1}{2},-\frac{1}{2};\frac{1}{2},-\frac{1}{2};\frac{1}{2},\frac{1}{2} \middle| S \middle| \frac{1}{2},-\frac{1}{2};\frac{1}{2},-\frac{1}{2};\frac{1}{2},-\frac{1}{2} \right\rangle \\ \left\langle \frac{1}{2},-\frac{1}{2};\frac{1}{2},-\frac{1}{2};\frac{1}{2},-\frac{1}{2} \middle| S \middle| \frac{1}{2},\frac{1}{2};\frac{1}{2},\frac{1}{2};\frac{1}{2},-\frac{1}{2} \right\rangle & \cdots\cdots & \left\langle \frac{1}{2},-\frac{1}{2};\frac{1}{2},-\frac{1}{2};\frac{1}{2},-\frac{1}{2} \middle| S \middle| \frac{1}{2},-\frac{1}{2};\frac{1}{2},-\frac{1}{2};\frac{1}{2},-\frac{1}{2} \right\rangle \end{bmatrix}_{8*8} \tag{4.17}$$

If these three electrons are expressed separately, S for the electrons are as in Eqs. (A.1) to (A.3), which express the spin states of the first to the third electrons, respectively. The dimensions of each spin state matrix are 8×8 because it describes a three-particle system. Moreover, each electron has two quantum numbers, which can be represented by an 8×8 matrix [130].

After calculating the matrix of each electron, the x, y, and z components of the angular momentum of each electron must be calculated to perform quantum measurements. The momentum in the direction of the z-axis, S_z, can be calculated from Eq. (4.4) [130]. The S_z matrices of the electrons in the quaternary system are expressed in Eqs. (A.1) to (A.3) as follows:

$$S_{1z} = \frac{\hbar}{2}\begin{bmatrix} 1 & 1 & 1 & 1 & 0 & 0 & 0 & 0 \\ 1 & 1 & 1 & 1 & 0 & 0 & 0 & 0 \\ 1 & 1 & 1 & 1 & 0 & 0 & 0 & 0 \\ 1 & 1 & 1 & 1 & 0 & 0 & 0 & 0 \\ 0 & 0 & 0 & 0 & -1 & -1 & -1 & -1 \\ 0 & 0 & 0 & 0 & -1 & -1 & -1 & -1 \\ 0 & 0 & 0 & 0 & -1 & -1 & -1 & -1 \\ 0 & 0 & 0 & 0 & -1 & -1 & -1 & -1 \end{bmatrix} \quad S_{2z} = \frac{\hbar}{2}\begin{bmatrix} 1 & 1 & 0 & 0 & 1 & 1 & 0 & 0 \\ 1 & 1 & 0 & 0 & 1 & 1 & 0 & 0 \\ 0 & 0 & -1 & -1 & 0 & 0 & -1 & -1 \\ 0 & 0 & -1 & -1 & 0 & 0 & -1 & -1 \\ 1 & 1 & 0 & 0 & 1 & 1 & 0 & 0 \\ 1 & 1 & 0 & 0 & 1 & 1 & 0 & 0 \\ 0 & 0 & -1 & -1 & 0 & 0 & -1 & -1 \\ 0 & 0 & -1 & -1 & 0 & 0 & -1 & -1 \end{bmatrix}$$

$$S_{3z} = \frac{\hbar}{2}\begin{bmatrix} 1 & 0 & 1 & 0 & 1 & 0 & 1 & 0 \\ 0 & -1 & 0 & -1 & 0 & -1 & 0 & -1 \\ 1 & 0 & 1 & 0 & 1 & 0 & 1 & 0 \\ 0 & -1 & 0 & -1 & 0 & -1 & 0 & -1 \\ 1 & 0 & 1 & 0 & 1 & 0 & 1 & 0 \\ 0 & -1 & 0 & -1 & 0 & -1 & 0 & -1 \\ 1 & 0 & 1 & 0 & 1 & 0 & 1 & 0 \\ 0 & -1 & 0 & -1 & 0 & -1 & 0 & -1 \end{bmatrix}$$

To calculate the momentum along the x and y axes, first, the raising (S_+) and lowering (S_-) operators are calculated for each electron. Equations (4.5) and (4.6) calculate S_+ and S_- for the quaternary system in the form of the following matrices.

$$S_{1+} = \hbar\begin{bmatrix} 0 & 0 & 0 & 0 & 1 & 1 & 1 & 1 \\ 0 & 0 & 0 & 0 & 1 & 1 & 1 & 1 \\ 0 & 0 & 0 & 0 & 1 & 1 & 1 & 1 \\ 0 & 0 & 0 & 0 & 1 & 1 & 1 & 1 \\ 0 & 0 & 0 & 0 & 0 & 0 & 0 & 0 \\ 0 & 0 & 0 & 0 & 0 & 0 & 0 & 0 \\ 0 & 0 & 0 & 0 & 0 & 0 & 0 & 0 \\ 0 & 0 & 0 & 0 & 0 & 0 & 0 & 0 \end{bmatrix} \quad S_{1-} = \hbar\begin{bmatrix} 0 & 0 & 0 & 0 & 0 & 0 & 0 & 0 \\ 0 & 0 & 0 & 0 & 0 & 0 & 0 & 0 \\ 0 & 0 & 0 & 0 & 0 & 0 & 0 & 0 \\ 0 & 0 & 0 & 0 & 0 & 0 & 0 & 0 \\ 1 & 1 & 1 & 1 & 0 & 0 & 0 & 0 \\ 1 & 1 & 1 & 1 & 0 & 0 & 0 & 0 \\ 1 & 1 & 1 & 1 & 0 & 0 & 0 & 0 \\ 1 & 1 & 1 & 1 & 0 & 0 & 0 & 0 \end{bmatrix}$$

$$S_{2+} = \hbar\begin{bmatrix} 0 & 0 & 1 & 1 & 0 & 0 & 1 & 1 \\ 0 & 0 & 1 & 1 & 0 & 0 & 1 & 1 \\ 0 & 0 & 0 & 0 & 0 & 0 & 0 & 0 \\ 0 & 0 & 0 & 0 & 0 & 0 & 0 & 0 \\ 0 & 0 & 1 & 1 & 0 & 0 & 1 & 1 \\ 0 & 0 & 1 & 1 & 0 & 0 & 1 & 1 \\ 0 & 0 & 0 & 0 & 0 & 0 & 0 & 0 \\ 0 & 0 & 0 & 0 & 0 & 0 & 0 & 0 \end{bmatrix} \quad S_{2-} = \hbar\begin{bmatrix} 0 & 0 & 0 & 0 & 0 & 0 & 0 & 0 \\ 0 & 0 & 0 & 0 & 0 & 0 & 0 & 0 \\ 1 & 1 & 0 & 0 & 1 & 1 & 0 & 0 \\ 1 & 1 & 0 & 0 & 1 & 1 & 0 & 0 \\ 0 & 0 & 0 & 0 & 0 & 0 & 0 & 0 \\ 0 & 0 & 0 & 0 & 0 & 0 & 0 & 0 \\ 1 & 1 & 0 & 0 & 1 & 1 & 0 & 0 \\ 1 & 1 & 0 & 0 & 1 & 1 & 0 & 0 \end{bmatrix}$$

$$S_{3+} = \hbar\begin{bmatrix} 0 & 1 & 0 & 1 & 0 & 1 & 0 & 1 \\ 0 & 0 & 0 & 0 & 0 & 0 & 0 & 0 \\ 0 & 1 & 0 & 1 & 0 & 1 & 0 & 1 \\ 0 & 0 & 0 & 0 & 0 & 0 & 0 & 0 \\ 0 & 1 & 0 & 1 & 0 & 1 & 0 & 1 \\ 0 & 0 & 0 & 0 & 0 & 0 & 0 & 0 \\ 0 & 1 & 0 & 1 & 0 & 1 & 0 & 1 \\ 0 & 0 & 0 & 0 & 0 & 0 & 0 & 0 \end{bmatrix} \quad S_{3-} = \hbar\begin{bmatrix} 0 & 0 & 0 & 0 & 0 & 0 & 0 & 0 \\ 1 & 0 & 1 & 0 & 1 & 0 & 1 & 0 \\ 0 & 0 & 0 & 0 & 0 & 0 & 0 & 0 \\ 1 & 0 & 1 & 0 & 1 & 0 & 1 & 0 \\ 0 & 0 & 0 & 0 & 0 & 0 & 0 & 0 \\ 1 & 0 & 1 & 0 & 1 & 0 & 1 & 0 \\ 0 & 0 & 0 & 0 & 0 & 0 & 0 & 0 \\ 1 & 0 & 1 & 0 & 1 & 0 & 1 & 0 \end{bmatrix}$$

By finding S_+ and S_- for each electron according to Eqs. (4.7) and (4.8), the momentum in the x and y directions can be determined. The value of the matrix is the momentum in the x-direction for each electron as follows:

$$S_{1x} = \frac{\hbar}{2}\begin{bmatrix} 0 & 0 & 0 & 0 & 1 & 1 & 1 & 1 \\ 0 & 0 & 0 & 0 & 1 & 1 & 1 & 1 \\ 0 & 0 & 0 & 0 & 1 & 1 & 1 & 1 \\ 0 & 0 & 0 & 0 & 1 & 1 & 1 & 1 \\ 1 & 1 & 1 & 1 & 0 & 0 & 0 & 0 \\ 1 & 1 & 1 & 1 & 0 & 0 & 0 & 0 \\ 1 & 1 & 1 & 1 & 0 & 0 & 0 & 0 \\ 1 & 1 & 1 & 1 & 0 & 0 & 0 & 0 \end{bmatrix} \quad S_{1y} = \frac{\hbar}{2}\begin{bmatrix} 0 & 0 & 0 & 0 & -i & -i & -i & -i \\ 0 & 0 & 0 & 0 & -i & -i & -i & -i \\ 0 & 0 & 0 & 0 & -i & -i & -i & -i \\ 0 & 0 & 0 & 0 & -i & -i & -i & -i \\ i & i & i & i & 0 & 0 & 0 & 0 \\ i & i & i & i & 0 & 0 & 0 & 0 \\ i & i & i & i & 0 & 0 & 0 & 0 \\ i & i & i & i & 0 & 0 & 0 & 0 \end{bmatrix}$$

$$S_{2x} = \frac{\hbar}{2}\begin{bmatrix} 0 & 0 & 1 & 1 & 0 & 0 & 1 & 1 \\ 0 & 0 & 1 & 1 & 0 & 0 & 1 & 1 \\ 1 & 1 & 0 & 0 & 1 & 1 & 0 & 0 \\ 1 & 1 & 0 & 0 & 1 & 1 & 0 & 0 \\ 0 & 0 & 1 & 1 & 0 & 0 & 1 & 1 \\ 0 & 0 & 1 & 1 & 0 & 0 & 1 & 1 \\ 1 & 1 & 0 & 0 & 1 & 1 & 0 & 0 \\ 1 & 1 & 0 & 0 & 1 & 1 & 0 & 0 \end{bmatrix} \quad S_{2y} = \frac{\hbar}{2}\begin{bmatrix} 0 & 0 & -i & -i & 0 & 0 & -i & -i \\ 0 & 0 & -i & -i & 0 & 0 & -i & -i \\ i & i & 0 & 0 & i & i & 0 & 0 \\ i & i & 0 & 0 & i & i & 0 & 0 \\ 0 & 0 & -i & -i & 0 & 0 & -i & -i \\ 0 & 0 & -i & -i & 0 & 0 & -i & -i \\ i & i & 0 & 0 & i & i & 0 & 0 \\ i & i & 0 & 0 & i & i & 0 & 0 \end{bmatrix}$$

$$S_{3x} = \frac{\hbar}{2}\begin{bmatrix} 0 & 1 & 0 & 1 & 0 & 1 & 0 & 1 \\ 1 & 0 & 1 & 0 & 1 & 0 & 1 & 0 \\ 0 & 1 & 0 & 1 & 0 & 1 & 0 & 1 \\ 1 & 0 & 1 & 0 & 1 & 0 & 1 & 0 \\ 0 & 1 & 0 & 1 & 0 & 1 & 0 & 1 \\ 1 & 0 & 1 & 0 & 1 & 0 & 1 & 0 \\ 0 & 1 & 0 & 1 & 0 & 1 & 0 & 1 \\ 1 & 0 & 1 & 0 & 1 & 0 & 1 & 0 \end{bmatrix} \quad S_{3y} = \frac{\hbar}{2}\begin{bmatrix} 0 & -i & 0 & -i & 0 & -i & 0 & -i \\ i & 0 & i & 0 & i & 0 & i & 0 \\ 0 & -i & 0 & -i & 0 & -i & 0 & -i \\ i & 0 & i & 0 & i & 0 & i & 0 \\ 0 & -i & 0 & -i & 0 & -i & 0 & -i \\ i & 0 & i & 0 & i & 0 & i & 0 \\ 0 & -i & 0 & -i & 0 & -i & 0 & -i \\ i & 0 & i & 0 & i & 0 & i & 0 \end{bmatrix}$$

After calculating the electron momentum matrices in the quaternary system, the total momentum can be calculated using the linear combination (algebraic sum) of S_x, S_y, and S_z of the electrons. Equations (4.18) to (4.20) express the momentum of the whole system [130].

$$S_x = S_{1x} + S_{2x} + S_{3x} \tag{4.18}$$

$$S_y = S_{1y} + S_{2y} + S_{3y} \tag{4.19}$$

$$S_z = S_{1z} + S_{2z} + S_{3z} \tag{4.20}$$

According to Eqs. (4.18) to (4.20), the matrices S_x, S_y, and S_z for a quaternary system or cell are in the following form [130]:

$$S_x = \frac{\hbar}{2}\begin{bmatrix} 0 & 1 & 1 & 2 & 1 & 2 & 2 & 3 \\ 1 & 0 & 2 & 1 & 2 & 1 & 3 & 2 \\ 1 & 2 & 0 & 1 & 2 & 3 & 1 & 2 \\ 2 & 1 & 1 & 0 & 3 & 2 & 2 & 1 \\ 1 & 2 & 2 & 3 & 0 & 1 & 1 & 2 \\ 2 & 1 & 3 & 2 & 1 & 0 & 2 & 1 \\ 2 & 3 & 1 & 2 & 1 & 2 & 0 & 1 \\ 3 & 2 & 2 & 1 & 2 & 1 & 1 & 0 \end{bmatrix} \quad S_y = \frac{\hbar}{2}\begin{bmatrix} 0 & -i & -i & -2i & -i & -2i & -2i & -3i \\ i & 0 & 0 & -i & 0 & -i & -i & -2i \\ i & 0 & 0 & -i & 0 & -i & -i & -2i \\ 2i & i & i & 0 & i & 0 & 0 & -i \\ i & 0 & 0 & -i & 0 & -i & -i & -2i \\ 2i & i & i & 0 & i & 0 & 0 & -i \\ 2i & i & i & 0 & i & 0 & 0 & -i \\ 3i & 2i & 2i & i & 2i & i & i & 0 \end{bmatrix}$$

$$S_z = \frac{\hbar}{2}\begin{bmatrix} 3 & 2 & 2 & 1 & 2 & 1 & 1 & 0 \\ 2 & 1 & 1 & 0 & 1 & 0 & 0 & -1 \\ 2 & 1 & 1 & 0 & 1 & 0 & 0 & -1 \\ 1 & 0 & 0 & -1 & 0 & -1 & -1 & -2 \\ 2 & 1 & 1 & 0 & 1 & 0 & 0 & -1 \\ 1 & 0 & 0 & -1 & 0 & -1 & -1 & -2 \\ 1 & 0 & 0 & -1 & 0 & -1 & -1 & -2 \\ 0 & -1 & -1 & -2 & -1 & -2 & -2 & -3 \end{bmatrix}$$

By removing $\frac{\hbar}{2}$ in these matrices, the three vectors δ_x, $\delta_{y,}$ and δ_z, which represent quasi-Pauli matrices in this book, are obtained. These matrices do not have all the properties of the Pauli matrices [130].

To calculate the polarization of the kth quantum cell in the z-axis direction, quantum measurements are made based on the δ_z matrix. Equation (4.9) is used for quantum measurement and calculating polarizations.

Before calculating the polarization, the input state for a quaternary system must be specified. Based on the dimensions of the δ_z matrix, the dimensions of the input vector for eight matrices are 8×1, as given in Table 4.3.

Using Eq. (4.9) and Table 4.3, the polarization of a quaternary (three-particle) cell is presented in Table 4.4.

According to Table 4.4, considering a three-particle system, the polarization set in QQCA is {3 −3, −1, −1, 1, −1, 1, 1, 3}. Several polarization values are duplicates for some inputs. The quaternary cell has four polarizations {−3, −1, 1, 3}, and a cell has to be designed to create these polarizations [130]. Figure 4.2 shows the polarization diagram in QQCA based on calculations.

Table 4.3 Vectors of quaternary input states.

Input state vector			
input state 1 $= \lvert 000\rangle = \begin{bmatrix}1\\0\\0\\0\\0\\0\\0\\0\end{bmatrix}$	input state 2 $= \lvert 001\rangle = \begin{bmatrix}0\\1\\0\\0\\0\\0\\0\\0\end{bmatrix}$	input state 3 $= \lvert 010\rangle = \begin{bmatrix}0\\0\\1\\0\\0\\0\\0\\0\end{bmatrix}$	input state 3 $= \lvert 011\rangle = \begin{bmatrix}0\\0\\0\\1\\0\\0\\0\\0\end{bmatrix}$
input state 4 $= \lvert 100\rangle = \begin{bmatrix}0\\0\\0\\0\\1\\0\\0\\0\end{bmatrix}$	input state 4 $= \lvert 101\rangle = \begin{bmatrix}0\\0\\0\\0\\0\\1\\0\\0\end{bmatrix}$	input state 4 $= \lvert 110\rangle = \begin{bmatrix}0\\0\\0\\0\\0\\0\\1\\0\end{bmatrix}$	input state 4 $= \lvert 111\rangle = \begin{bmatrix}0\\0\\0\\0\\0\\0\\0\\1\end{bmatrix}$

Table 4.4 Polarization of a three-particle system in QQCA.

Polarization in QQCA			
$P_{\|000\rangle} = -3$	$P_{\|001\rangle} = -1$	$P_{\|010\rangle} = -1$	$P_{\|011\rangle} = 1$
$P_{\|100\rangle} = -1$	$P_{\|101\rangle} = 1$	$P_{\|110\rangle} = 1$	$P_{\|111\rangle} = 3$

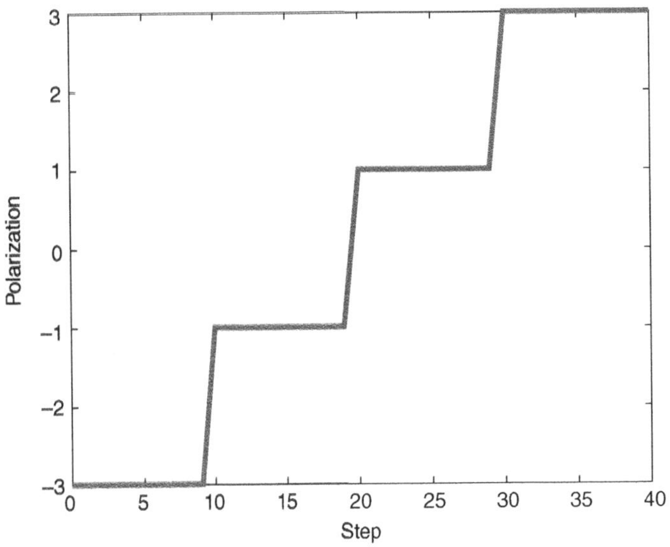

Figure 4.2 Polarization diagram in QQCA based on calculations.

As calculated, the polarization set is $\{-3, -1, 1, 3\}$, which can be considered equivalent to {A, B, C, D}. In the quaternary logic, A or polarization -3 is equal to level 0, B or polarization -1 is equivalent to level 1/3, C or polarization 1 is equivalent to level 2/3, and D or polarization 3 is equivalent to level 1. For creating polarizations -1 and 1, which are duplicates of several input states, only one state that produces these polarizations is considered in the calculations [130]. This study considers polarizations $-3, -1, 1$, and 3 as input states [130].

4.2.3.2 Hamiltonian Matrix in Quaternary QCA

In the following, the Hamiltonian matrix is presented as a quaternary system. To calculate this matrix, the Hartree–Fock approximation in Eq. (4.10) is used [129–132].

In Eq. (4.10), γ_j is the amount of tunneling energy, δ_x is the quaternary quasi-Pauli matrix in the x-direction, δ_z is the quaternary quasi-Pauli matrix in the z-direction, E_k is the kink energy, and the parameter P_j is the value of polarization in QQCA. By inserting any of the parameters expressed in the Hamiltonian approximation, the Hamiltonian matrix in QQCA can be expressed as follows [130]:

$$\hat{H}_j = \begin{bmatrix} -\frac{3}{2}E_kP_j & -E_kP_j - \gamma_j & -E_kP_j - \gamma_j & -\frac{1}{2}E_kP_j - \gamma_j & -E_kP_j - \gamma_j & -\frac{1}{2}E_kP_j - 2\gamma_j & -\frac{1}{2}E_kP_j - 2\gamma_j & -3\gamma_j \\ -E_kP_j - \gamma_j & -\frac{1}{2}E_kP_j & -\frac{1}{2}E_kP_j - 2\gamma_j & -\gamma_j & -\frac{1}{2}E_kP_j - 2\gamma_j & -\gamma_j & -3\gamma_j & \frac{1}{2}E_kP_j - 2\gamma_j \\ -E_kP_j - \gamma_j & -\frac{1}{2}E_kP_j - 2\gamma_j & -\frac{1}{2}E_kP_j & -\gamma_j & -\frac{1}{2}E_kP_j - 2\gamma_j & -3\gamma_j & -\gamma_j & \frac{1}{2}E_kP_j - 2\gamma_j \\ -\frac{1}{2}E_kP_j - 2\gamma_j & -\gamma_j & -\gamma_j & \frac{1}{2}E_kP_j & -3\gamma_i & \frac{1}{2}E_kP_j - 2\gamma_j & \frac{1}{2}E_kP_j - 2\gamma_j & E_kP_j - \gamma_j \\ -E_kP_j - \gamma_j & -\frac{1}{2}E_kP_j - 2\gamma_j & -\frac{1}{2}E_kP_j - 2\gamma_j & -3\gamma_i & -\frac{1}{2}E_kP_j & -\gamma_i & -\gamma_i & \frac{1}{2}E_kP_j - 2\gamma_j \\ -\frac{1}{2}E_kP_j - 2\gamma_j & -\gamma_i & -3\gamma_i & \frac{1}{2}E_kP_j - 2\gamma_j & -\gamma_i & \frac{1}{2}E_kP_j & \frac{1}{2}E_kP_j - 2\gamma_j & E_kP_j - \gamma_j \\ -\frac{1}{2}E_kP_j - 2\gamma_j & -3\gamma_j & -\gamma_j & \frac{1}{2}E_kP_j - 2\gamma_j & -\gamma_j & \frac{1}{2}E_kP_j - 2\gamma_j & \frac{1}{2}E_kP_j & E_kP_j - \gamma_j \\ -3\gamma_i & \frac{1}{2}E_kP_j - 2\gamma_j & \frac{1}{2}E_kP_j - 2\gamma_j & E_kP_j - \gamma_j & \frac{1}{2}E_kP_j - 2\gamma_j & E_kP_j - \gamma_j & E_kP_j - \gamma_j & \frac{3}{2}E_kP_i \end{bmatrix}$$

In the next section, we will present a new cellular model for QQCA that is fully stable based on the standard structure of bQCA gates for different gate input states.

4.2.4 Polarization and Hamiltonian Matrix in Quinary QCA

4.2.4.1 Polarization in Quinary QCA

In this section, the polarization in quinary logic is examined. If the quinary logic is considered for a four-particle system, the system has four spins and eight quantum numbers, and the quantum number of the four particles can create different states [131] as given in Table 4.5.

After determining the quantum states in a four-particle system, the polarizations of the quinary system can be determined. Each polarization can be considered at a logical level.

A quinary quantum dot cellular automata (QuQCA) system is a four-particle system in which the four particles are electrons. The spin value of each electron is ½, and electron quantum numbers are +1/2 and −1/2. The spin state matrix for a system with four electrons is in the form of the S matrix in Eq. (4.21). Only some states of S are given for the sake of brevity.

Table 4.5 Spin states of a four-electron system.

First particle	Second particle	Third particle	Fourth particle
1/2	1/2	1/2	1/2
1/2	1/2	1/2	−1/2
1/2	1/2	−1/2	1/2
1/2	1/2	−1/2	−1/2
1/2	1/2	1/2	1/2
1/2	−1/2	1/2	−1/2
1/2	−1/2	−1/2	1/2
1/2	−1/2	−1/2	−1/2
−1/2	1/2	1/2	1/2
−1/2	1/2	1/2	−1/2
−1/2	1/2	−1/2	1/2
−1/2	1/2	−1/2	−1/2
−1/2	−1/2	1/2	1/2
−1/2	−1/2	1/2	−1/2
−1/2	−1/2	−1/2	1/2
−1/2	−1/2	−1/2	−1/2

$$S=\begin{bmatrix}
\left\langle \frac{1}{2},\frac{1}{2};\frac{1}{2},\frac{1}{2};\frac{1}{2},\frac{1}{2};\frac{1}{2},\frac{1}{2}\middle| S \middle|\frac{1}{2},\frac{1}{2};\frac{1}{2},\frac{1}{2};\frac{1}{2},\frac{1}{2};\frac{1}{2},\frac{1}{2}\right\rangle & \cdots & \cdots & \left\langle \frac{1}{2},\frac{1}{2};\frac{1}{2},\frac{1}{2};\frac{1}{2},\frac{1}{2};\frac{1}{2},\frac{1}{2}\middle| S \middle|\frac{1}{2},-\frac{1}{2};\frac{1}{2},-\frac{1}{2};\frac{1}{2},-\frac{1}{2};\frac{1}{2},-\frac{1}{2}\right\rangle \\
\left\langle \frac{1}{2},\frac{1}{2};\frac{1}{2},\frac{1}{2};\frac{1}{2},\frac{1}{2};\frac{1}{2},-\frac{1}{2}\middle| S \middle|\frac{1}{2},\frac{1}{2};\frac{1}{2},\frac{1}{2};\frac{1}{2},\frac{1}{2};\frac{1}{2},\frac{1}{2}\right\rangle & \cdots & \cdots & \left\langle \frac{1}{2},\frac{1}{2};\frac{1}{2},\frac{1}{2};\frac{1}{2},\frac{1}{2};\frac{1}{2},-\frac{1}{2}\middle| S \middle|\frac{1}{2},-\frac{1}{2};\frac{1}{2},-\frac{1}{2};\frac{1}{2},-\frac{1}{2};\frac{1}{2},-\frac{1}{2}\right\rangle \\
 & & \iddots & \\
\vdots & \ddots & & \vdots \\
 & \iddots & & \\
 & \ddots & & \\
 & \iddots & & \\
\vdots & & & \vdots \\
 & & \ddots & \\
 & \iddots & & \\
\left\langle \frac{1}{2},-\frac{1}{2};\frac{1}{2},-\frac{1}{2};\frac{1}{2},-\frac{1}{2};\frac{1}{2},\frac{1}{2}\middle| S \middle|\frac{1}{2},\frac{1}{2};\frac{1}{2},\frac{1}{2};\frac{1}{2},\frac{1}{2};\frac{1}{2},\frac{1}{2}\right\rangle & \cdots & \cdots & \left\langle \frac{1}{2},-\frac{1}{2};\frac{1}{2},-\frac{1}{2};\frac{1}{2},-\frac{1}{2};\frac{1}{2},\frac{1}{2}\middle| S \middle|\frac{1}{2},-\frac{1}{2};\frac{1}{2},-\frac{1}{2};\frac{1}{2},-\frac{1}{2};\frac{1}{2},-\frac{1}{2}\right\rangle \\
\left\langle \frac{1}{2},-\frac{1}{2};\frac{1}{2},-\frac{1}{2};\frac{1}{2},-\frac{1}{2};\frac{1}{2},-\frac{1}{2}\middle| S \middle|\frac{1}{2},\frac{1}{2};\frac{1}{2},\frac{1}{2};\frac{1}{2},\frac{1}{2};\frac{1}{2},\frac{1}{2}\right\rangle & \cdots & \cdots & \left\langle \frac{1}{2},-\frac{1}{2};\frac{1}{2},-\frac{1}{2};\frac{1}{2},-\frac{1}{2};\frac{1}{2},-\frac{1}{2}\middle| S \middle|\frac{1}{2},-\frac{1}{2};\frac{1}{2},-\frac{1}{2};\frac{1}{2},-\frac{1}{2};\frac{1}{2},-\frac{1}{2}\right\rangle
\end{bmatrix}_{16\times 16} \tag{4.21}$$

Equation (4.21) expresses the spin and quantum states of a four-electron system. The states of these four electrons are separated, as shown in Eqs. (A.4) to (A.7). Equation (A.4) expresses the states of the first electron, Eq. (A.5) shows the states of the second electron, Eq. (A.6) expresses the states of the third electron, and Eq. (A.7) gives the states of the fourth electron. Each of the spin state matrices is 16×16 because each electron has two quantum numbers, and four electrons are used in the structure of the QuQCA system [131].

By knowing the state matrix of each electron, the momenta of each particle in the x, y, and z directions can be calculated to allow quantum measurements [131]. The momentum in the z-direction is calculated using Eq. (4.4). The S_z matrix for each electron in the QuQCA system is as follows:

$$S_{1z} = \frac{\hbar}{2}\begin{bmatrix} 1 & 1 & 1 & 1 & 1 & 1 & 1 & 1 & 0 & 0 & 0 & 0 & 0 & 0 & 0 & 0 \\ 1 & 1 & 1 & 1 & 1 & 1 & 1 & 1 & 0 & 0 & 0 & 0 & 0 & 0 & 0 & 0 \\ 1 & 1 & 1 & 1 & 1 & 1 & 1 & 1 & 0 & 0 & 0 & 0 & 0 & 0 & 0 & 0 \\ 1 & 1 & 1 & 1 & 1 & 1 & 1 & 1 & 0 & 0 & 0 & 0 & 0 & 0 & 0 & 0 \\ 1 & 1 & 1 & 1 & 1 & 1 & 1 & 1 & 0 & 0 & 0 & 0 & 0 & 0 & 0 & 0 \\ 1 & 1 & 1 & 1 & 1 & 1 & 1 & 1 & 0 & 0 & 0 & 0 & 0 & 0 & 0 & 0 \\ 1 & 1 & 1 & 1 & 1 & 1 & 1 & 1 & 0 & 0 & 0 & 0 & 0 & 0 & 0 & 0 \\ 1 & 1 & 1 & 1 & 1 & 1 & 1 & 1 & 0 & 0 & 0 & 0 & 0 & 0 & 0 & 0 \\ 0 & 0 & 0 & 0 & 0 & 0 & 0 & 0 & -1 & -1 & -1 & -1 & -1 & -1 & -1 & -1 \\ 0 & 0 & 0 & 0 & 0 & 0 & 0 & 0 & -1 & -1 & -1 & -1 & -1 & -1 & -1 & -1 \\ 0 & 0 & 0 & 0 & 0 & 0 & 0 & 0 & -1 & -1 & -1 & -1 & -1 & -1 & -1 & -1 \\ 0 & 0 & 0 & 0 & 0 & 0 & 0 & 0 & -1 & -1 & -1 & -1 & -1 & -1 & -1 & -1 \\ 0 & 0 & 0 & 0 & 0 & 0 & 0 & 0 & -1 & -1 & -1 & -1 & -1 & -1 & -1 & -1 \\ 0 & 0 & 0 & 0 & 0 & 0 & 0 & 0 & -1 & -1 & -1 & -1 & -1 & -1 & -1 & -1 \\ 0 & 0 & 0 & 0 & 0 & 0 & 0 & 0 & -1 & -1 & -1 & -1 & -1 & -1 & -1 & -1 \\ 0 & 0 & 0 & 0 & 0 & 0 & 0 & 0 & -1 & -1 & -1 & -1 & -1 & -1 & -1 & -1 \end{bmatrix}_{16\times16}$$

$$S_{2z} = \frac{\hbar}{2}\begin{bmatrix} 1 & 1 & 1 & 1 & 0 & 0 & 0 & 0 & 1 & 1 & 1 & 1 & 0 & 0 & 0 & 0 \\ 1 & 1 & 1 & 1 & 0 & 0 & 0 & 0 & 1 & 1 & 1 & 1 & 0 & 0 & 0 & 0 \\ 1 & 1 & 1 & 1 & 0 & 0 & 0 & 0 & 1 & 1 & 1 & 1 & 0 & 0 & 0 & 0 \\ 1 & 1 & 1 & 1 & 0 & 0 & 0 & 0 & 1 & 1 & 1 & 1 & 0 & 0 & 0 & 0 \\ 0 & 0 & 0 & 0 & -1 & -1 & -1 & -1 & 0 & 0 & 0 & 0 & -1 & -1 & -1 & -1 \\ 0 & 0 & 0 & 0 & -1 & -1 & -1 & -1 & 0 & 0 & 0 & 0 & -1 & -1 & -1 & -1 \\ 0 & 0 & 0 & 0 & -1 & -1 & -1 & -1 & 0 & 0 & 0 & 0 & -1 & -1 & -1 & -1 \\ 0 & 0 & 0 & 0 & -1 & -1 & -1 & -1 & 0 & 0 & 0 & 0 & -1 & -1 & -1 & -1 \\ 1 & 1 & 1 & 1 & 0 & 0 & 0 & 0 & 1 & 1 & 1 & 1 & 0 & 0 & 0 & 0 \\ 1 & 1 & 1 & 1 & 0 & 0 & 0 & 0 & 1 & 1 & 1 & 1 & 0 & 0 & 0 & 0 \\ 1 & 1 & 1 & 1 & 0 & 0 & 0 & 0 & 1 & 1 & 1 & 1 & 0 & 0 & 0 & 0 \\ 1 & 1 & 1 & 1 & 0 & 0 & 0 & 0 & 1 & 1 & 1 & 1 & 0 & 0 & 0 & 0 \\ 0 & 0 & 0 & 0 & -1 & -1 & -1 & -1 & 0 & 0 & 0 & 0 & -1 & -1 & -1 & -1 \\ 0 & 0 & 0 & 0 & -1 & -1 & -1 & -1 & 0 & 0 & 0 & 0 & -1 & -1 & -1 & -1 \\ 0 & 0 & 0 & 0 & -1 & -1 & -1 & -1 & 0 & 0 & 0 & 0 & -1 & -1 & -1 & -1 \\ 0 & 0 & 0 & 0 & -1 & -1 & -1 & -1 & 0 & 0 & 0 & 0 & -1 & -1 & -1 & -1 \end{bmatrix}_{16\times16}$$

$$S_{3z} = \frac{\hbar}{2}\begin{bmatrix} 1 & 1 & 0 & 0 & 1 & 1 & 0 & 0 & 1 & 1 & 0 & 0 & 1 & 1 & 0 & 0 \\ 1 & 1 & 0 & 0 & 1 & 1 & 0 & 0 & 1 & 1 & 0 & 0 & 1 & 1 & 0 & 0 \\ 0 & 0 & -1 & -1 & 0 & 0 & -1 & -1 & 0 & 0 & -1 & -1 & 0 & 0 & -1 & -1 \\ 0 & 0 & -1 & -1 & 0 & 0 & -1 & -1 & 0 & 0 & -1 & -1 & 0 & 0 & -1 & -1 \\ 1 & 1 & 0 & 0 & 1 & 1 & 0 & 0 & 1 & 1 & 0 & 0 & 1 & 1 & 0 & 0 \\ 1 & 1 & 0 & 0 & 1 & 1 & 0 & 0 & 1 & 1 & 0 & 0 & 1 & 1 & 0 & 0 \\ 0 & 0 & -1 & -1 & 0 & 0 & -1 & -1 & 0 & 0 & -1 & -1 & 0 & 0 & -1 & -1 \\ 0 & 0 & -1 & -1 & 0 & 0 & -1 & -1 & 0 & 0 & -1 & -1 & 0 & 0 & -1 & -1 \\ 1 & 1 & 0 & 0 & 1 & 1 & 0 & 0 & 1 & 1 & 0 & 0 & 1 & 1 & 0 & 0 \\ 1 & 1 & 0 & 0 & 1 & 1 & 0 & 0 & 1 & 1 & 0 & 0 & 1 & 1 & 0 & 0 \\ 0 & 0 & -1 & -1 & 0 & 0 & -1 & -1 & 0 & 0 & -1 & -1 & 0 & 0 & -1 & -1 \\ 0 & 0 & -1 & -1 & 0 & 0 & -1 & -1 & 0 & 0 & -1 & -1 & 0 & 0 & -1 & -1 \\ 1 & 1 & 0 & 0 & 1 & 1 & 0 & 0 & 1 & 1 & 0 & 0 & 1 & 1 & 0 & 0 \\ 1 & 1 & 0 & 0 & 1 & 1 & 0 & 0 & 1 & 1 & 0 & 0 & 1 & 1 & 0 & 0 \\ 0 & 0 & -1 & -1 & 0 & 0 & -1 & -1 & 0 & 0 & -1 & -1 & 0 & 0 & -1 & -1 \\ 0 & 0 & -1 & -1 & 0 & 0 & -1 & -1 & 0 & 0 & -1 & -1 & 0 & 0 & -1 & -1 \end{bmatrix}_{16\times16}$$

$$S_{4z} = \frac{\hbar}{2}\begin{bmatrix} 1 & 0 & 1 & 0 & 1 & 0 & 1 & 0 & 1 & 0 & 1 & 0 & 1 & 0 & 1 & 0 \\ 0 & -1 & 0 & -1 & 0 & -1 & 0 & -1 & 0 & -1 & 0 & -1 & 0 & -1 & 0 & -1 \\ 1 & 0 & 1 & 0 & 1 & 0 & 1 & 0 & 1 & 0 & 1 & 0 & 1 & 0 & 1 & 0 \\ 0 & -1 & 0 & -1 & 0 & -1 & 0 & -1 & 0 & -1 & 0 & -1 & 0 & -1 & 0 & -1 \\ 1 & 0 & 1 & 0 & 1 & 0 & 1 & 0 & 1 & 0 & 1 & 0 & 1 & 0 & 1 & 0 \\ 0 & -1 & 0 & -1 & 0 & -1 & 0 & -1 & 0 & -1 & 0 & -1 & 0 & -1 & 0 & -1 \\ 1 & 0 & 1 & 0 & 1 & 0 & 1 & 0 & 1 & 0 & 1 & 0 & 1 & 0 & 1 & 0 \\ 0 & -1 & 0 & -1 & 0 & -1 & 0 & -1 & 0 & -1 & 0 & -1 & 0 & -1 & 0 & -1 \\ 1 & 0 & 1 & 0 & 1 & 0 & 1 & 0 & 1 & 0 & 1 & 0 & 1 & 0 & 1 & 0 \\ 0 & -1 & 0 & -1 & 0 & -1 & 0 & -1 & 0 & -1 & 0 & -1 & 0 & -1 & 0 & -1 \\ 1 & 0 & 1 & 0 & 1 & 0 & 1 & 0 & 1 & 0 & 1 & 0 & 1 & 0 & 1 & 0 \\ 0 & -1 & 0 & -1 & 0 & -1 & 0 & -1 & 0 & -1 & 0 & -1 & 0 & -1 & 0 & -1 \\ 1 & 0 & 1 & 0 & 1 & 0 & 1 & 0 & 1 & 0 & 1 & 0 & 1 & 0 & 1 & 0 \\ 0 & -1 & 0 & -1 & 0 & -1 & 0 & -1 & 0 & -1 & 0 & -1 & 0 & -1 & 0 & -1 \\ 1 & 0 & 1 & 0 & 1 & 0 & 1 & 0 & 1 & 0 & 1 & 0 & 1 & 0 & 1 & 0 \\ 0 & -1 & 0 & -1 & 0 & -1 & 0 & -1 & 0 & -1 & 0 & -1 & 0 & -1 & 0 & -1 \end{bmatrix}_{16\times16}$$

To calculate momenta along the x and y axes, we must first calculate the raising operator (S_+) and the lowering operator (S_-) for each electron using Eqs. (4.5) and (4.6), respectively [131]. Accordingly, the S_+ and S_- of each electron are obtained as follows:

$$S_{1+} = \hbar\begin{bmatrix} 0&0&0&0&0&0&0&0&1&1&1&1&1&1&1&1 \\ 0&0&0&0&0&0&0&0&1&1&1&1&1&1&1&1 \\ 0&0&0&0&0&0&0&0&1&1&1&1&1&1&1&1 \\ 0&0&0&0&0&0&0&0&1&1&1&1&1&1&1&1 \\ 0&0&0&0&0&0&0&0&1&1&1&1&1&1&1&1 \\ 0&0&0&0&0&0&0&0&1&1&1&1&1&1&1&1 \\ 0&0&0&0&0&0&0&0&1&1&1&1&1&1&1&1 \\ 0&0&0&0&0&0&0&0&1&1&1&1&1&1&1&1 \\ 0&0&0&0&0&0&0&0&0&0&0&0&0&0&0&0 \\ 0&0&0&0&0&0&0&0&0&0&0&0&0&0&0&0 \\ 0&0&0&0&0&0&0&0&0&0&0&0&0&0&0&0 \\ 0&0&0&0&0&0&0&0&0&0&0&0&0&0&0&0 \\ 0&0&0&0&0&0&0&0&0&0&0&0&0&0&0&0 \\ 0&0&0&0&0&0&0&0&0&0&0&0&0&0&0&0 \\ 0&0&0&0&0&0&0&0&0&0&0&0&0&0&0&0 \\ 0&0&0&0&0&0&0&0&0&0&0&0&0&0&0&0 \end{bmatrix}_{16\times16} \quad S_{1-} = \hbar\begin{bmatrix} 0&0&0&0&0&0&0&0&0&0&0&0&0&0&0&0 \\ 0&0&0&0&0&0&0&0&0&0&0&0&0&0&0&0 \\ 0&0&0&0&0&0&0&0&0&0&0&0&0&0&0&0 \\ 0&0&0&0&0&0&0&0&0&0&0&0&0&0&0&0 \\ 0&0&0&0&0&0&0&0&0&0&0&0&0&0&0&0 \\ 0&0&0&0&0&0&0&0&0&0&0&0&0&0&0&0 \\ 0&0&0&0&0&0&0&0&0&0&0&0&0&0&0&0 \\ 0&0&0&0&0&0&0&0&0&0&0&0&0&0&0&0 \\ 1&1&1&1&1&1&1&1&0&0&0&0&0&0&0&0 \\ 1&1&1&1&1&1&1&1&0&0&0&0&0&0&0&0 \\ 1&1&1&1&1&1&1&1&0&0&0&0&0&0&0&0 \\ 1&1&1&1&1&1&1&1&0&0&0&0&0&0&0&0 \\ 1&1&1&1&1&1&1&1&0&0&0&0&0&0&0&0 \\ 1&1&1&1&1&1&1&1&0&0&0&0&0&0&0&0 \\ 1&1&1&1&1&1&1&1&0&0&0&0&0&0&0&0 \\ 1&1&1&1&1&1&1&1&0&0&0&0&0&0&0&0 \end{bmatrix}_{16\times16}$$

$$S_{2+} = \hbar\begin{bmatrix} 0&0&0&0&1&1&1&1&0&0&0&0&1&1&1&1 \\ 0&0&0&0&1&1&1&1&0&0&0&0&1&1&1&1 \\ 0&0&0&0&1&1&1&1&0&0&0&0&1&1&1&1 \\ 0&0&0&0&1&1&1&1&0&0&0&0&1&1&1&1 \\ 0&0&0&0&0&0&0&0&0&0&0&0&0&0&0&0 \\ 0&0&0&0&0&0&0&0&0&0&0&0&0&0&0&0 \\ 0&0&0&0&0&0&0&0&0&0&0&0&0&0&0&0 \\ 0&0&0&0&0&0&0&0&0&0&0&0&0&0&0&0 \\ 0&0&0&0&1&1&1&1&0&0&0&0&1&1&1&1 \\ 0&0&0&0&1&1&1&1&0&0&0&0&1&1&1&1 \\ 0&0&0&0&1&1&1&1&0&0&0&0&1&1&1&1 \\ 0&0&0&0&1&1&1&1&0&0&0&0&1&1&1&1 \\ 0&0&0&0&0&0&0&0&0&0&0&0&0&0&0&0 \\ 0&0&0&0&0&0&0&0&0&0&0&0&0&0&0&0 \\ 0&0&0&0&0&0&0&0&0&0&0&0&0&0&0&0 \\ 0&0&0&0&0&0&0&0&0&0&0&0&0&0&0&0 \end{bmatrix}_{16\times16} \quad S_{2-} = \hbar\begin{bmatrix} 0&0&0&0&0&0&0&0&0&0&0&0&0&0&0&0 \\ 0&0&0&0&0&0&0&0&0&0&0&0&0&0&0&0 \\ 0&0&0&0&0&0&0&0&0&0&0&0&0&0&0&0 \\ 0&0&0&0&0&0&0&0&0&0&0&0&0&0&0&0 \\ 1&1&1&1&0&0&0&0&1&1&1&1&0&0&0&0 \\ 1&1&1&1&0&0&0&0&1&1&1&1&0&0&0&0 \\ 1&1&1&1&0&0&0&0&1&1&1&1&0&0&0&0 \\ 1&1&1&1&0&0&0&0&1&1&1&1&0&0&0&0 \\ 0&0&0&0&0&0&0&0&0&0&0&0&0&0&0&0 \\ 0&0&0&0&0&0&0&0&0&0&0&0&0&0&0&0 \\ 0&0&0&0&0&0&0&0&0&0&0&0&0&0&0&0 \\ 0&0&0&0&0&0&0&0&0&0&0&0&0&0&0&0 \\ 1&1&1&1&0&0&0&0&1&1&1&1&0&0&0&0 \\ 1&1&1&1&0&0&0&0&1&1&1&1&0&0&0&0 \\ 1&1&1&1&0&0&0&0&1&1&1&1&0&0&0&0 \\ 1&1&1&1&0&0&0&0&1&1&1&1&0&0&0&0 \end{bmatrix}_{16\times16}$$

$$S_{3+} = \hbar\begin{bmatrix} 0&0&1&1&0&0&1&1&0&0&1&1&0&0&1&1 \\ 0&0&1&1&0&0&1&1&0&0&1&1&0&0&1&1 \\ 0&0&0&0&0&0&0&0&0&0&0&0&0&0&0&0 \\ 0&0&0&0&0&0&0&0&0&0&0&0&0&0&0&0 \\ 0&0&1&1&0&0&1&1&0&0&1&1&0&0&1&1 \\ 0&0&1&1&0&0&1&1&0&0&1&1&0&0&1&1 \\ 0&0&0&0&0&0&0&0&0&0&0&0&0&0&0&0 \\ 0&0&0&0&0&0&0&0&0&0&0&0&0&0&0&0 \\ 0&0&1&1&0&0&1&1&0&0&1&1&0&0&1&1 \\ 0&0&1&1&0&0&1&1&0&0&1&1&0&0&1&1 \\ 0&0&0&0&0&0&0&0&0&0&0&0&0&0&0&0 \\ 0&0&0&0&0&0&0&0&0&0&0&0&0&0&0&0 \\ 0&0&1&1&0&0&1&1&0&0&1&1&0&0&1&1 \\ 0&0&1&1&0&0&1&1&0&0&1&1&0&0&1&1 \\ 0&0&0&0&0&0&0&0&0&0&0&0&0&0&0&0 \\ 0&0&0&0&0&0&0&0&0&0&0&0&0&0&0&0 \end{bmatrix}_{16\times16} \quad S_{3-} = \hbar\begin{bmatrix} 0&0&0&0&0&0&0&0&0&0&0&0&0&0&0&0 \\ 0&0&0&0&0&0&0&0&0&0&0&0&0&0&0&0 \\ 1&1&0&0&1&1&0&0&1&1&0&0&1&1&0&0 \\ 1&1&0&0&1&1&0&0&1&1&0&0&1&1&0&0 \\ 0&0&0&0&0&0&0&0&0&0&0&0&0&0&0&0 \\ 0&0&0&0&0&0&0&0&0&0&0&0&0&0&0&0 \\ 1&1&0&0&1&1&0&0&1&1&0&0&1&1&0&0 \\ 1&1&0&0&1&1&0&0&1&1&0&0&1&1&0&0 \\ 0&0&0&0&0&0&0&0&0&0&0&0&0&0&0&0 \\ 0&0&0&0&0&0&0&0&0&0&0&0&0&0&0&0 \\ 1&1&0&0&1&1&0&0&1&1&0&0&1&1&0&0 \\ 1&1&0&0&1&1&0&0&1&1&0&0&1&1&0&0 \\ 0&0&0&0&0&0&0&0&0&0&0&0&0&0&0&0 \\ 0&0&0&0&0&0&0&0&0&0&0&0&0&0&0&0 \\ 1&1&0&0&1&1&0&0&1&1&0&0&1&1&0&0 \\ 1&1&0&0&1&1&0&0&1&1&0&0&1&1&0&0 \end{bmatrix}_{16\times16}$$

$$S_{4+} = \hbar \begin{bmatrix} 0&1&0&1&0&1&0&1&0&1&0&1&0&1&0&1 \\ 0&0&0&0&0&0&0&0&0&0&0&0&0&0&0&0 \\ 0&1&0&1&0&1&0&1&0&1&0&1&0&1&0&1 \\ 0&0&0&0&0&0&0&0&0&0&0&0&0&0&0&0 \\ 0&1&0&1&0&1&0&1&0&1&0&1&0&1&0&1 \\ 0&0&0&0&0&0&0&0&0&0&0&0&0&0&0&0 \\ 0&1&0&1&0&1&0&1&0&1&0&1&0&1&0&1 \\ 0&0&0&0&0&0&0&0&0&0&0&0&0&0&0&0 \\ 0&1&0&1&0&1&0&1&0&1&0&1&0&1&0&1 \\ 0&0&0&0&0&0&0&0&0&0&0&0&0&0&0&0 \\ 0&1&0&1&0&1&0&1&0&1&0&1&0&1&0&1 \\ 0&0&0&0&0&0&0&0&0&0&0&0&0&0&0&0 \\ 0&1&0&1&0&1&0&1&0&1&0&1&0&1&0&1 \\ 0&0&0&0&0&0&0&0&0&0&0&0&0&0&0&0 \\ 0&1&0&1&0&1&0&1&0&1&0&1&0&1&0&1 \\ 0&0&0&0&0&0&0&0&0&0&0&0&0&0&0&0 \end{bmatrix}_{16\times16} \quad S_{4-} = \hbar \begin{bmatrix} 0&0&0&0&0&0&0&0&0&0&0&0&0&0&0&0 \\ 1&0&1&0&1&0&1&0&1&0&1&0&1&0&1&0 \\ 0&0&0&0&0&0&0&0&0&0&0&0&0&0&0&0 \\ 1&0&1&0&1&0&1&0&1&0&1&0&1&0&1&0 \\ 0&0&0&0&0&0&0&0&0&0&0&0&0&0&0&0 \\ 1&0&1&0&1&0&1&0&1&0&1&0&1&0&1&0 \\ 0&0&0&0&0&0&0&0&0&0&0&0&0&0&0&0 \\ 1&0&1&0&1&0&1&0&1&0&1&0&1&0&1&0 \\ 0&0&0&0&0&0&0&0&0&0&0&0&0&0&0&0 \\ 1&0&1&0&1&0&1&0&1&0&1&0&1&0&1&0 \\ 0&0&0&0&0&0&0&0&0&0&0&0&0&0&0&0 \\ 1&0&1&0&1&0&1&0&1&0&1&0&1&0&1&0 \\ 0&0&0&0&0&0&0&0&0&0&0&0&0&0&0&0 \\ 1&0&1&0&1&0&1&0&1&0&1&0&1&0&1&0 \\ 0&0&0&0&0&0&0&0&0&0&0&0&0&0&0&0 \\ 1&0&1&0&1&0&1&0&1&0&1&0&1&0&1&0 \end{bmatrix}_{16\times16}$$

After finding S_+ and S_- for each electron in a quinary system, we can use Eqs. (4.7) and (4.8) to measure the momentum in the x-direction (S_x) and y-direction (S_y) as follows:

$$S_{1x} = \frac{\hbar}{2} \begin{bmatrix} 0&0&0&0&0&0&0&0&1&1&1&1&1&1&1&1 \\ 0&0&0&0&0&0&0&0&1&1&1&1&1&1&1&1 \\ 0&0&0&0&0&0&0&0&1&1&1&1&1&1&1&1 \\ 0&0&0&0&0&0&0&0&1&1&1&1&1&1&1&1 \\ 0&0&0&0&0&0&0&0&1&1&1&1&1&1&1&1 \\ 0&0&0&0&0&0&0&0&1&1&1&1&1&1&1&1 \\ 0&0&0&0&0&0&0&0&1&1&1&1&1&1&1&1 \\ 0&0&0&0&0&0&0&0&1&1&1&1&1&1&1&1 \\ 1&1&1&1&1&1&1&1&0&0&0&0&0&0&0&0 \\ 1&1&1&1&1&1&1&1&0&0&0&0&0&0&0&0 \\ 1&1&1&1&1&1&1&1&0&0&0&0&0&0&0&0 \\ 1&1&1&1&1&1&1&1&0&0&0&0&0&0&0&0 \\ 1&1&1&1&1&1&1&1&0&0&0&0&0&0&0&0 \\ 1&1&1&1&1&1&1&1&0&0&0&0&0&0&0&0 \\ 1&1&1&1&1&1&1&1&0&0&0&0&0&0&0&0 \\ 1&1&1&1&1&1&1&1&0&0&0&0&0&0&0&0 \end{bmatrix} \quad S_{1y} = \frac{\hbar}{2} \begin{bmatrix} 0&0&0&0&0&0&0&0&-i&-i&-i&-i&-i&-i&-i&-i \\ 0&0&0&0&0&0&0&0&-i&-i&-i&-i&-i&-i&-i&-i \\ 0&0&0&0&0&0&0&0&-i&-i&-i&-i&-i&-i&-i&-i \\ 0&0&0&0&0&0&0&0&-i&-i&-i&-i&-i&-i&-i&-i \\ 0&0&0&0&0&0&0&0&-i&-i&-i&-i&-i&-i&-i&-i \\ 0&0&0&0&0&0&0&0&-i&-i&-i&-i&-i&-i&-i&-i \\ 0&0&0&0&0&0&0&0&-i&-i&-i&-i&-i&-i&-i&-i \\ 0&0&0&0&0&0&0&0&-i&-i&-i&-i&-i&-i&-i&-i \\ i&i&i&i&i&i&i&i&0&0&0&0&0&0&0&0 \\ i&i&i&i&i&i&i&i&0&0&0&0&0&0&0&0 \\ i&i&i&i&i&i&i&i&0&0&0&0&0&0&0&0 \\ i&i&i&i&i&i&i&i&0&0&0&0&0&0&0&0 \\ i&i&i&i&i&i&i&i&0&0&0&0&0&0&0&0 \\ i&i&i&i&i&i&i&i&0&0&0&0&0&0&0&0 \\ i&i&i&i&i&i&i&i&0&0&0&0&0&0&0&0 \\ i&i&i&i&i&i&i&i&0&0&0&0&0&0&0&0 \end{bmatrix}$$

$$S_{2x} = \frac{\hbar}{2} \begin{bmatrix} 0&0&0&0&1&1&1&1&0&0&0&0&1&1&1&1 \\ 0&0&0&0&1&1&1&1&0&0&0&0&1&1&1&1 \\ 0&0&0&0&1&1&1&1&0&0&0&0&1&1&1&1 \\ 0&0&0&0&1&1&1&1&0&0&0&0&1&1&1&1 \\ 1&1&1&1&0&0&0&0&1&1&1&1&0&0&0&0 \\ 1&1&1&1&0&0&0&0&1&1&1&1&0&0&0&0 \\ 1&1&1&1&0&0&0&0&1&1&1&1&0&0&0&0 \\ 1&1&1&1&0&0&0&0&1&1&1&1&0&0&0&0 \\ 0&0&0&0&1&1&1&1&0&0&0&0&1&1&1&1 \\ 0&0&0&0&1&1&1&1&0&0&0&0&1&1&1&1 \\ 0&0&0&0&1&1&1&1&0&0&0&0&1&1&1&1 \\ 0&0&0&0&1&1&1&1&0&0&0&0&1&1&1&1 \\ 1&1&1&1&0&0&0&0&1&1&1&1&0&0&0&0 \\ 1&1&1&1&0&0&0&0&1&1&1&1&0&0&0&0 \\ 1&1&1&1&0&0&0&0&1&1&1&1&0&0&0&0 \\ 1&1&1&1&0&0&0&0&1&1&1&1&0&0&0&0 \end{bmatrix} \quad S_{2y} = \frac{\hbar}{2} \begin{bmatrix} 0&0&0&0&-i&-i&-i&-i&0&0&0&0&-i&-i&-i&-i \\ 0&0&0&0&-i&-i&-i&-i&0&0&0&0&-i&-i&-i&-i \\ 0&0&0&0&-i&-i&-i&-i&0&0&0&0&-i&-i&-i&-i \\ 0&0&0&0&-i&-i&-i&-i&0&0&0&0&-i&-i&-i&-i \\ i&i&i&i&0&0&0&0&i&i&i&i&0&0&0&0 \\ i&i&i&i&0&0&0&0&i&i&i&i&0&0&0&0 \\ i&i&i&i&0&0&0&0&i&i&i&i&0&0&0&0 \\ i&i&i&i&0&0&0&0&i&i&i&i&0&0&0&0 \\ 0&0&0&0&-i&-i&-i&-i&0&0&0&0&-i&-i&-i&-i \\ 0&0&0&0&-i&-i&-i&-i&0&0&0&0&-i&-i&-i&-i \\ 0&0&0&0&-i&-i&-i&-i&0&0&0&0&-i&-i&-i&-i \\ 0&0&0&0&-i&-i&-i&-i&0&0&0&0&-i&-i&-i&-i \\ i&i&i&i&0&0&0&0&i&i&i&i&0&0&0&0 \\ i&i&i&i&0&0&0&0&i&i&i&i&0&0&0&0 \\ i&i&i&i&0&0&0&0&i&i&i&i&0&0&0&0 \\ i&i&i&i&0&0&0&0&i&i&i&i&0&0&0&0 \end{bmatrix}$$

$$S_{3x} = \frac{\hbar}{2}\begin{bmatrix} 0&0&1&1&0&0&1&1&0&0&1&1&0&0&1&1\\ 0&0&1&1&0&0&1&1&0&0&1&1&0&0&1&1\\ 1&1&0&0&1&1&0&0&1&1&0&0&1&1&0&0\\ 1&1&0&0&1&1&0&0&1&1&0&0&1&1&0&0\\ 0&0&1&1&0&0&1&1&0&0&1&1&0&0&1&1\\ 0&0&1&1&0&0&1&1&0&0&1&1&0&0&1&1\\ 1&1&0&0&1&1&0&0&1&1&0&0&1&1&0&0\\ 1&1&0&0&1&1&0&0&1&1&0&0&1&1&0&0\\ 0&0&1&1&0&0&1&1&0&0&1&1&0&0&1&1\\ 0&0&1&1&0&0&1&1&0&0&1&1&0&0&1&1\\ 1&1&0&0&1&1&0&0&1&1&0&0&1&1&0&0\\ 1&1&0&0&1&1&0&0&1&1&0&0&1&1&0&0\\ 0&0&1&1&0&0&1&1&0&0&1&1&0&0&1&1\\ 0&0&1&1&0&0&1&1&0&0&1&1&0&0&1&1\\ 1&1&0&0&1&1&0&0&1&1&0&0&1&1&0&0\\ 1&1&0&0&1&1&0&0&1&1&0&0&1&1&0&0 \end{bmatrix} \quad S_{3y} = \frac{\hbar}{2}\begin{bmatrix} 0&0&-i&-i&0&0&-i&-i&0&0&-i&-i&0&0&-i&-i\\ 0&0&-i&-i&0&0&-i&-i&0&0&-i&-i&0&0&-i&-i\\ i&i&0&0&i&i&0&0&i&i&0&0&i&i&0&0\\ i&i&0&0&i&i&0&0&i&i&0&0&i&i&0&0\\ 0&0&-i&-i&0&0&-i&-i&0&0&-i&-i&0&0&-i&-i\\ 0&0&-i&-i&0&0&-i&-i&0&0&-i&-i&0&0&-i&-i\\ i&i&0&0&i&i&0&0&i&i&0&0&i&i&0&0\\ i&i&0&0&i&i&0&0&i&i&0&0&i&i&0&0\\ 0&0&-i&-i&0&0&-i&-i&0&0&-i&-i&0&0&-i&-i\\ 0&0&-i&-i&0&0&-i&-i&0&0&-i&-i&0&0&-i&-i\\ i&i&0&0&i&i&0&0&i&i&0&0&i&i&0&0\\ i&i&0&0&i&i&0&0&i&i&0&0&i&i&0&0\\ 0&0&-i&-i&0&0&-i&-i&0&0&-i&-i&0&0&-i&-i\\ 0&0&-i&-i&0&0&-i&-i&0&0&-i&-i&0&0&-i&-i\\ i&i&0&0&i&i&0&0&i&i&0&0&i&i&0&0\\ i&i&0&0&i&i&0&0&i&i&0&0&i&i&0&0 \end{bmatrix}$$

$$S_{4x} = \frac{\hbar}{2}\begin{bmatrix} 0&1&0&1&0&1&0&1&0&1&0&1&0&1&0&1\\ 1&0&1&0&1&0&1&0&1&0&1&0&1&0&1&0\\ 0&1&0&1&0&1&0&1&0&1&0&1&0&1&0&1\\ 1&0&1&0&1&0&1&0&1&0&1&0&1&0&1&0\\ 0&1&0&1&0&1&0&1&0&1&0&1&0&1&0&1\\ 1&0&1&0&1&0&1&0&1&0&1&0&1&0&1&0\\ 0&1&0&1&0&1&0&1&0&1&0&1&0&1&0&1\\ 1&0&1&0&1&0&1&0&1&0&1&0&1&0&1&0\\ 0&1&0&1&0&1&0&1&0&1&0&1&0&1&0&1\\ 1&0&1&0&1&0&1&0&1&0&1&0&1&0&1&0\\ 0&1&0&1&0&1&0&1&0&1&0&1&0&1&0&1\\ 1&0&1&0&1&0&1&0&1&0&1&0&1&0&1&0\\ 0&1&0&1&0&1&0&1&0&1&0&1&0&1&0&1\\ 1&0&1&0&1&0&1&0&1&0&1&0&1&0&1&0\\ 0&1&0&1&0&1&0&1&0&1&0&1&0&1&0&1\\ 1&0&1&0&1&0&1&0&1&0&1&0&1&0&1&0 \end{bmatrix} \quad S_{4y} = \frac{\hbar}{2}\begin{bmatrix} 0&-i&0&-i&0&-i&0&-i&0&-i&0&-i&0&-i&0&-i\\ i&0&i&0&i&0&i&0&i&0&i&0&i&0&i&0\\ 0&-i&0&-i&0&-i&0&-i&0&-i&0&-i&0&-i&0&-i\\ i&0&i&0&i&0&i&0&i&0&i&0&i&0&i&0\\ 0&-i&0&-i&0&-i&0&-i&0&-i&0&-i&0&-i&0&-i\\ i&0&i&0&i&0&i&0&i&0&i&0&i&0&i&0\\ 0&-i&0&-i&0&-i&0&-i&0&-i&0&-i&0&-i&0&-i\\ i&0&i&0&i&0&i&0&i&0&i&0&i&0&i&0\\ 0&-i&0&-i&0&-i&0&-i&0&-i&0&-i&0&-i&0&-i\\ i&0&i&0&i&0&i&0&i&0&i&0&i&0&i&0\\ 0&-i&0&-i&0&-i&0&-i&0&-i&0&-i&0&-i&0&-i\\ i&0&i&0&i&0&i&0&i&0&i&0&i&0&i&0\\ 0&-i&0&-i&0&-i&0&-i&0&-i&0&-i&0&-i&0&-i\\ i&0&i&0&i&0&i&0&i&0&i&0&i&0&i&0\\ 0&-i&0&-i&0&-i&0&-i&0&-i&0&-i&0&-i&0&-i\\ i&0&i&0&i&0&i&0&i&0&i&0&i&0&i&0 \end{bmatrix}$$

After calculating the momentum matrices for each electron in the quinary system, we can use Eqs. (4.22) to (4.24) as the linear combination of S_x, S_y, and S_z of each electron to determine the momentum of the whole system [131].

$$S_x = S_{1x} + S_{2x} + S_{3x} + S_{4x} \tag{4.22}$$

$$S_y = S_{1y} + S_{2y} + S_{3y} + S_{4y} \tag{4.23}$$

$$S_z = S_{1z} + S_{2z} + S_{3z} + S_{4z} \tag{4.24}$$

According to Eqs. (4.22) to (4.24), the momentum matrices in the quinary system are [131]:

$$S_x = \frac{\hbar}{2}\begin{bmatrix}
0 & 1 & 1 & 2 & 1 & 2 & 2 & 3 & 1 & 2 & 2 & 3 & 2 & 3 & 3 & 4 \\
1 & 0 & 2 & 1 & 2 & 1 & 3 & 2 & 2 & 1 & 3 & 2 & 3 & 2 & 4 & 3 \\
1 & 2 & 0 & 1 & 2 & 3 & 1 & 2 & 2 & 3 & 1 & 2 & 3 & 4 & 2 & 3 \\
2 & 1 & 1 & 0 & 3 & 2 & 2 & 1 & 3 & 2 & 2 & 1 & 4 & 3 & 3 & 2 \\
1 & 2 & 2 & 3 & 0 & 1 & 1 & 2 & 2 & 3 & 3 & 4 & 1 & 2 & 2 & 3 \\
2 & 1 & 3 & 2 & 1 & 0 & 2 & 1 & 3 & 2 & 4 & 3 & 2 & 1 & 3 & 2 \\
2 & 3 & 1 & 2 & 1 & 2 & 0 & 1 & 3 & 4 & 2 & 3 & 2 & 3 & 1 & 2 \\
3 & 2 & 2 & 1 & 2 & 1 & 1 & 0 & 4 & 3 & 3 & 2 & 3 & 2 & 2 & 1 \\
1 & 2 & 2 & 3 & 2 & 3 & 3 & 4 & 0 & 1 & 1 & 2 & 1 & 2 & 2 & 3 \\
2 & 1 & 3 & 2 & 3 & 2 & 4 & 3 & 1 & 0 & 2 & 1 & 2 & 1 & 3 & 2 \\
2 & 3 & 1 & 2 & 3 & 4 & 2 & 3 & 1 & 2 & 0 & 1 & 2 & 3 & 1 & 2 \\
3 & 2 & 2 & 1 & 4 & 3 & 3 & 2 & 2 & 1 & 1 & 0 & 3 & 2 & 2 & 1 \\
2 & 3 & 3 & 4 & 1 & 2 & 2 & 3 & 1 & 2 & 2 & 3 & 0 & 1 & 1 & 2 \\
3 & 2 & 4 & 3 & 2 & 1 & 3 & 2 & 2 & 1 & 3 & 2 & 1 & 0 & 2 & 1 \\
3 & 4 & 2 & 3 & 2 & 3 & 1 & 2 & 2 & 3 & 1 & 2 & 1 & 2 & 0 & 1 \\
4 & 3 & 3 & 2 & 3 & 2 & 2 & 1 & 3 & 2 & 2 & 1 & 2 & 1 & 1 & 0
\end{bmatrix}$$

$$S_y = \frac{\hbar}{2}\begin{bmatrix}
0 & -i & -i & -2i & -i & -2i & -2i & -3i & -i & -2i & -2i & -3i & -2i & -3i & -3i & -4i \\
i & 0 & 0 & -i & 0 & -i & -i & -2i & 0 & -i & -i & -2i & -i & -2i & -2i & -3i \\
i & 0 & 0 & -i & 0 & -i & -i & -2i & 0 & -i & -i & -2i & -i & -2i & -2i & -3i \\
2i & i & i & 0 & i & 0 & 0 & -i & i & 0 & 0 & -i & 0 & -i & -i & -2i \\
i & 0 & 0 & -i & 0 & -i & -i & -2i & 0 & -i & -i & -2i & -i & -2i & -2i & -3i \\
2i & i & i & 0 & i & 0 & 0 & -i & i & 0 & 0 & -i & 0 & -i & -i & -2i \\
2i & i & i & 0 & i & 0 & 0 & -i & i & 0 & 0 & -i & 0 & -i & -i & -2i \\
3i & 2i & 2i & i & 2i & i & i & 0 & 2i & i & i & 0 & i & 0 & 0 & -i \\
i & 0 & 0 & -i & 0 & -i & -i & -2i & 0 & -i & -i & -2i & -i & -2i & -2i & -3i \\
2i & i & i & 0 & i & 0 & 0 & -i & i & 0 & 0 & -i & 0 & -i & -i & -2i \\
2i & i & i & 0 & i & 0 & 0 & -i & i & 0 & 0 & -i & 0 & -i & -i & -2i \\
3i & 2i & 2i & i & 2i & i & i & 0 & 2i & i & i & 0 & i & 0 & 0 & -i \\
2i & i & i & 0 & i & 0 & 0 & -i & i & 0 & 0 & -i & 0 & -i & -i & -2i \\
3i & 2i & 2i & i & 2i & i & i & 0 & 2i & i & i & 0 & i & 0 & 0 & -i \\
3i & 2i & 2i & i & 2i & i & i & 0 & 2i & i & i & 0 & i & 0 & 0 & -i \\
4i & 3i & 3i & 2i & 3i & 2i & 2i & i & 3i & 2i & 2i & i & 2i & i & i & 0
\end{bmatrix}$$

$$
S_z = \frac{\hbar}{2}\begin{bmatrix}
4 & 3 & 3 & 2 & 3 & 2 & 2 & 1 & 3 & 2 & 2 & 1 & 2 & 1 & 1 & 0 \\
3 & 2 & 2 & 1 & 2 & 1 & 1 & 0 & 2 & 1 & 1 & 0 & 1 & 0 & 0 & -1 \\
3 & 2 & 2 & 1 & 2 & 1 & 1 & 0 & 2 & 1 & 1 & 0 & 1 & 0 & 0 & -1 \\
2 & 1 & 1 & 0 & 1 & 0 & 0 & -1 & 1 & 0 & 0 & -1 & 0 & -1 & -1 & -2 \\
3 & 2 & 2 & 1 & 2 & 1 & 1 & 0 & 2 & 1 & 1 & 0 & 1 & 0 & 0 & -1 \\
2 & 1 & 1 & 0 & 1 & 0 & 0 & -1 & 1 & 0 & 0 & -1 & 0 & -1 & -1 & -2 \\
2 & 1 & 1 & 0 & 1 & 0 & 0 & -1 & 1 & 0 & 0 & -1 & 0 & -1 & -1 & -2 \\
1 & 0 & 0 & -1 & 0 & -1 & -1 & -2 & 0 & -1 & -1 & -2 & -1 & -2 & -2 & -3 \\
3 & 2 & 2 & 1 & 2 & 1 & 1 & 0 & 2 & 1 & 1 & 0 & 1 & 0 & 0 & -1 \\
2 & 1 & 1 & 0 & 1 & 0 & 0 & -1 & 1 & 0 & 0 & -1 & 0 & -1 & -1 & -2 \\
2 & 1 & 1 & 0 & 1 & 0 & 0 & -1 & 1 & 0 & 0 & -1 & 0 & -1 & -1 & -2 \\
1 & 0 & 0 & -1 & 0 & -1 & -1 & -2 & 0 & -1 & -1 & -2 & -1 & -2 & -2 & -3 \\
2 & 1 & 1 & 0 & 1 & 0 & 0 & -1 & 1 & 0 & 0 & -1 & 0 & -1 & -1 & -2 \\
1 & 0 & 0 & -1 & 0 & -1 & -1 & -2 & 0 & -1 & -1 & -2 & -1 & -2 & -2 & -3 \\
1 & 0 & 0 & -1 & 0 & -1 & -1 & -2 & 0 & -1 & -1 & -2 & -1 & -2 & -2 & -3 \\
0 & -1 & -1 & -2 & -1 & -2 & -2 & -3 & -1 & -2 & -2 & -3 & -2 & -3 & -3 & -4
\end{bmatrix}
$$

By removing $\hbar/2$ from S_x, S_y, and S_z, the parameters σ_x, σ_y, and σ_z are obtained as quasi-Pauli matrices. These matrices are called so because they do not have all the properties of the Pauli matrices [131]. To calculate polarization in a quinary system, quantum measurements must be made in the z-direction as given in Eq. (4.9). To calculate polarizations, we must first determine the input states for a quinary system with four particles. Table 4.6 shows the input states of the quinary system.

Using Eq. (4.9) and performing quantum measurements, the polarizations in the quinary system are obtained as given in Table 4.7 [131].

According to Table 4.7, several states are repeated for some inputs. Therefore, a cell must be designed to produce the polarizations {−4, −2, 0, 2, 4} [131]. Figure 4.3 shows the polarization diagram in QuQCA based on calculations.

According to the calculations, the polarizations {−4, −2, 0, 2, 4} in the QuQCA cell are considered equivalent to {A, B, C, D, E} [131]. In quinary logic, A or polarization −4 is equal to logical zero, B

Table 4.6 Input states of the quinary system.

Input State			
input state 1 = $\lvert 0000\rangle = \begin{bmatrix}1\\0\\0\\0\\0\\0\\0\\0\\0\\0\\0\\0\\0\\0\\0\\0\end{bmatrix}$	input state 2 = $\lvert 0001\rangle = \begin{bmatrix}0\\1\\0\\0\\0\\0\\0\\0\\0\\0\\0\\0\\0\\0\\0\\0\end{bmatrix}$	input state 3 = $\lvert 0010\rangle = \begin{bmatrix}0\\0\\1\\0\\0\\0\\0\\0\\0\\0\\0\\0\\0\\0\\0\\0\end{bmatrix}$	input state 4 = $\lvert 0011\rangle = \begin{bmatrix}0\\0\\0\\1\\0\\0\\0\\0\\0\\0\\0\\0\\0\\0\\0\\0\end{bmatrix}$

Table 4.6 (Continued)

Input State			
input state 5 $= \|0100\rangle = \begin{bmatrix}0\\0\\0\\0\\1\\0\\0\\0\\0\\0\\0\\0\\0\\0\\0\\0\end{bmatrix}$	input state 6 $= \|0101\rangle = \begin{bmatrix}0\\0\\0\\0\\0\\1\\0\\0\\0\\0\\0\\0\\0\\0\\0\\0\end{bmatrix}$	input state 7 $= \|0110\rangle = \begin{bmatrix}0\\0\\0\\0\\0\\0\\1\\0\\0\\0\\0\\0\\0\\0\\0\\0\end{bmatrix}$	input state 8 $= \|0111\rangle = \begin{bmatrix}0\\0\\0\\0\\0\\0\\0\\1\\0\\0\\0\\0\\0\\0\\0\\0\end{bmatrix}$
input state 9 $= \|1000\rangle = \begin{bmatrix}0\\0\\0\\0\\0\\0\\0\\0\\1\\0\\0\\0\\0\\0\\0\\0\end{bmatrix}$	input state 10 $= \|1001\rangle = \begin{bmatrix}0\\0\\0\\0\\0\\0\\0\\0\\0\\1\\0\\0\\0\\0\\0\\0\end{bmatrix}$	input state 11 $= \|1010\rangle = \begin{bmatrix}0\\0\\0\\0\\0\\0\\0\\0\\0\\0\\1\\0\\0\\0\\0\\0\end{bmatrix}$	input state 12 $= \|1011\rangle = \begin{bmatrix}0\\0\\0\\0\\0\\0\\0\\0\\0\\0\\0\\1\\0\\0\\0\\0\end{bmatrix}$
input state 13 $= \|1100\rangle = \begin{bmatrix}0\\0\\0\\0\\0\\0\\0\\0\\0\\0\\0\\0\\1\\0\\0\\0\end{bmatrix}$	input state 13 $= \|1101\rangle = \begin{bmatrix}0\\0\\0\\0\\0\\0\\0\\0\\0\\0\\0\\0\\0\\1\\0\\0\end{bmatrix}$	input state 13 $= \|1110\rangle = \begin{bmatrix}0\\0\\0\\0\\0\\0\\0\\0\\0\\0\\0\\0\\0\\0\\1\\0\end{bmatrix}$	input state 14 $= \|1110\rangle = \begin{bmatrix}0\\0\\0\\0\\0\\0\\0\\0\\0\\0\\0\\0\\0\\0\\0\\1\end{bmatrix}$

Table 4.7 Polarizations of the quinary system.

Polarizations in QuQCA							
$P_{\|0000\rangle} = -4$	$P_{\|0001\rangle} = -2$	$P_{\|0010\rangle} = -2$	$P_{\|0011\rangle} = 0$	$P_{\|0100\rangle} = -2$	$P_{\|0101\rangle} = 0$	$P_{\|0110\rangle} = 0$	$P_{\|0111\rangle} = 2$
$P_{\|1000\rangle} = -2$	$P_{\|1001\rangle} = 0$	$P_{\|1010\rangle} = 0$	$P_{\|1011\rangle} = 2$	$P_{\|1100\rangle} = 0$	$P_{\|1101\rangle} = 2$	$P_{\|1110\rangle} = 2$	$P_{\|1111\rangle} = 4$

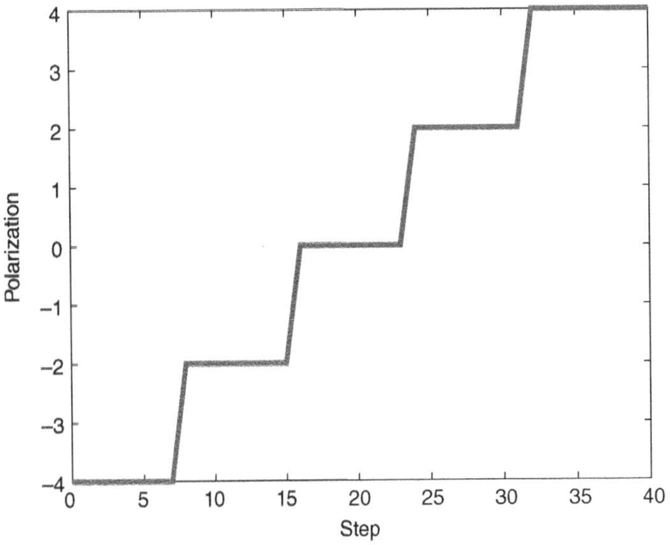

Figure 4.3 Polarization diagram in QuQCA.

or polarization −2 is equal to logical level 1/4, C or polarization 0 is equal to 2/4, D or polarization 2 is equal to level 3/4, and E or polarization 4 is equal to level 1. Only one state is considered for duplicate polarizations. Therefore, here, for polarization −4, the input state is equal to |0000>; for polarization −2, the input state is equal to |0001>; for polarization 0, the input state is equal to |1001>; for polarization 2, the input state is equal to |1110>, and for polarization 4, the input state is equal to |1111>. Accordingly, all polarizations related to the quinary system can be calculated [131].

As stated, each quinary system consists of four particles. If the electrons are placed inside a cell as a layer with more than two electrons, the logic gates become unstable, leading to incorrect outputs of the quinary gates. Therefore, to prevent instability, the proposed cell structure should be such that the constructed basic gates have the correct output. In Chapter 5, the structure of the QuQCA cell will be presented, and it will be shown that the cell has an accurate and stable response to the various inputs of the logic gates.

4.2.4.2 Hamiltonian Matrix in Quinary QCA

In the following, the Hamiltonian matrix of a quinary system is presented. To calculate the Hamiltonian matrix, the Hartree–Fock approximation in Eq. (4.10) is used [129–132].

In Eq. (4.10), γ_j is the amount of tunneling energy, δ_x is the quinary quasi-Pauli matrix in the x-direction, δ_z is the quinary quasi-Pauli matrix in the z-direction, E_k is equal to the kink energy, and P_j is the quinary polarization value. Applying these parameters to the Hartree–Fock Hamiltonian matrix in the quinary system gives [131]:

$$
\widehat{H}(j) =
\begin{bmatrix}
-2E_kP_j & -\frac{3E_k}{2}P_j-\gamma_j & -\frac{3E_k}{2}P_j-\gamma_j & -E_kP_j-2\gamma_j & -\frac{3E_k}{2}P_j-\gamma_j & -E_kP_j-2\gamma_j & -E_kP_j-2\gamma_j & -\frac{E_k}{2}P_j-3\gamma_j & -\frac{3E_k}{2}P_j-\gamma_j & -E_kP_j-2\gamma_j & -E_kP_j-2\gamma_j & -\frac{E_k}{2}P_j-3\gamma_j & -E_kP_j-2\gamma_j & -\frac{E_k}{2}P_j-3\gamma_j & -\frac{E_k}{2}P_j-3\gamma_j & -4\gamma_j \\
-\frac{3E_k}{2}P_j-\gamma_j & -E_kP_j & -E_kP_j-2\gamma_j & -\frac{E_k}{2}P_j-\gamma_j & -E_kP_j-2\gamma_j & -\frac{E_k}{2}P_j-\gamma_j & -\frac{E_k}{2}P_j-3\gamma_j & -2\gamma_j & -E_kP_j-2\gamma_j & -\frac{E_k}{2}P_j-\gamma_j & -\frac{E_k}{2}P_j-3\gamma_j & -2\gamma_j & -\frac{E_k}{2}P_j-3\gamma_j & -2\gamma_j & -4\gamma_j & \frac{E_k}{2}P_j-3\gamma_j \\
-\frac{3E_k}{2}P_j-\gamma_j & -E_kP_j-2\gamma_j & -E_kP_j & -\frac{E_k}{2}P_j-\gamma_j & -E_kP_j-2\gamma_j & -\frac{E_k}{2}P_j-3\gamma_j & -\frac{E_k}{2}P_j-\gamma_j & -2\gamma_j & -E_kP_j-2\gamma_j & -\frac{E_k}{2}P_j-3\gamma_j & -\frac{E_k}{2}P_j-\gamma_j & -2\gamma_j & -\frac{E_k}{2}P_j-3\gamma_j & -4\gamma_j & -2\gamma_j & \frac{E_k}{2}P_j-3\gamma_j \\
-E_kP_j-2\gamma_j & -\frac{E_k}{2}P_j-\gamma_j & -\frac{E_k}{2}P_j-\gamma_j & 0 & -\frac{E_k}{2}P_j-3\gamma_j & -2\gamma_j & -2\gamma_j & \frac{E_k}{2}P_j-\gamma_j & -\frac{E_k}{2}P_j-3\gamma_j & -2\gamma_j & -2\gamma_j & \frac{E_k}{2}P_j-\gamma_j & -4\gamma_j & \frac{E_k}{2}P_j-3\gamma_j & \frac{E_k}{2}P_j-3\gamma_j & E_kP_j-2\gamma_j \\
-\frac{3E_k}{2}P_j-\gamma_j & -E_kP_j-2\gamma_j & -E_kP_j-2\gamma_j & -\frac{E_k}{2}P_j-3\gamma_j & -E_kP_j & -\frac{E_k}{2}P_j-\gamma_j & -\frac{E_k}{2}P_j-\gamma_j & -2\gamma_j & -E_kP_j-2\gamma_j & -\frac{E_k}{2}P_j-3\gamma_j & -\frac{E_k}{2}P_j-3\gamma_j & -4\gamma_j & -\frac{E_k}{2}P_j-\gamma_j & -2\gamma_j & -2\gamma_j & \frac{E_k}{2}P_j-3\gamma_j \\
-E_kP_j-2\gamma_j & -\frac{E_k}{2}P_j-\gamma_j & -\frac{E_k}{2}P_j-3\gamma_j & -2\gamma_j & -\frac{E_k}{2}P_j-\gamma_j & 0 & -2\gamma_j & \frac{E_k}{2}P_j-\gamma_j & -\frac{E_k}{2}P_j-3\gamma_j & -2\gamma_j & -4\gamma_j & \frac{E_k}{2}P_j-3\gamma_j & -2\gamma_j & \frac{E_k}{2}P_j-\gamma_j & \frac{E_k}{2}P_j-3\gamma_j & E_kP_j-2\gamma_j \\
-E_kP_j-2\gamma_j & -\frac{E_k}{2}P_j-3\gamma_j & -\frac{E_k}{2}P_j-\gamma_j & -2\gamma_j & -\frac{E_k}{2}P_j-\gamma_j & -2\gamma_j & 0 & \frac{E_k}{2}P_j-\gamma_j & -\frac{E_k}{2}P_j-3\gamma_j & -4\gamma_j & -2\gamma_j & \frac{E_k}{2}P_j-3\gamma_j & -2\gamma_j & \frac{E_k}{2}P_j-3\gamma_j & \frac{E_k}{2}P_j-\gamma_j & \frac{E_k}{2}P_j-\gamma_j \\
-\frac{E_k}{2}P_j-3\gamma_j & -2\gamma_j & -2\gamma_j & \frac{E_k}{2}P_j-\gamma_j & -2\gamma_j & \frac{E_k}{2}P_j-\gamma_j & \frac{E_k}{2}P_j-\gamma_j & E_kP_j & -4\gamma_j & \frac{E_k}{2}P_j-3\gamma_j & \frac{E_k}{2}P_j-3\gamma_j & E_kP_j-2\gamma_j & \frac{E_k}{2}P_j-3\gamma_j & E_kP_j-2\gamma_j & E_kP_j-2\gamma_j & \frac{3E_k}{2}P_j-\gamma_j \\
-\frac{3E_k}{2}P_j-\gamma_j & -E_kP_j-2\gamma_j & -E_kP_j-2\gamma_j & -\frac{E_k}{2}P_j-3\gamma_j & -E_kP_j-2\gamma_j & -\frac{E_k}{2}P_j-3\gamma_j & -\frac{E_k}{2}P_j-3\gamma_j & -4\gamma_j & -E_kP_j & -\frac{E_k}{2}P_j-\gamma_j & -\frac{E_k}{2}P_j-\gamma_j & -2\gamma_j & -\frac{E_k}{2}P_j-\gamma_j & -2\gamma_j & -2\gamma_j & \frac{E_k}{2}P_j-3\gamma_j \\
-E_kP_j-2\gamma_j & -\frac{E_k}{2}P_j-\gamma_j & -\frac{E_k}{2}P_j-3\gamma_j & -2\gamma_j & -\frac{E_k}{2}P_j-3\gamma_j & -2\gamma_j & -4\gamma_j & \frac{E_k}{2}P_j-3\gamma_j & -\frac{E_k}{2}P_j-\gamma_j & 0 & -2\gamma_j & \frac{E_k}{2}P_j-\gamma_j & -2\gamma_j & \frac{E_k}{2}P_j-\gamma_j & \frac{E_k}{2}P_j-3\gamma_j & E_kP_j-2\gamma_j \\
-E_kP_j-2\gamma_j & -\frac{E_k}{2}P_j-3\gamma_j & \frac{E_k}{2}P_j-\gamma_j & -2\gamma_j & -\frac{E_k}{2}P_j-3\gamma_j & -4\gamma_j & -2\gamma_j & \frac{E_k}{2}P_j-3\gamma_j & -\frac{E_k}{2}P_j-\gamma_j & -2\gamma_j & 0 & \frac{E_k}{2}P_j-\gamma_j & -2\gamma_j & \frac{E_k}{2}P_j-3\gamma_j & \frac{E_k}{2}P_j-\gamma_j & E_kP_j-2\gamma_j \\
-\frac{E_k}{2}P_j-3\gamma_j & -2\gamma_j & -2\gamma_j & \frac{E_k}{2}P_j-\gamma_j & -4\gamma_j & \frac{E_k}{2}P_j-3\gamma_j & \frac{E_k}{2}P_j-3\gamma_j & E_kP_j-2\gamma_j & -2\gamma_j & \frac{E_k}{2}P_j-\gamma_j & \frac{E_k}{2}P_j-\gamma_j & E_kP_j & \frac{E_k}{2}P_j-3\gamma_j & E_kP_j-2\gamma_j & E_kP_j-2\gamma_j & \frac{3E_k}{2}P_j-\gamma_j \\
-E_kP_j-2\gamma_j & -\frac{E_k}{2}P_j-3\gamma_j & -\frac{E_k}{2}P_j-3\gamma_j & -4\gamma_j & -\frac{E_k}{2}P_j-\gamma_j & -2\gamma_j & -2\gamma_j & \frac{E_k}{2}P_j-3\gamma_j & -\frac{E_k}{2}P_j-\gamma_j & -2\gamma_j & -2\gamma_j & \frac{E_k}{2}P_j-3\gamma_j & 0 & \frac{E_k}{2}P_j-\gamma_j & \frac{E_k}{2}P_j-\gamma_j & E_kP_j-2\gamma_j \\
-\frac{E_k}{2}P_j-3\gamma_j & -2\gamma_j & -4\gamma_j & \frac{E_k}{2}P_j-3\gamma_j & -2\gamma_j & \frac{E_k}{2}P_j-\gamma_j & \frac{E_k}{2}P_j-3\gamma_j & E_kP_j-2\gamma_j & -2\gamma_j & \frac{E_k}{2}P_j-\gamma_j & \frac{E_k}{2}P_j-3\gamma_j & E_kP_j-2\gamma_j & \frac{E_k}{2}P_j-\gamma_j & E_kP_j & E_kP_j-2\gamma_j & \frac{3E_k}{2}P_j-\gamma_j \\
-\frac{E_k}{2}P_j-3\gamma_j & -4\gamma_j & -2\gamma_j & \frac{E_k}{2}P_j-3\gamma_j & -2\gamma_j & \frac{E_k}{2}P_j-3\gamma_j & \frac{E_k}{2}P_j-\gamma_j & E_kP_j-2\gamma_j & -2\gamma_j & \frac{E_k}{2}P_j-3\gamma_j & \frac{E_k}{2}P_j-\gamma_j & E_kP_j-2\gamma_j & \frac{E_k}{2}P_j-\gamma_j & E_kP_j-2\gamma_j & E_kP_j & \frac{3E_k}{2}P_j-\gamma_j \\
-4\gamma_j & \frac{E_k}{2}P_j-3\gamma_j & \frac{E_k}{2}P_j-3\gamma_j & E_kP_j-2\gamma_j & \frac{E_k}{2}P_j-3\gamma_j & E_kP_j-2\gamma_j & E_kP_j-2\gamma_j & \frac{3E_k}{2}P_j-\gamma_j & \frac{E_k}{2}P_j-3\gamma_j & E_kP_j-2\gamma_j & E_kP_j-2\gamma_j & \frac{3E_k}{2}P_j-\gamma_j & E_kP_j-2\gamma_j & \frac{3E_k}{2}P_j-\gamma_j & \frac{3E_k}{2}P_j-\gamma_j & 2E_kP_j
\end{bmatrix}_{16\times16}
$$

5

Multi-Valued Cells Based on Polarization Calculations

5.1 Ternary QCA

5.1.1 Proposed Ternary QCA

A ternary quantum dot cellular automata (TQCA) cell is an enhanced version of a binary quantum dot cellular automata (bQCA) cell. The technology consists of eight potential wells positioned on a circle at angles separated by 45° (i.e., at 0°, 45°, 90°, 135°, 180°, 225°, 270°, and 315°). According to the positions of wells in TQCA [93, 101, 102, 104, 108, 109], four states are created. Here, the structure presented in Ref. [109] is used with changes in the cell dimensions. The polarizations of each cell were obtained in the previous chapter. The number of polarization states is equal to the number of states as reported in Refs. [93, 101, 103, 104, 108]. However, the obtained values are different because the two-electron structure is used here. All the states that the two electrons can take are considered. Here, we use polarizations {−1, 0, 1}, instead of {0, −2, 0, 0, 2} or {A, C, D, B}. Using this concept, we design AND, OR, and NOT gates and simulate them. Figure 5.1 shows the calculated polarizations.

The proposed cell structure in Ref. [129] is the same as in Ref. [101]; only the distance from the well to the cell wall is changed to 1.5 nm, and the distance from each cell to the next cell is 0.5 nm [129]. Figure 5.2 shows the proposed cell, and Figure 5.3 shows the distance between the two cells.

5.1.2 Energy Calculations in Proposed TQCA

Generally, energy in TQCA, like in bQCA, is divided into two categories: intracellular energy and extracellular energy [109, 129]. Intracellular energy includes internal electrostatic energy, ground state energy, kink energy, and quantum tunneling [129]. Intracellular energies are constant values that do not change with cell polarization. Extracellular energy contains external electrostatic energy that affects the adjacent cell, and the interaction resulting from this energy is the operation principle of quantum dot cellular automata (QCA) cells [101, 129–132]. The following is a review of each of the mentioned energies.

5.1.2.1 Internal Electrostatic Energy

Internal electrostatic energy in a single cell is due to the electrostatic force between the two electrons in the cell. Equation (5.1) calculates the internal energy of a cell in which q_i and q_j are the

Computing with Multi-Value Logic in Quantum Dot Cellular Automata, First Edition.
Reza Sabbaghi-Nadooshan, Reza Akbari-Hasanjani, Leila Dehbozorgi, Majid Haghparast, and Hamid Reza Akbari-Hasanjani.

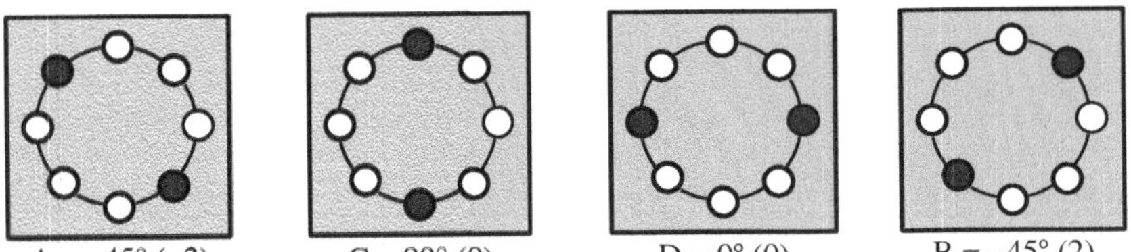

Figure 5.1 Polarizations of the ternary model.

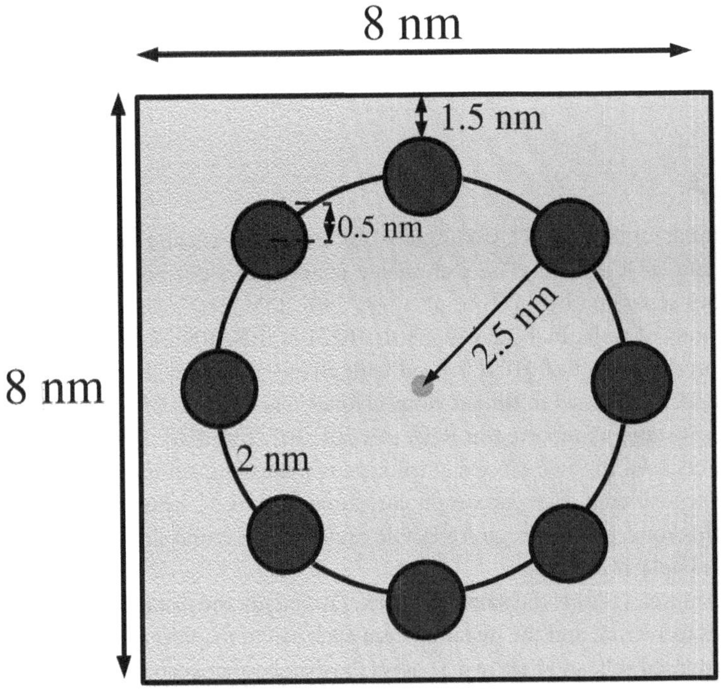

Figure 5.2 TQCA cell proposed in Ref. [129]. Source: Akbari-Hasanjani et al. [129]/with permission of Elsevier.

charge of electrons, d_{ij} is the distance between electrons in each cell, ε_0 is the vacuum permeability, and ε_r is the permeability of the material used in the cell [109, 129].

$$E_{\text{internal}} = \frac{1}{4\pi\varepsilon_0\varepsilon_r}\sum_{i=0}^{m}\sum_{j=0}^{n}\frac{q_i q_j}{d_{ij}} \tag{5.1}$$

According to Figure 5.1, the internal electrostatic energy for each cell is equal to:

$$E_{\text{internal}} = 4.6115 \times 10^{-20}\ (j)$$

5.1.2.2 Ground State Energy

Ground state energy occurs when no clock enters the circuit and electrons are in the ground state. To calculate the state energy of each electron, the time-independent Schrodinger wave equation (5.2) is used as follows [109, 129]:

$$E_n = \frac{n^2\pi^2\hbar^2}{2mL^2} \tag{5.2}$$

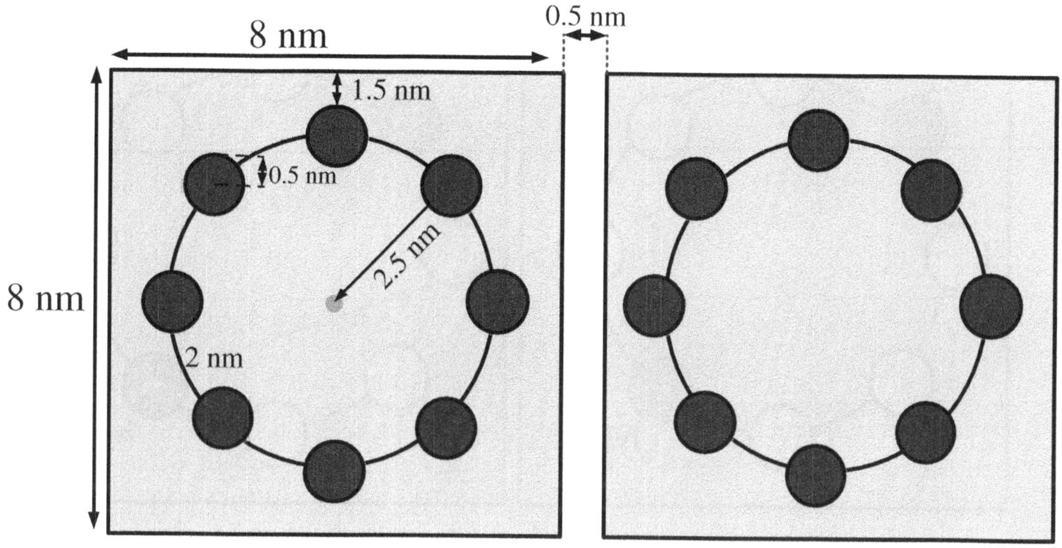

Figure 5.3 Distance between items of two proposed TQCA cells.

where $\hbar$ is the reduced Planck constant, m is the mass of the electron, and L is the width of the well. To calculate the ground state energy ($n = 1$) of an electron, the state energy of the electron is equal to:

$$E_{G-S} = 6.025 \times 10^{-20}\ (j)$$

5.1.2.3 Kinetic Energy

This energy is related to the movement of electrons and depends on both the ambient temperature and the rotation of the electrons. The final value of kinetic energy is obtained from the algebraic sum of these two factors as given in Eq. (5.3) [109, 129].

$$E_{\text{kinetic}} = \frac{1}{2}K_B T + K_B T = \frac{3}{2}K_B T \tag{5.3}$$

In Eq. (5.3), K_B is the Boltzmann constant, and T is the Kelvin temperature. Kinetic energy for each cell is equal to:

$$E_{\text{kinetic}} = \frac{3}{2}K_B T = 6.171 \times 10^{-21}\ (\text{j})$$

The radius of the cell proposed in the present work is equal to that reported in Refs. [105, 109]. The distance between the two wells is equal to 2 nm, which is the same as the height of the barrier between the two potential wells. Moreover, the radius of each well is 0.5 nm, and tunneling rarely occurs. Therefore, it is assumed that the wave function of each electron in one well is not related to those in other wells, and the wave function of each electron is distinct [129].

5.1.2.4 External Electrostatic Energy

External electrostatic energy stems from the operation of QCA-based cells. The external electrostatic field changes the polarization of cells adjacent to another cell. Based on the location of two

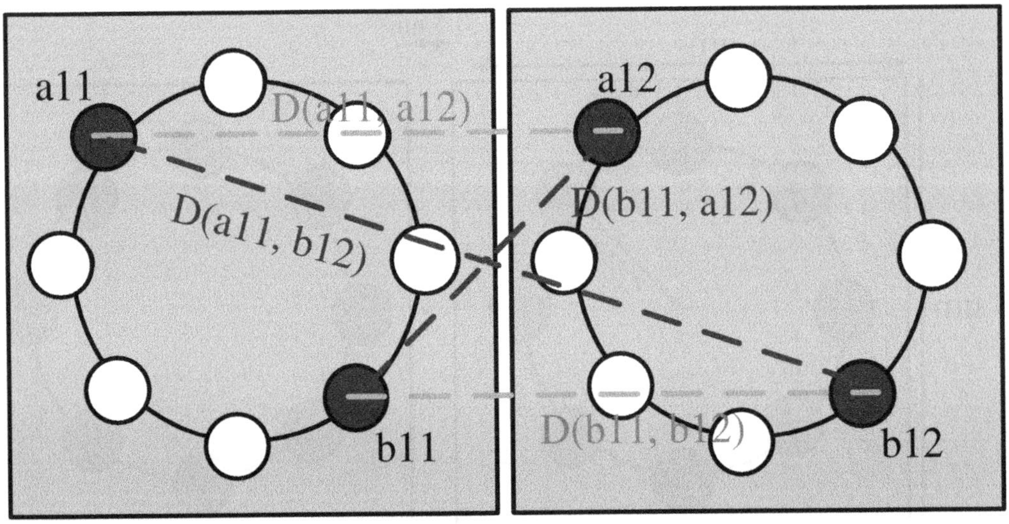

Figure 5.4 Distances between two ternary cells to calculate the external electrostatic energy. Source: Adapted from Refs. [130, 131].

moving electrons in a cell, the force given in Equation (5.4) is applied by each electron in a cell to the electrons in adjacent cells [93, 109, 129, 130].

$$E_{ij} = \frac{1}{4\pi\varepsilon_0\varepsilon_r}\sum_{i=0}^{m}\sum_{j=0}^{n}\frac{q_i q_j}{d_{ij}} \tag{5.4}$$

To calculate the external electrostatic energy between two cells, we follow the technique presented in Refs. [103, 129]. There are two electrons in each cell; the distance from each electron to the first cell to the electrons of the second cell must be initially calculated based on the distance of the pairs (a1 and a2), (b1 and b2), (a1 and b2), and (a2 and b1). Then, using Eq. (5.4), the external electrostatic energy between the potential well pairs a1 and a2, b1 and b2, a1 and b2, and a2 and b1 are calculated as E_{e1}, E_{e2}, E_{e3}, and E_{e4}, respectively. In the next step, the external electrostatic energies denoted by the same letter (E_{e1} and E_{e2}) are summed, and the sum of the external electrostatic energies with the same letter is calculated again. To calculate the final energy (E_t), we subtract the amount of external electrostatic energy of the pairs with the same letter from the electrostatic energy of the pairs with different letters. Equation (5.5) calculates the final electrostatic energy of each state. Figure 5.4 shows the distances between two ternary cells used to calculate the external electrostatic energy [130, 131].

$$E_t = (E_{e1} + E_{e2}) - (E_{e3} + E_{e4}) \tag{5.5}$$

5.2 Quaternary QCA

5.2.1 Proposed QQCA

As mentioned in the previous chapter, a quaternary quantum dot cellular automata (QQCA) cell needs three particles to create four distinct states. The presence of three electrons in a single-layer cell causes instability in the gates designed with this cell. Therefore, a structure that can create QQCA polarization and is always stable must be designed [130].

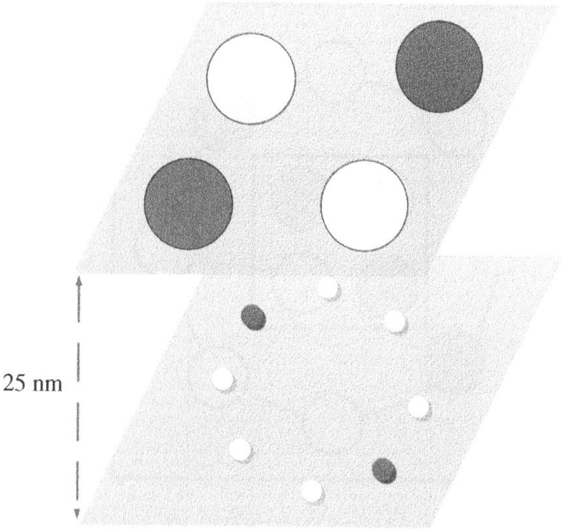

Figure 5.5 Shows the proposed QQCA model. Source: Adapted from Akbari-Hasanjani and Sabbaghi-Nadooshan [130].

The polarizations in QQCA are calculated as {−3, −1, −1, 3} [130]. The proposed cell uses the algebraic sum of bQCA and TQCA polarizations in two layers at a distance of 25 nm from each other. The distance of 25 nm is chosen to achieve maximum reduction in the electrostatic interaction between the two layers that are assumed to be isolated. Figure 5.5 shows the proposed QQCA model [130].

As mentioned, in the proposed QQCA model, two layers are used; the first layer is based on TQCA, and the second layer is based on bQCA. The specifications of the layers used for the proposed QQCA model are given in Figure 5.6 [130]. In this figure, the wells are located on a circle with a radius of 2.5 nm and a distance of 1.5 nm from the cell wall [130]. Also, in the first layer (ternary cell), each well is located at an angle of 45° from the other well, and the first well is at 0°. In the second layer (bQCA cell), the angular distance from each well to the next well is 90°, the distance from the wells to the cell wall is 1.5 nm, and the radius of the potential wells is 0.5 nm [130].

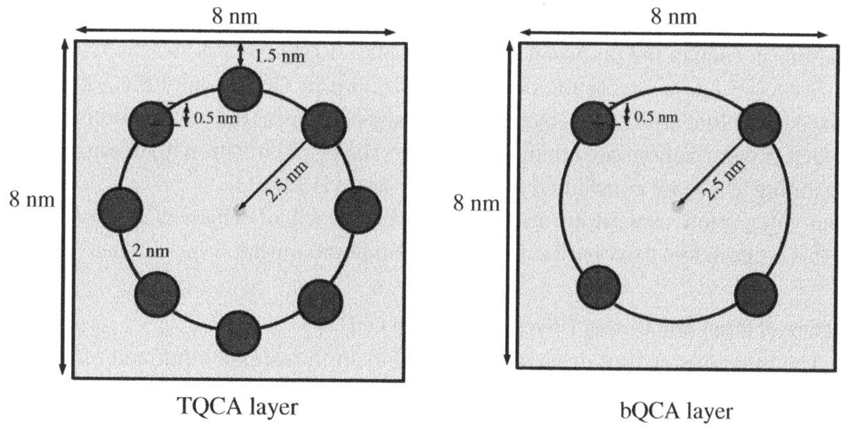

Figure 5.6 Specifications of the used layers. Source: Adapted from Akbari-Hasanjani and Sabbaghi-Nadooshan [130].

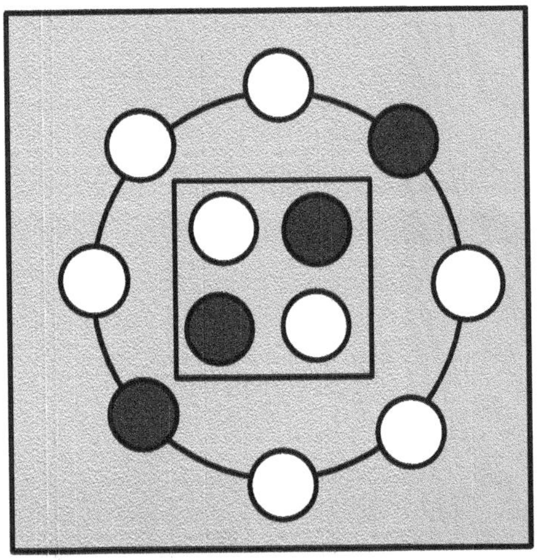

Figure 5.7 QQCA cell symbol. Source: Adapted from Akbari-Hasanjani and Sabbaghi-Nadooshan [130].

The proposed model is a three-dimensional structure, so it is difficult to represent this model in the design of basic gates and quaternary circuits. Figure 5.7 is used to represent the QQCA cell symbol [130].

To construct each of the QQCA polarizations, we have [130]:

To obtain polarization −3 in QQCA, the polarizations −2 in the TQCA layer and −1 in the bQCA layer are used. To obtain polarization −1 in QQCA, the polarizations −2 in the TQCA layer and +1 in the bQCA layer are used. To obtain polarization +1 in QQCA, +2 in the TQCA layer and −1 in the bQCA layer are used. To obtain polarization +3 in QQCA, polarizations +2 and +1 in TQCA and bQCA layers are used, respectively. Also, the dimensions of bQCA and TQCA cells in each layer equal 8 nm × 8 nm. Table 5.1 shows the polarizations of the QQCA cell.

5.2.2 Integration in Input Applied to Proposed QQCA Cell

Each proposed QQCA cell consists of two layers. The first layer contains a ternary quantum cell, and the second layer contains a binary cell [130]. Also, the input is applied to each cell using single-electron transistors (SETs). For this reason, it is possible to apply integrated input to the proposed cell structure using digital circuits. Integration is concerned with the application of the calculated polarizations in each of the proposed QQCA cells [130–133].

As described in Section 5.2.1, to create the desired polarization in QQCA, only the −2 and 2 polarizations are used in the first layer. Then, by combining (algebraic sum) the polarizations of the layers, the desired QQCA polarizations are obtained. In the first layer, polarization 0 corresponding to 0° and 0 corresponding to 90° are considered intracellular states [130–133].

Input and output integration uses input and output drives, each of which is discussed in Section 5.2.2.1. In this section, two drives are suggested for input and output.

5.2.2.1 First Structure of Input and Output Drives in Proposed Cell

Input Drive Design The input and output drives are intended to apply a single input and receive a single output based on polarization. An input drive for QQCA uses two 2*4 decoders (DEC) with one Enable. The input of the bQCA layer is used as the decoder input so that the polarization of the TQCA layer can be created at its input, and the provided drive is evaluated as a block. In the proposed input drive, the first decoder creates polarization −2, and the second decoder creates

Table 5.1 QQCA cell polarizations.

Polarization	Cell Model	Symbol Cell
(−3): A	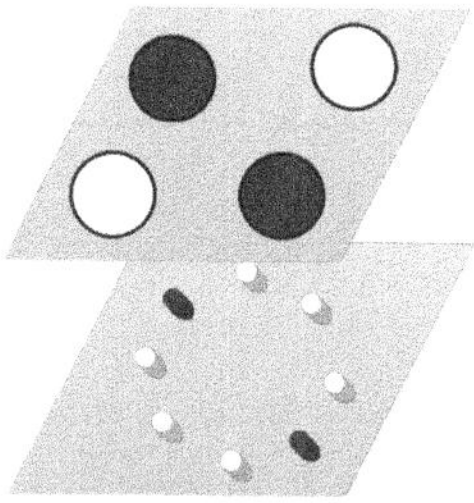	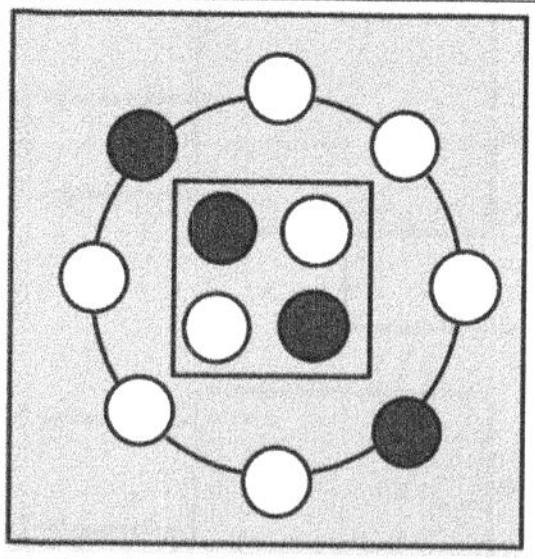
(−1): B	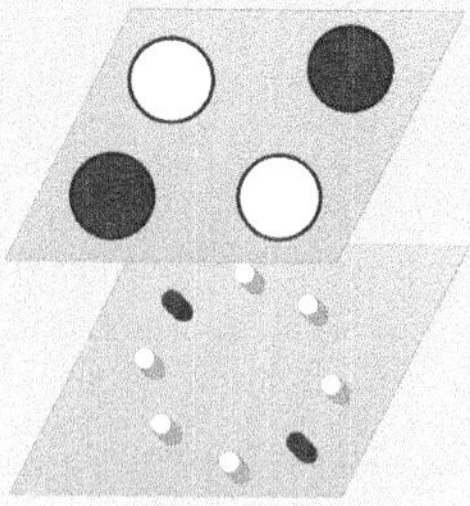	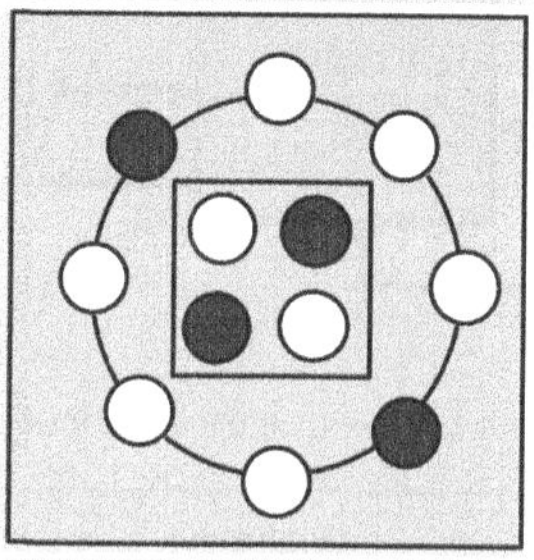
(+1): C	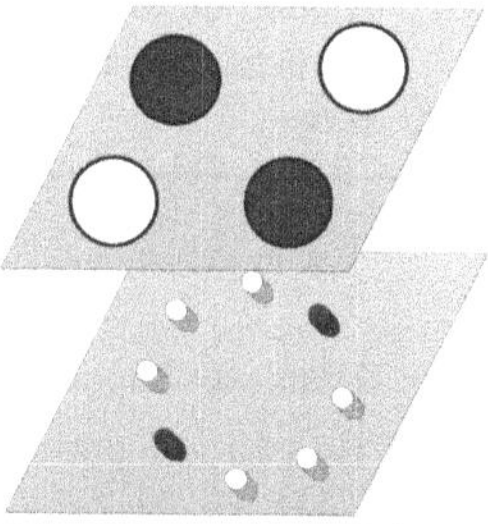	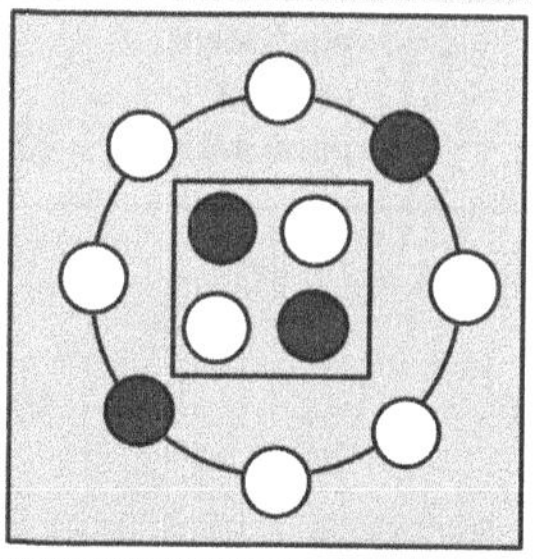
(+3): D	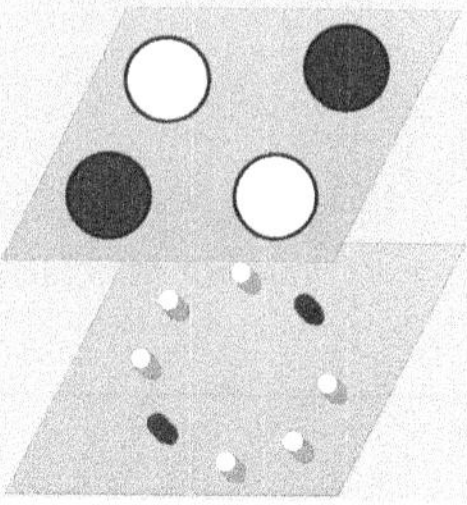	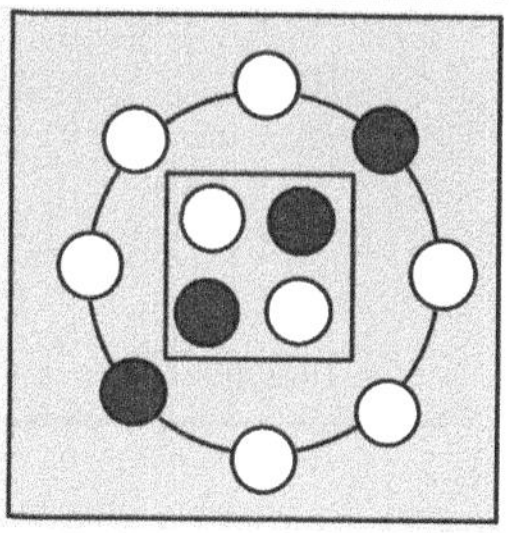

Source: Adapted from Akbari-Hasanjani and Sabbaghi-Nadooshan [130]

polarization 2. Decoder 1 is activated if the Enable base is high in the proposed circuit, and decoder 2 is activated if the Enable is high.

The noteworthy point in designing based on this method is that in applying binary inputs that use SETs, the transistors related to polarizations −1 and +1 are never activated simultaneously. These transistors and the way of applying the input to them are the basis of applying the input layer to the ternary layer or SETs. Figure 5.8 shows the schematic of the input drive for input integration in a QQCA cell [133].

Table 5.2 presents the input drive truth table. Here, the input to the binary layer SETs determines the polarization in the ternary layer.

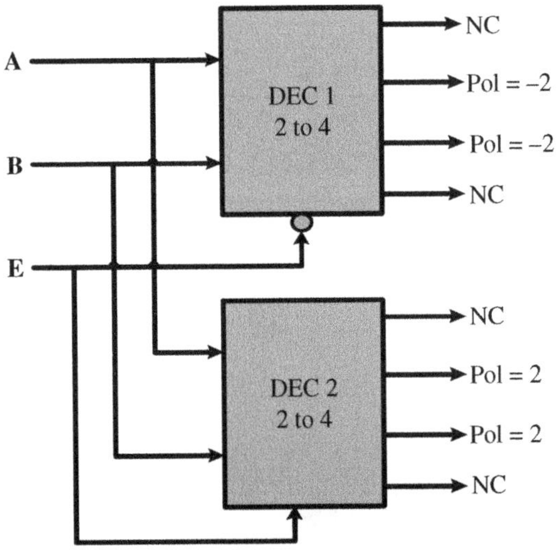

Figure 5.8 Input drive block in QQCA. Source: Akbari-Hasanjani et al. [133]/IEEE/CC-BY 4.0.

Table 5.2 Truth table for the quaternary input drive.

Enable	Input binary		Quantum equivalent of input polarization	Output	Connection to transistor corresponding to ternary layer	Quantum equivalent of quaternary
0	A	B	Illegal	Decoder 1	—	—
	0	0		—		
0	0	1	–1	D1	Transistor of polarization –2	–3
0	1	0	1	D2	Transistor of polarization –2	–1
0	1	1	Illegal	—	—	
1	0	0	Illegal	Decoder 2	—	—
				—		
1	0	1	–1	D1	Transistor of polarization 2	1
1	1	0	1	D2	Transistor of polarization 2	3
1	1	1	Illegal	—	—	—

Structure of First Output Drive The purpose of designing an output drive in QQCA is to provide a single output. In the proposed drive, if there is some charge in the SET channel of TQCA and bQCA, it is considered equal to 1, while in the absence of charge, it is assumed to be 0.

Two encoders (ENC) and a decoder (DEC) are used for integration. In the bQCA layer, two SETs have opposite states to show two polarizations. The two SETs are never active together. In the TQCA layer, four SETs are used to achieve the required polarizations. These transistors are never activated together. In Figure 5.9, the first proposed model for the output drive is shown [133].

Table 5.3 shows the truth table of the output drive. In this table, an illegal state occurs if the two inputs, B and C, of the decoder in output are the same. An electron corresponds to each polarization, and two or more polarizations never occur simultaneously.

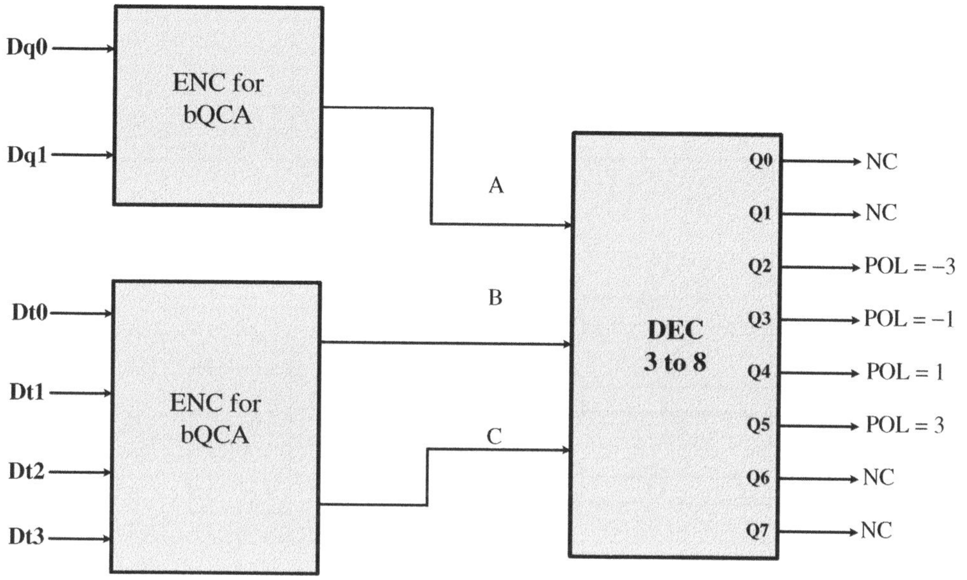

Figure 5.9 The first proposed output drive. Source: Akbari-Hasanjani et al. [133]/IEEE/CC-BY 4.0.

Table 5.3 Truth table of the first structure of the quaternary output drive.

C	B	A	bQCA polarization	TQCA polarization	Q7	Q6	Q5	Q4	Q3	Q2	Q1	Q0	Output polarization of QQCA
0	0	0	−1	Illegal	0	0	0	0	0	0	0	1	Illegal
0	0	1	1	Illegal	0	0	0	0	0	0	1	0	Illegal
0	1	0	−1	−2	0	0	0	0	0	1	0	0	−3
0	1	1	1	−2	0	0	0	0	1	0	0	0	−1
1	0	0	−1	2	0	0	0	1	0	0	0	0	1
1	0	1	1	2	0	0	1	0	0	0	0	0	3
1	1	0	−1	Illegal	0	1	0	0	0	0	0	0	Illegal
1	1	1	1	Illegal	1	0	0	0	0	0	0	0	Illegal

5.2.2.2 Second Structure of Input and Output Drives

Second Structure of Input Drive In the proposed method, the second input drive uses QCA cells and is designed using a circuit shown in Figure 5.10. Using the binary layer input and an Enable, this method determines the ternary layer state based on the QQCA input polarization [130]. A bQCA to TQCA converter is used in the second input drive design investigated in [132].

In the input drive structure provided for both inputs, a bQCA to TQCA converter is used to create polarizations −2 and +2 in the ternary layer. All gates of the proposed structure must be of TQCA type. Moreover, due to the ternary structure of the circuit, a 1*3 decoder is used. The decoder line j1 is not used because the produced polarization is undesirable [130]. Table 5.4 shows the truth table of the second structure of the input drive [130].

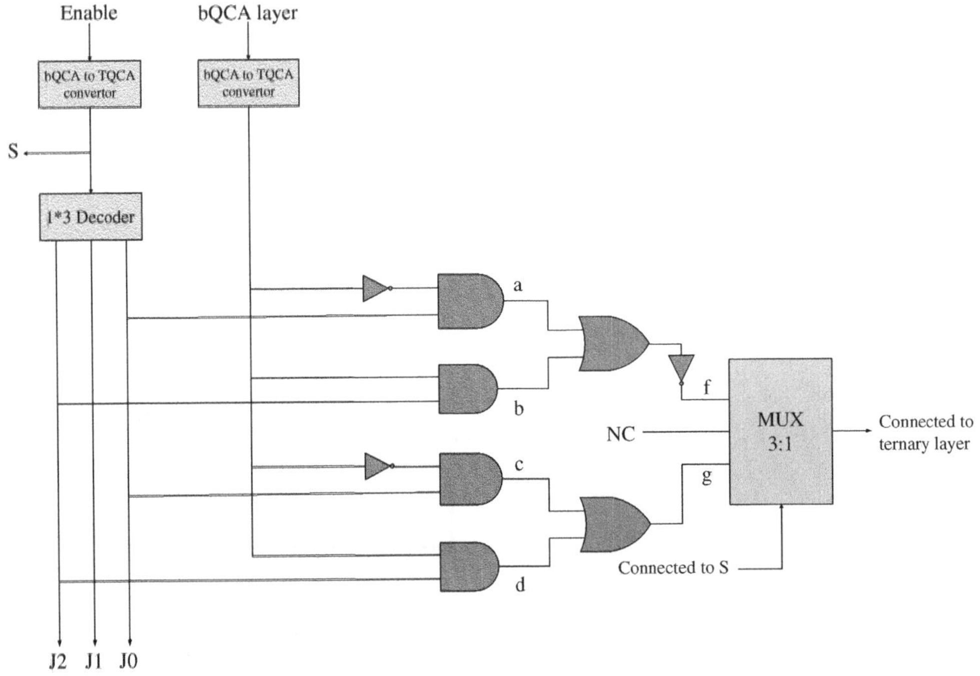

Figure 5.10 Second structure of the input drive. Source: Akbari-Hasanjani and Sabbaghi-Nadooshan [130]/with permission of John Wiley & Sons.

Table 5.4 The truth table of the second structure of the quaternary input drive.

Enable (Pol)	bQCA input (Pol)	S(Pol)	A	J2	J1	J0	d	C	b	A	f	g	Output of MUX
0 (−1)	0 (−1)	0 (−2)	0 (−2)	0	0	2	−2	−2	−2	2	−2	−2	−2
0 (−1)	1 (+1)	0 (−2)	2 (+2)	0	0	2	−2	−2	2	−2	−2	−2	−2
1 (+1)	0 (−1)	2 (+2)	0 (−2)	2	0	0	−2	2	−2	−2	2	2	2
1 (+1)	1 (+1)	2 (+2)	2 (+2)	2	0	0	2	−2	−2	−2	2	2	2

Source: Akbari-Hasanjani and Sabbaghi-Nadooshan [130]/with permission of John Wiley & Sons.

Structure of Second Output Drive In the proposed method, a TQCA to bQCA converter is used to create a single output, which converts the output of the ternary layer into bQCA [135]. In the design of the second output drive, the TQCA to bQCA converter, which is provided in [135], is used.

After using the converter, the QQCA output cell information is converted to binary information. Now, the output value can be determined from a bQCA-based collector, and the output is shown in Table 5.5 [130].

In Table 5.5, the states marked with * do not occur since, in the QQCA structure, polarization 0 does not occur in the TQCA layer. This is because 0 is prevented in the TQCA layer, or in other words, 0 events are eliminated [130].

Table 5.5 The truth table for determination of quaternary output using bQCA adder.

bQCA layer input	TQCA layer input	QCA Pol for input adder	TQCA Pol before conversion to binary		Carry (*t*)	Sum	Equivalent polarization
		Carry(*t*−1)	b	A			
−1	−2	−1	−1	−1	0	0	−3
−1	0	−1	−1	1	—	—	*
−1	0	−1	1	−1	—	—	*
−1	2	−1	1	1	1	0	+1
1	−2	1	−1	−1	0	1	−1
1	0	1	−1	1	—	—	*
1	0	1	1	−1	—	—	*
1	2	1	1	1	1	1	3

Source: Akbari-Hasanjani and Sabbaghi-Nadooshan [130]/with permission of John Wiley & Sons.

5.2.3 Energy in Proposed QQCA Cell

Energy in QQCA cells, similar to bQCA and TQCA cells, consists of both intracellular and extracellular energies. Two layers are used to generate QQCA polarization in the proposed cell. The first layer contains a TQCA cell, and the second one contains a bQCA cell. Intracellular energy is constant and unchangeable with the changes in polarization. The types of this energy are internal electrostatic energy, ground state energy, kink energy, and quantum tunneling energy. Extracellular energy is the external electrostatic energy that affects adjacent cells. A QCA-based cell operates based on the external electrostatic energy between adjacent cells.

5.2.3.1 Internal Electrostatic Energy

In a QQCA cell, Eqs. (5.6) to (5.8) are used for the two layers to calculate the internal electrostatic energy between the electrons of each layer. The values of the internal electrostatic energy are algebraically summed [130].

$$E_{\text{internal}_{\text{TQCA}}} = \frac{1}{4\pi\varepsilon_0\varepsilon_r}\sum_{i=0}^{m}\sum_{j=0}^{n}\frac{q_i q_j}{d_{ij_{\text{TQCA}}}} \tag{5.6}$$

$$E_{\text{internal}_{\text{bQCA}}} = \frac{1}{4\pi\varepsilon_0\varepsilon_r}\sum_{i=0}^{m}\sum_{j=0}^{n}\frac{q_i q_j}{d_{ij_{\text{bQCA}}}} \tag{5.7}$$

$$E_{\text{internal}} = E_{\text{internal}_{\text{TQCA}}} + E_{\text{internal}_{\text{bQCA}}} \tag{5.8}$$

5.2.3.2 Ground State Energy

Ground state energy occurs when no clock enters the cell and electrons are in the ground state. To calculate the ground state energy, the time-independent Schrodinger equation is used, and for the proposed QQCA, the energy of each layer is calculated separately and summed algebraically. Equations (5.9) to (5.11) calculate the ground state energy in QQCA [130].

$$E_{\text{G-S}_{\text{TQCA}}} = \frac{n^2\pi^2\hbar^2}{2mL_{\text{TQCA}}{}^2} \tag{5.9}$$

$$E_{\text{G-S}_{\text{bQCA}}} = \frac{n^2\pi^2\hbar^2}{2mL_{\text{bQCA}}{}^2} \tag{5.10}$$

$$E_{\text{G-S}_{\text{total}}} = E_{\text{G-S}_{\text{TQCA}}} + E_{\text{G-S}_{\text{bQCA}}} \tag{5.11}$$

5.2.3.3 Kinetic Energy

Kinetic energy is related to the movement of electrons inside a cell. This energy depends on the ambient temperature. Since there are two layers in the proposed quaternary cell, the kinetic energy in each layer must be calculated and summed algebraically [130]. Equations (5.12) to (5.14) calculate the kinetic energy in the proposed quaternary cell [130].

$$E_{\text{kinetic}_{\text{TQCA}}} = \frac{1}{2}K_{\text{B}}T_{\text{TQCA}} + K_{\text{B}}T_{\text{TQCA}} = \frac{3}{2}K_{\text{B}}T_{\text{TQCA}} \tag{5.12}$$

$$E_{\text{kinetic}_{\text{bQCA}}} = \frac{1}{2}K_{\text{B}}T_{\text{bQCA}} + K_{\text{B}}T_{\text{bQCA}} = \frac{3}{2}K_{\text{B}}T_{\text{bQCA}} \tag{5.13}$$

$$E_{\text{kinetic}_{\text{Total}}} = E_{\text{kinetic}_{\text{TQCA}}} + E_{\text{kinetic}_{\text{bQCA}}} \tag{5.14}$$

Because both layers are created on the same plate, the ambient temperatures of the layers are the same ($T_{\text{TQCA}} = T_{\text{bQCA}} = T$), so Eq. (5.14) can be written in the form of Eq. (5.15) [130].

$$E_{\text{kinetic}_{\text{Total}}} = 3K_{\text{B}}T \tag{5.15}$$

In Eqs. (5.12) to (5.15), K_{B} is the Boltzmann constant, and T is the ambient temperature in degrees Kelvin.

5.2.3.4 External Electrostatic Energy

Electrostatic energy is the basis of the work of QCA-based cells. This energy causes a change in the polarization of adjacent cells without transmitting current [129–132]. Considering the position of electrons in a cell, the external electrostatic energy is obtained using Eq. (5.16) [130].

$$E_{ij} = \frac{1}{4\pi\varepsilon_0\varepsilon_r}\sum_{i=0}^{m}\sum_{j=0}^{n}\frac{q_i q_j}{d_{ij}} \tag{5.16}$$

In the proposed cell, two layers are used to prevent electrostatic interactions of the layers, and the distance between the isolated layers is 25 nm. To calculate the electrostatic energy between two cells, the method used in Ref. [80] is employed with the only difference that instead of one layer, there are two layers in each cell [129–131].

To calculate the external electrostatic energy, first, the distance of each electron from the first cell to the second cell in the first layer (ternary) is calculated, i.e., for the pairs (a12, a11), (b12, b11), (b12, a11), and (b11, a12). Then, using Eq. (3.16), the external electrostatic energy between the electron pairs (a12, a11), (b12, b11), (b12, a11), and (b11, a12) is calculated as E_{e11}, E_{e12}, E_{e13}, and E_{e14}, respectively. In the next step, the electrostatic energies from the pairs with the same letter (E_{e11} and E_{e12}) are added together. Then, electrostatic energies from the pairs with different letters (E_{e13} and E_{e14}) are added algebraically. To calculate the final electrostatic energy of the ternary layer, the previously obtained values of external electrostatic energy are subtracted from each other. Equation (5.17) calculates the final energy value of the first layer, and Figure 5.11 shows the distances considered for calculations [130].

$$E_{t_{\text{TQCAlayer}}} = (E_{\text{e11}} + E_{\text{e12}}) - (E_{\text{e13}} + E_{\text{e14}}) \tag{5.17}$$

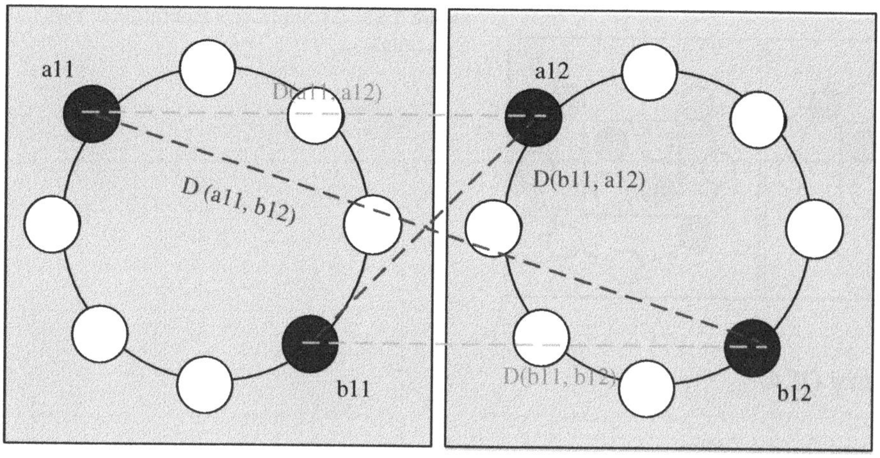

Figure 5.11 Distances between two ternary cells in the first layer. Source: Akbari-Hasanjani and Sabbaghi-Nadooshan [130]/with permission of John Wiley & Sons.

Figure 5.12 Distances between two QQCA cells in the second layer.

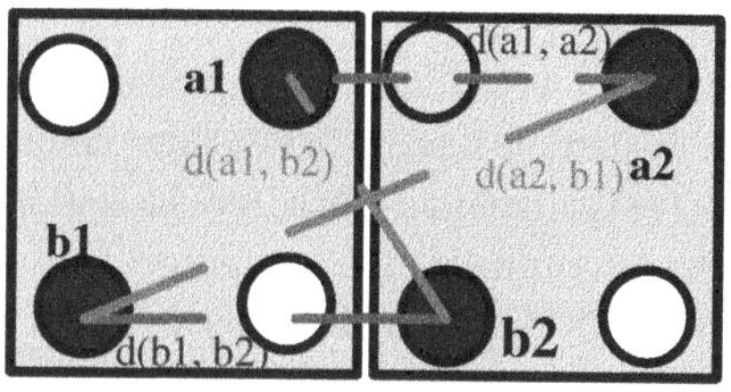

The calculations for the second layer are similar to those performed for the first layer. First, the distances of the electrons in the second layer of the first cell from the electrons of the second layer of the second cell, represented by the pairs (a22, a21), (b22, b21), (b22, a21), and (b21, a22), are calculated. Then, Eq. (3.16) is used to calculate the external electrostatic energy between the electron pairs (a22, a21), (b22, b21), (b22, a21), and (b21, a22), which are denoted by E_{e21}, E_{e22}, E_{e23}, and E_{e24}, respectively. To calculate the electrostatic energy of the second (binary) layer ($E_{\text{total binary layer}}$), the electrostatic energy of the pairs with the same letter is subtracted from those with different letters. Finally, the magnitude (absolute value) of the electrostatic energy of the binary layer is obtained [54]. Figure 5.12 shows how to calculate the distance between the electrons of two cells, which are represented by the same and different letters. Equation (5.18) expresses the final energy value of the second (binary) layer.

$$E_{t_{\text{bQCAlayer}}} = (E_{e21} + E_{e22}) - (E_{e23} + E_{e24}) \tag{5.18}$$

In the final step for calculating the electrostatic energy in a QQCA cell is to obtain the algebraic sum of $|E_{t_{\text{TQCA layer}}}|$ and $|E_{t_{\text{bQCA layer}}}|$ [54]. Equation (5.19) shows the amount of external electrostatic energy of two QQCA cells. Figure 5.13 is a symbolic representation of two QQCA cells [130].

$$|E_{t_{\text{QQCA}}}| = |E_{t_{\text{TQCA layer}}}| + |E_{t_{\text{bQCA layer}}}| \tag{5.19}$$

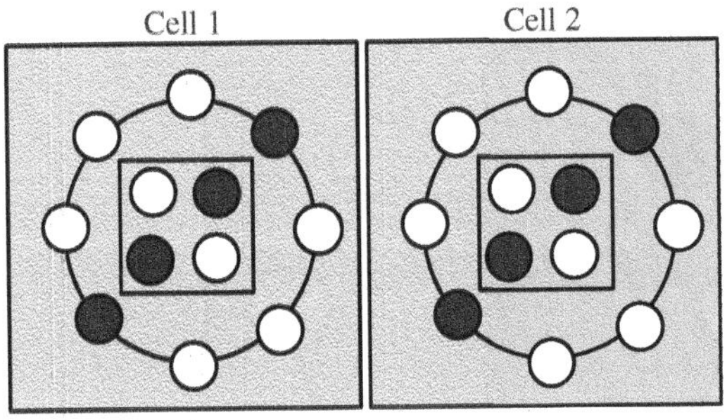

Figure 5.13 Symbolic representation of two QCA cells.

5.3 Quinary QCA

5.3.1 Proposed QuQCA Cells

A quinary quantum dot cellular automata (QuQCA) cell is a four-particle system (with four electrons), which produces five distinct states [131]. The presence of more than two electrons in a layer destabilizes the logic gates, so a cell should be designed with the ability to produce stable logic gates [131].

Chapter 4 calculates the polarization values as {−4, −2, 0, 2, −4} using quantum measurement in a four-particle system. A two-layer cell is presented to design stable logic gates with the proposed cell [131]. To obtain QuQCA polarizations, the algebraic sum of the polarizations of the layers has to be determined [131]. In Figure 5.14, the proposed QuQCA cell model consists of two layers, each with a TQCA cell. In the proposed QuQCA cell, the layers are isolated at a distance of 25 nm so that the electrostatic interaction between the layers can be ignored [131].

Figure 5.15 shows the characteristics of the ternary layers used in the proposed QuQCA cell, as reported by Akbari-Hasanjani and Sabbaghi-Nadooshan [131]. In each ternary layer, the potential wells are located on a circle with a radius of 2.5 nm and at a distance of 1.5 nm from the cell wall. Also, in the ternary cell, the angular distance from each well to the next well is 45°, the first well is located at 0°, and the radius of potential wells is 0.5 nm [131].

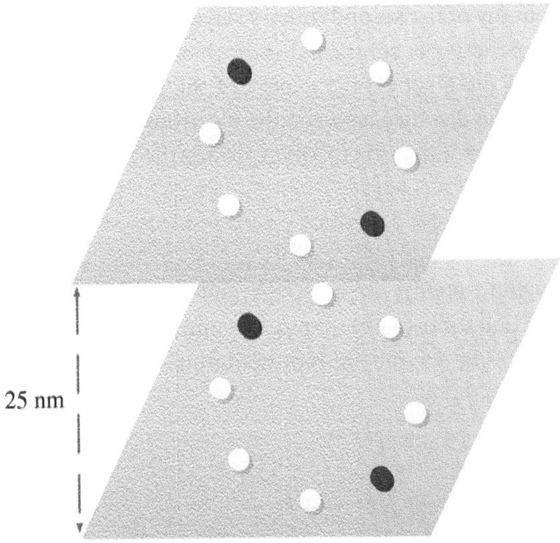

Figure 5.14 Proposed QuQCA model. Source: Akbari-Hasanjani and Sabbaghi-Nadooshan [131]/with permission of John Wiley & Sons.

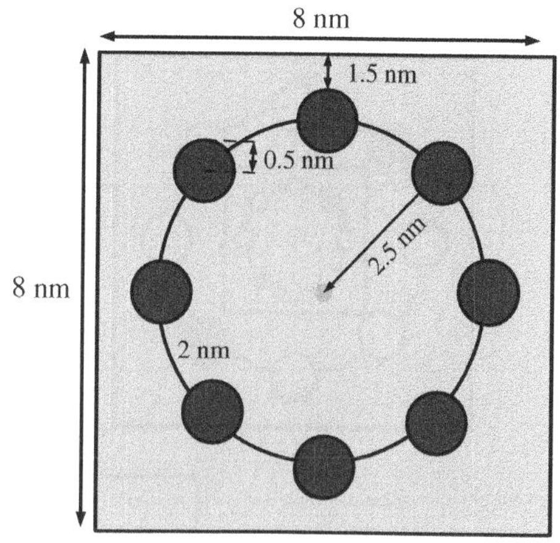

Figure 5.15 Specifications of the ternary layer. Source: Akbari-Hasanjani and Sabbaghi-Nadooshan [131]/with permission of John Wiley & Sons.

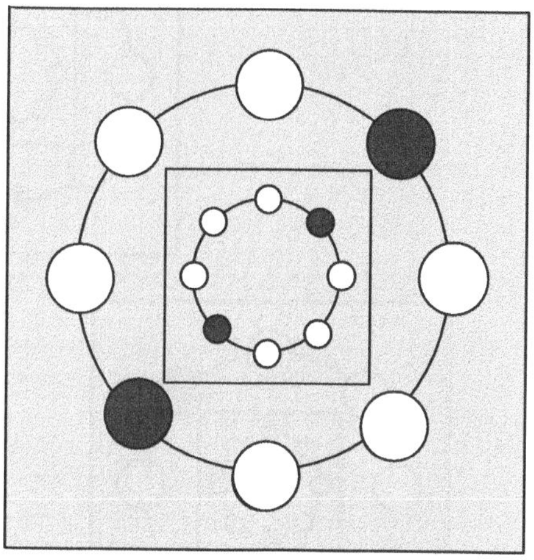

Figure 5.16 QuQCA cell symbol. Source: Akbari-Hasanjani and Sabbaghi-Nadooshan [131]/with permission of John Wiley & Sons.

The proposed model is a three-dimensional structure made of two layers of ternary cells, so visually presenting this model of logic gates is difficult. Figure 5.16 shows the QuQCA symbol [131].

The polarization of each layer of the proposed QuQCA cell can be created as follows:

To obtain polarization −4 in QuQCA, the polarizations of both TQCA layers have to be −2. To obtain polarization −2, polarizations −2 in the first ternary layer and 0 (90°) in the second layer are used. To obtain polarization 0 in QuQCA, polarizations 0 (90°) in both ternary layers are used. To obtain polarization +2, the polarizations +2 in the first layer and 0 (90°) in the second layer are used. Polarization +2 in both ternary layers obtains polarization 4 in QuQCA [131]. Table 5.6 shows how ternary cells construct polarizations in QuQCA.

5.3.2 Integration in Input to QuQCA Cells

The proposed QuQCA cell comprises two layers of TQCA cells [131]. SETs are used to apply input to TQCA layers. Therefore, integrated inputs can be applied to QuQCA cell structure using digital circuits. The integration aims to apply input polarization or receive output polarization calculated in QuQCA [131].

As described in Chapter 4, polarizations {−2, 0 (90°), +2{ can be constructed in QuQCA by algebraically summing the polarizations of both existing ternary layers. The polarization value of the

Table 5.6 Polarizations in QuQCA.

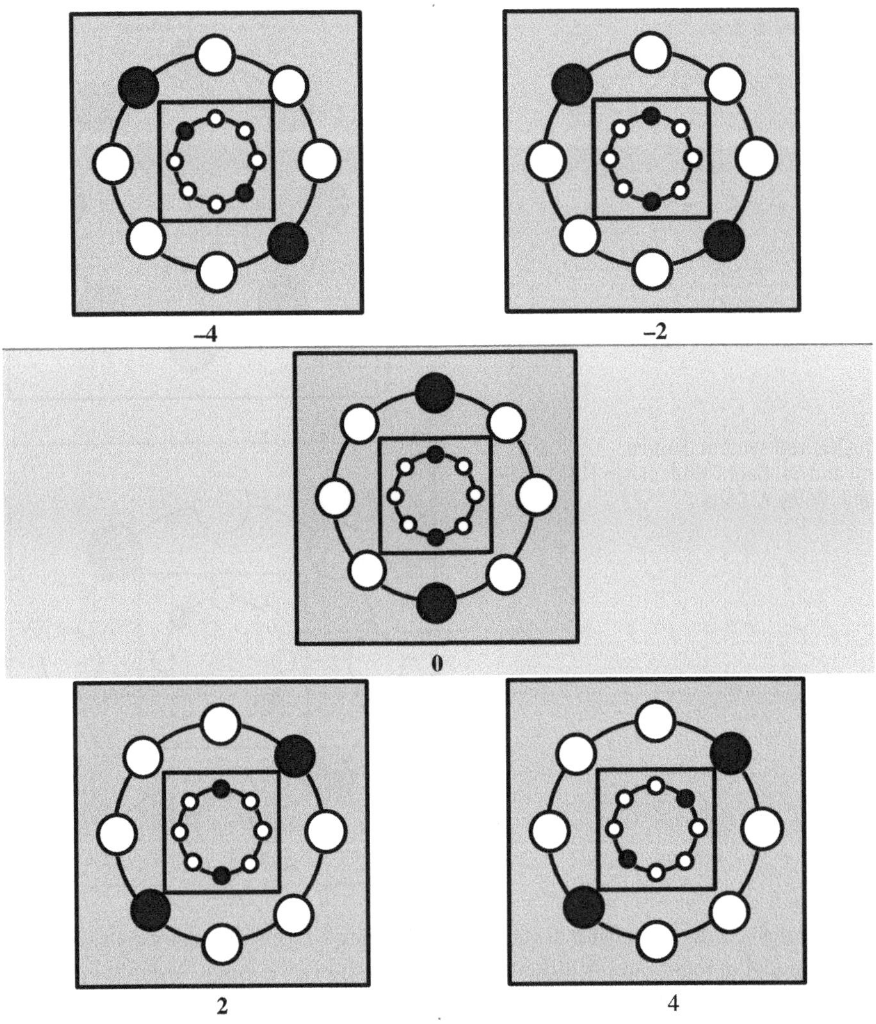

Source: Akbari-Hasanjani and Sabbaghi-Nadooshan [131]/with permission of John Wiley & Sons.

proposed QuQCA is generated, and 0 (0°) in each layer is considered an intracellular state [131]. The input and output drives for integrating input and output in QuQCA are examined in Sections 5.3.2.1 and 5.3.2.2.

5.3.2.1 First Structure of Input and Output Drive

Design of Input Drives The purpose of the input and output drive is to apply input and receive output based on the calculated QuQCA polarization. In designing the input drive in QuQCA, two 3×8 decoders can be used based on how the polarization is produced, and the polarization of one of the layers can be considered the input of the decoders.

As mentioned, QuQCA polarization uses a combination of two-layer ternary polarization. In the proposed method, the first layer SETs are used as the input of the decoders. In this design, the presence of electrons in the SET channel is considered logic 1, and the absence of electrons in the channel is considered 0 [131]. To produce QuQCA polarization, we need input modes.

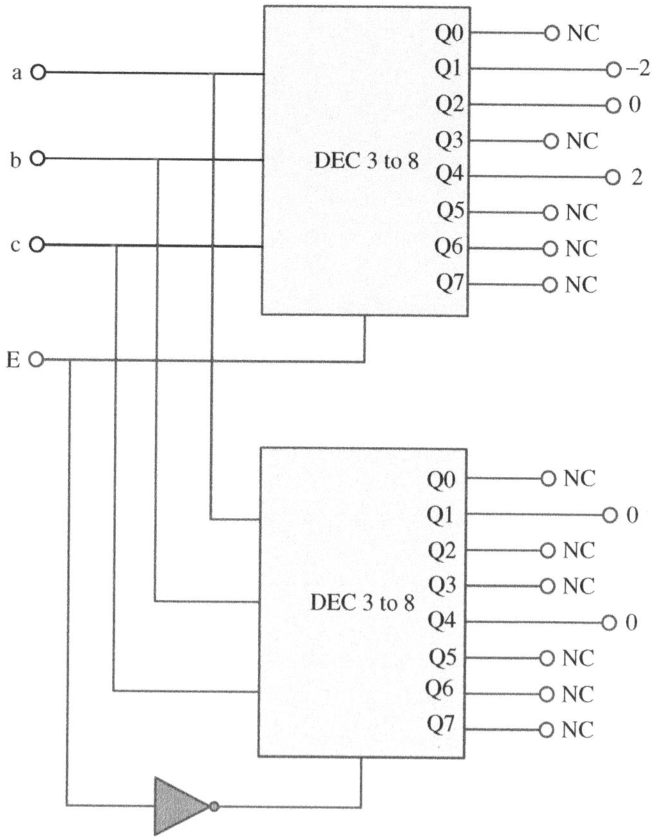

Figure 5.17 Schematic of the input drive using decoder. Source: Akbari-Hasanjani and Sabbaghi-Nadooshan [131]/with permission of John Wiley & Sons.

According to the input modes in QuQCA, the schematic structure of the input drive to integrate the QuQCA cell input is shown in Figure 5.17 [131].

Table 5.7 shows the truth table of the input drive. In this study, the input of the first ternary layer transistors is used to determine the polarization of the second ternary layer [131].

First Structure of Output Drive In the design of the output drive, the purpose is to obtain QuQCA polarization. In the design of the output drive, two 3×8 encoders and one 6×64 decoder are used. In QuQCA cells, each layer uses four SETs to produce polarization. Transistors are never activated together, and the simultaneous activation of more than one SET is an illegal state in QCA-based cells. Figure 5.18 shows the output drive model [131]. In the proposed output drive model shown in Figure 5.18, the output cell of each layer is connected to one encoder.

The above mentioned structure has a high level of complexity, so the structure in Figure 5.19 is presented to integrate the output. In Figure 5.19, five AND gates are used for integration. The inputs of these gates are the states of the output cell layers, or in other words, the SET states of the output cell layers. Table 5.8 shows the output drive states using the AND gate, where inputs a, b, and c are the output polarization states of the first layer and a′, b′, and c′ are the output polarization states of the second layer. Moreover, the SETs of each layer are connected to the input of the AND gates [131].

Table 5.7 Truth table of five-value input drive.

E	b	c	a	Connected to SET in second layer	Q15	Q14	Q13	Q12	Q11	Q10	Q9	Q8	Q7	Q6	Q5	Q4	Q3	Q2	Q1	Q0
0	0	0	0	Illegal	0	0	0	0	0	0	0	0	0	0	0	0	0	0	0	1
0	0	0	1	−2	0	0	0	0	0	0	0	0	0	0	0	0	0	0	1	0
0	0	1	0	0	0	0	0	0	0	0	0	0	0	0	0	0	0	1	0	0
0	0	1	1	Illegal	0	0	0	0	0	0	0	0	0	0	0	0	1	0	0	0
0	1	0	0	2	0	0	0	0	0	0	0	0	0	0	0	1	0	0	0	0
0	1	0	1	Illegal	0	0	0	0	0	0	0	0	0	0	1	0	0	0	0	0
0	1	1	0	Illegal	0	0	0	0	0	0	0	0	0	1	0	0	0	0	0	0
0	1	1	1	Illegal	0	0	0	0	0	0	0	0	1	0	0	0	0	0	0	0
1	0	0	0	Illegal	0	0	0	0	0	0	0	1	0	0	0	0	0	0	0	0
1	0	0	1	0	0	0	0	0	0	0	1	0	0	0	0	0	0	0	0	0
1	0	1	0	0	0	0	0	0	0	1	0	0	0	0	0	0	0	0	0	0
1	0	1	1	Illegal	0	0	0	0	1	0	0	0	0	0	0	0	0	0	0	0
1	1	0	0	0	0	0	0	1	0	0	0	0	0	0	0	0	0	0	0	0
1	1	0	1	Illegal	0	0	1	0	0	0	0	0	0	0	0	0	0	0	0	0
1	1	1	0	Illegal	0	1	0	0	0	0	0	0	0	0	0	0	0	0	0	0
1	1	1	1	Illegal	1	0	0	0	0	0	0	0	0	0	0	0	0	0	0	0

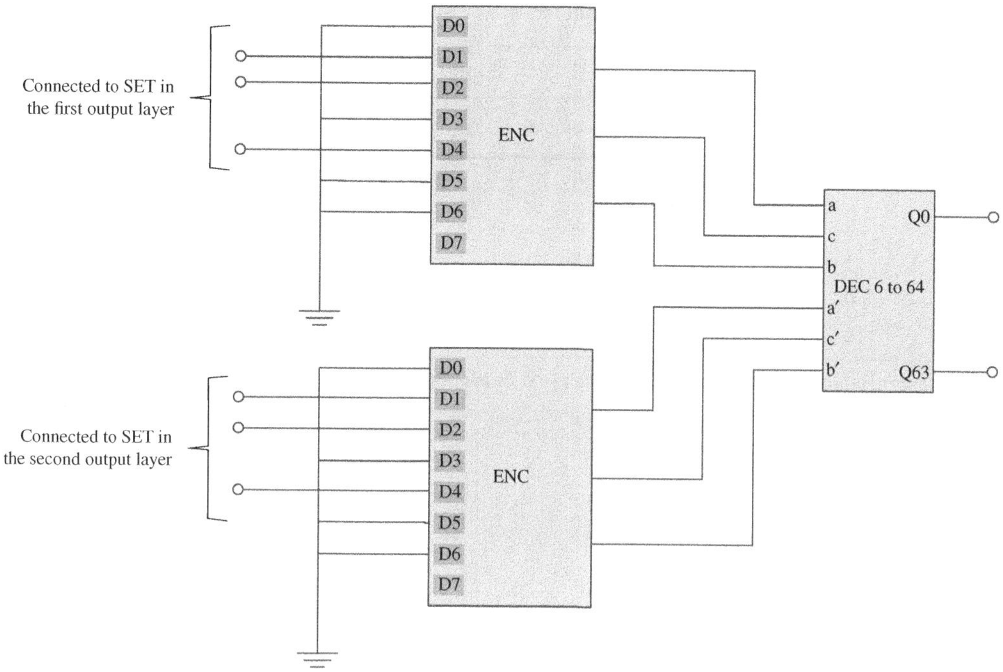

Figure 5.18 Schematic of the first output drive. Source: Akbari-Hasanjani and Sabbaghi-Nadooshan [131]/with permission of John Wiley & Sons.

Figure 5.19 Simplified output drive model [131].

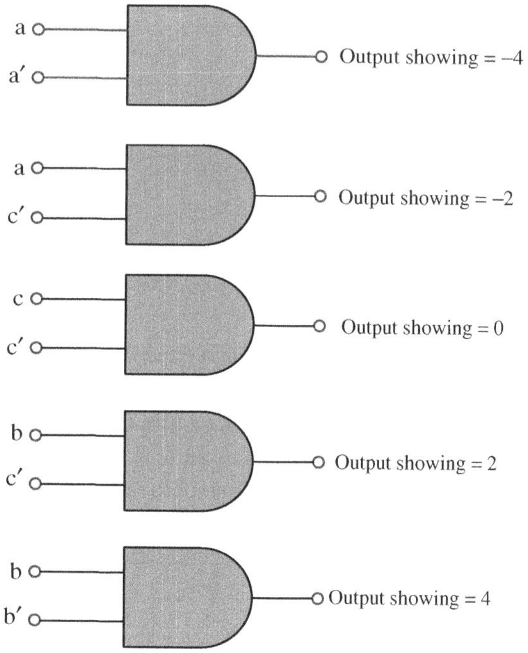

Table 5.8 Input states for the simplified output model in the quinary system.

SET in first layer			SET in second layer			Polarization output
a	c	b	a′	c′	b′	
1	0	0	1	0	0	−4
1	0	0	0	1	0	−2
0	1	0	0	1	0	0
0	0	1	0	1	0	2
0	0	1	0	0	1	4

Source: Akbari-Hasanjani and Sabbaghi-Nadooshan [131]/with permission of John Wiley & Sons.

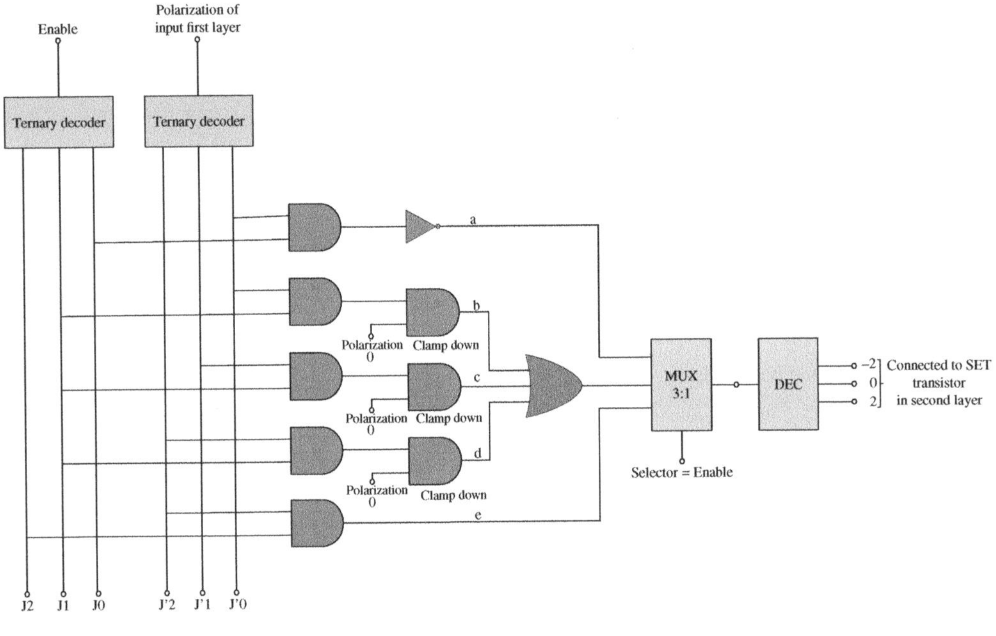

Figure 5.20 Second input drive in QuQCA.

5.3.2.2 Second Structure of Input and Output Drive

Second Structure of Input Drive In the second design of the input drive, the TQCA cell can be used for implementation. The circuit structure of the input drive is schematically shown in Figure 5.20. In the proposed circuit, an Enable and the input cell state of the first layer are used as the inputs of the proposed circuit, and the output of the input drive circuit is connected to the second layer. In other words, the output of the proposed circuit produces the polarization of the second layer. Table 5.9 shows the truth table of the second structure of the input drive.

Output Structure of Second Drive In this method, a half-adder is used to integrate the output. Using the output of the semiconductor circuit, the output states of the QuQCA-based circuit can be understood. The semiconductor used in this study is presented in Ref. [101] as shown in Figure 5.21. Table 5.10 shows the output states produced for QuQCA.

Table 5.9 Truth table of the second input drive in QuQCA.

Enable	Polarization of first input layer	Result of Z
−2	−2	a = −2 MUX Selector is Enable Z = −2
0	−2	b = 0 MUX Selector is Enable Z = 0
0	0	c = 0 MUX Selector is Enable Z = 0
0	2	d = 0 MUX Selector is Enable Z = 0
2	2	e = 2 MUX Selector is Enable Z = 2

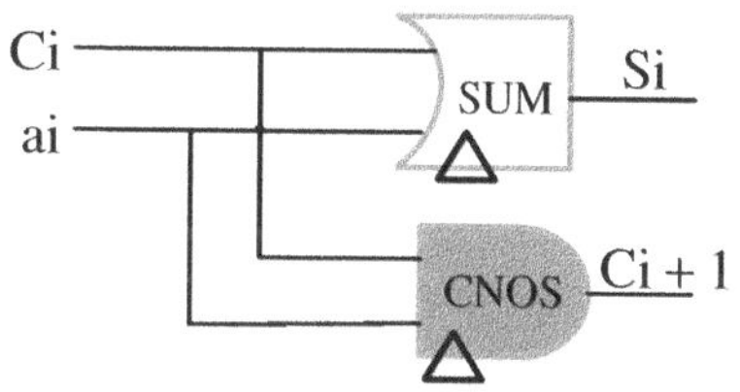

Figure 5.21 TQCA-based half-adder circuit. Source: Adapted from Mohaghegh et al. [101].

Table 5.10 QuQCA output truth table.

Polarization of output ternary in first layer	Polarization of output ternary in second layer	Sum	Carry	Equivalent polarization
−2	−2	2	−2	−4
−2	0	−2	0	−2
−2	2	0	0	0
0	−2	−2	0	−2
0	0	0	0	0
0	2	2	0	2
2	−2	0	0	0
2	0	2	0	2
2	2	−2	2	4

5.3.3 Energy in QuQCA Cell

Energy in QuQCA-based cells consists of both intracellular and extracellular energies. Intracellular energy is a constant value and is a function of cell size. Extracellular energy is a function of the distance between cells, or in other words, the positions of electrons relative to each other in each cell. Types of intracellular energy include internal electrostatic energy, ground state energy, kink energy, and tunneling energy [131]. Extracellular energy is described by only external electrostatic energy, which affects the adjacent cells [131].

5.3.3.1 Internal Electrostatic Energy

In the proposed QuQCA cell, two layers of TQCA cells are used. In Equation (5.20), the amount of external electrostatic energy is calculated according to Coulomb's law for the ternary cell.

Because two layers of ternary cells are used in the QuQCA cell, we have to algebraically sum the electrostatic energy of the layers [131]. Equation (5.21) expresses the electrostatic energy of the proposed QuQCA cell [131].

$$E_{\text{internal}_{\text{TQCA}}} = \frac{1}{4\pi\varepsilon_0\varepsilon_r}\sum_{i=0}^{m}\sum_{j=0}^{n}\frac{q_i q_j}{d_{ij_{\text{TQCA}}}} \tag{5.20}$$

$$E_{\text{internal(PQCA)}} = E_{\text{internal}_{\text{TQCA(layer1)}}} + E_{\text{internal}_{\text{TQCA(layer2)}}} \tag{5.21}$$

5.3.3.2 Ground State Energy

In QCA technology, ground state energy occurs when no clock enters the cell and electrons are in the ground state. The amount of state energy is obtained by solving the time-independent Schrodinger equation, in which two ternary layers are used in the proposed QuQCA cell [131]. Equation (5.22) shows the amount of state energy in the ternary cell [131]. In Equation (5.23), the total state energy is obtained from the algebraic sum of the state energies of the TQCA cells [131].

$$E_{\text{G-S}_{\text{TQCA}}} = \frac{n^2\pi^2\hbar^2}{2mL_{\text{TQCA}}{}^2} \tag{5.22}$$

$$E_{\text{G-S}_{\text{total}}} = E_{\text{G-S}_{\text{TQCA(layer1)}}} + E_{\text{G-S}_{\text{TQCA(layer2)}}} = 2E_{\text{G-S}_{\text{TQCA}}} \tag{5.23}$$

5.3.3.3 Kinetic Energy

Kinetic energy corresponds to the motion of electrons within a cell and is a function of ambient temperature. In a QuQCA cell, the amount of kinetic energy is obtained as an algebraic sum of kinetic energies of the layers [131]. Equation (5.24) gives the amount of kinetic energy in the ternary cell, and Eq. (5.25) gives the total kinetic energy in the QuQCA cell [131].

$$E_{\text{kinetic}_{\text{TQCA}}} = \frac{1}{2}K_B T_{\text{TQCA}} + K_B T_{\text{TQCA}} = \frac{3}{2}K_B T_{\text{TQCA}} \tag{5.24}$$

$$E_{\text{kinetic}_{\text{Total}}} = E_{\text{kinetic}_{\text{TQCA(layer1)}}} + E_{\text{kinetic}_{\text{TQCA(layer2)}}} \tag{5.25}$$

In the proposed QuQCA cell, both layers of TQCA are isolated on a plate, and the ambient temperatures of the layers are equal ($T_{\text{TQCA1}} = T_{\text{TQCA2}} = T$), so Eq. (5.25) can be rewritten in the form of Eq. (5.26) [131].

$$E_{\text{kinetic}_{\text{Total}}} = 2E_{\text{kinetic}_{\text{TQCA}}} = 3K_B T \tag{5.26}$$

5.3.3.4 External Electrostatic Energy

The operation of QCA cells is based on electrostatic energy. Moreover, the electrostatic interactions of cells change the polarization of adjacent cells [129–132] without transferring current. Equation (5.27) expresses the amount of external electrostatic energy obtained from Coulomb's law [131].

$$E_{ij} = \frac{1}{4\pi\varepsilon_0\varepsilon_r}\sum_{i=0}^{m}\sum_{j=0}^{n}\frac{q_i q_j}{d_{ij}} \tag{5.27}$$

In the proposed QuQCA cell, two ternary layers are used to prevent the electrostatic effect of the ternary cells on each other. The distance between the isolated layers is 25 nm [131]. The technique in Ref. [109] is used to calculate the electrostatic energy between the layers.

To calculate the electrostatic energy in the proposed QuQCA, first, the electrostatic energy of the first layer between the two cells and then the electrostatic energy of the second layer between

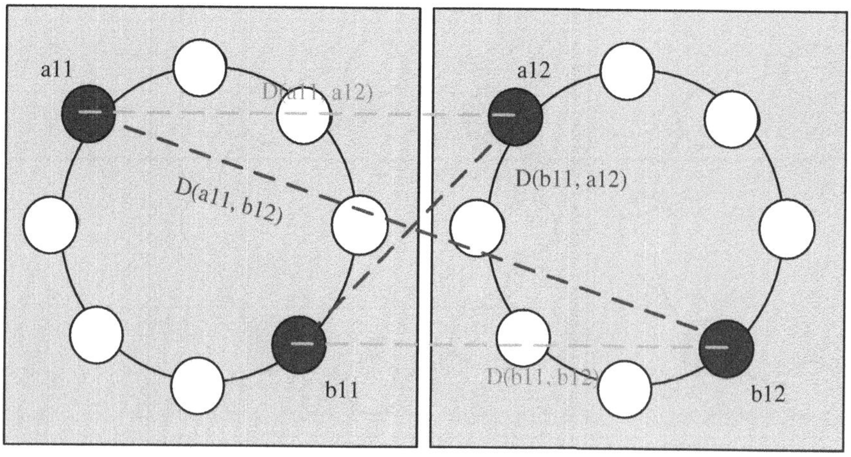

Figure 5.22 Distances between electrons in the first layer.

the two cells are calculated. Finally, the electrostatic energies of the two layers are summed algebraically [131].

In calculating the external electrostatic energy between adjacent cells, first, the distance of electrons between two adjacent cells is calculated in each layer. The distances between the electrons in the first layer of two adjacent cells are denoted by the pairs (a11, a12), (b12, b11), (b12, a11), and (b11, a12) [131]. Then, using Eq. (5.27), the external electrostatic energies for the pairs (a11, a12), (b12, b11), (b12, a11), and (b11, a12), i.e., respectively, E_{e11}, E_{e12}, E_{e13}, and Ee_{14} can be calculated.

After calculating the electrostatic energy of each of the mentioned states, the energies of the pairs with the same letter (E_{e11}, E_{e12}) are added together. Then, the electrostatic energies of the pairs with different letters (E_{e13}, E_{e14}) are added together. Equation (5.28) shows the total external electrostatic energy of the first layer, and Figure 5.22 shows the distances between the electrons in the first layer [131].

$$E_{t_{\text{TQCAlayer(layer1)}}} = (E_{e11} + E_{e12}) - (E_{e13} + E_{e14}) \tag{5.28}$$

To calculate the external electrostatic energy of the second layer, in a similar way for the first layer, the distances of electrons in the cells denoted by the pairs (a22, a21), (b22, b21), (b22, a21), and (b21, a22) are determined. Then, Eq. (5.27) is used to calculate the external electrostatic energies between the electron pairs (a22, a21), (b22, b21), (b22, a21), and (b21, a22) as E_{e21}, E_{e22}, E_{e23}, and E_{e24}, respectively [131]. Next, the electrostatic energies of the pairs with the same letter (E_{e21}, E_{e22}) and those with different letters (E_{e23}, E_{e24}) are added together [131]. Subsequently, the total external electrostatic energy of the second layer is obtained from Eq. (5.29) by subtracting the electrostatic energies calculated in the previous step. Figure 5.23 shows the distance between the electrons in the second layer [131].

$$E_{t_{\text{TQCAlayer(layer2)}}} = (E_{e21} + E_{e22}) - (E_{e23} + E_{e24}) \tag{5.29}$$

To calculate the total electrostatic energy in an adjacent QuQCA cell, the value of $|E_{t_{\text{TQCA1}}}|$ is algebraically summed with the value of $|E_{t_{\text{TQCA2}}}|$. Equation (5.30) gives the total electrostatic energy of two adjacent QuQCA cells [131]. Figure 5.24 shows the symbolic representation of two adjacent cells in QuQCA [131].

$$|E_{t_{\text{PQCA}}}| = |E_{t_{\text{TQCA (layer1)}}}| + |E_{t_{\text{bQCA (layer2)}}}| \tag{5.30}$$

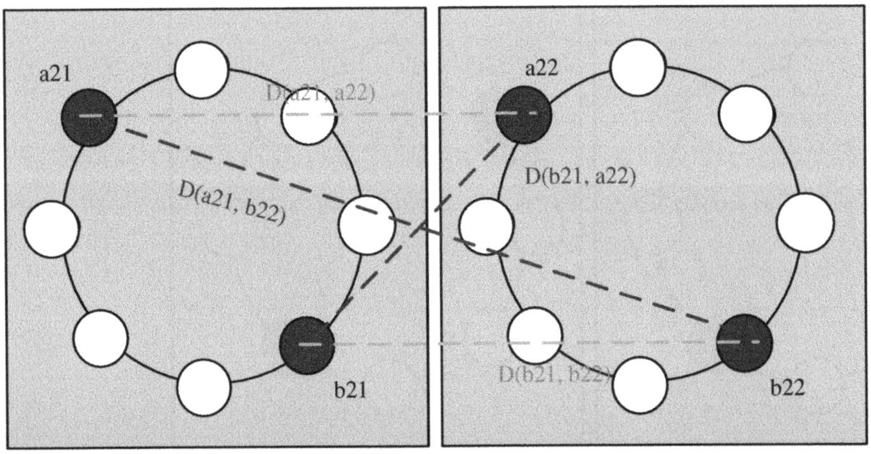

Figure 5.23 Distances between electrons in the second layer. Source: Akbari-Hasanjani and Sabbaghi-Nadooshan [131]/with permission of John Wiley & Sons.

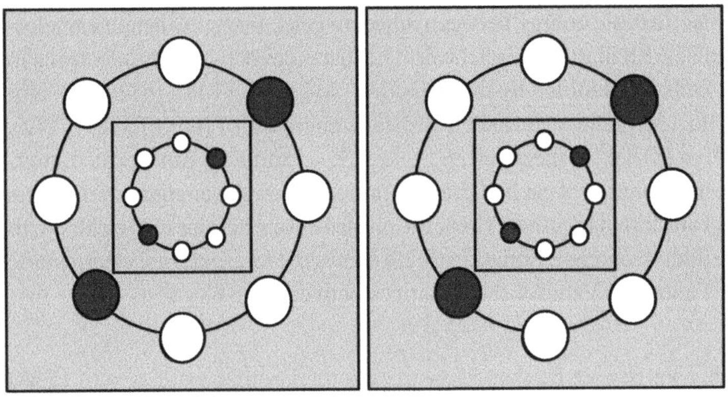

Figure 5.24 Symbolic representation of two adjacent cells in QuQCA. Source: Akbari-Hasanjani and Sabbaghi-Nadooshan [131]/with permission of John Wiley & Sons.

5.4 Hypothesis Based on *n*-Value QCA Cell

In this section, we present n-valued cells according to previous calculations. With n-valued cells, natural variables can be controlled based on the concept of logic or fuzzy control introduced by Zadeh. Moreover, the QCA polarization and n-valued cell structure are discussed without including quantum computations.

Binary, ternary, quaternary, and quinary QCA cells are single-particle, two-particle, three-particle, and four-particle systems. According to quantum calculations, higher-level QCA cells such as senary, septenary, and octonary cells require five, six, and seven particles, respectively; this trend is true for larger n-valued systems [131]. Equation (5.31) expresses the number of particles needed for an n-valued QCA system or cell [131].

$$n = m - 1 \tag{5.31}$$

where n is the number of particles and m is the QCA cell level. The polarization value of an n-valued cell satisfies Equation (5.32) [131].

$$-n \leq \text{Polarization} \leq +n \tag{5.32}$$

Table 5.11 shows the characteristics of cell structures from two values (binary cell) to eight values (Octonary cell).

In the proposed structure for n-level cells, the fuzzy concept can be implemented using cells based on QCA technology. Using these cells, a membership function is provided for the input and output of a system, and a processor is used for calculations [131].

Table 5.11 Specifications of n-valued cell structures.

Value	Number of particles	Polarization	Number of layers		Cell structure	Value graph
			Binary	Ternary		
2	1	{−1,1}	1	0	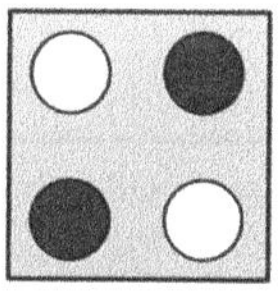	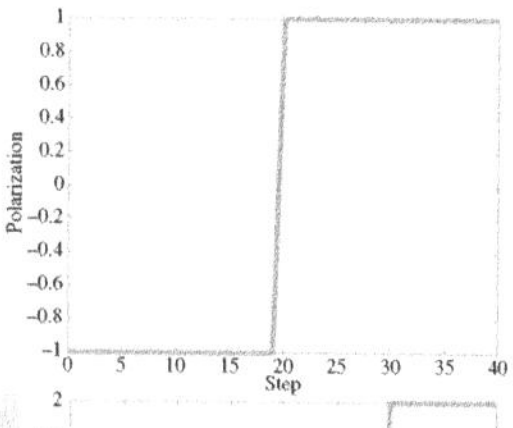
3	2	{−2,0,0,2}	0	1	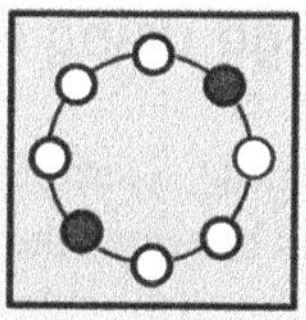	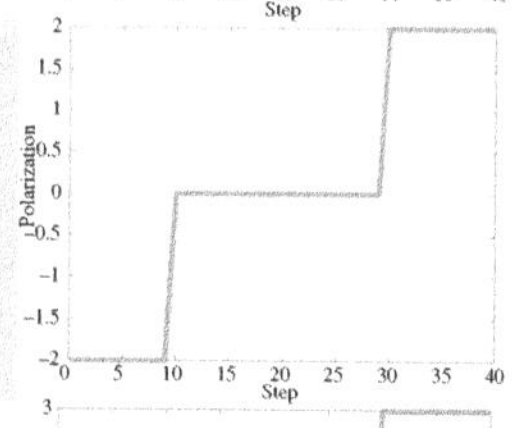
4	3	{−3, −1,1,3}	1	1	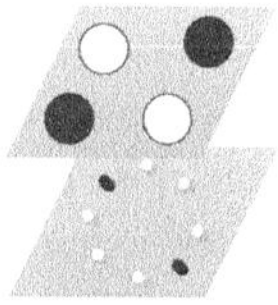	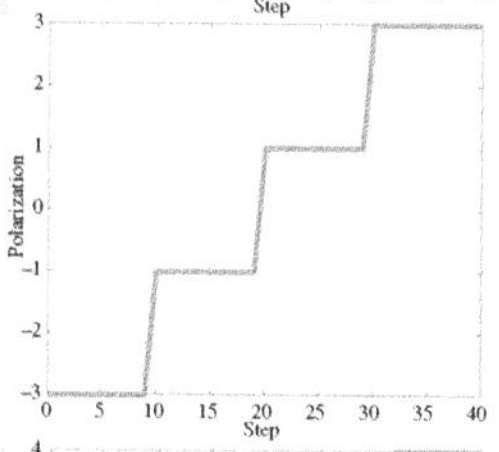
5	4	{−4, −2,0,2,4}	0	2	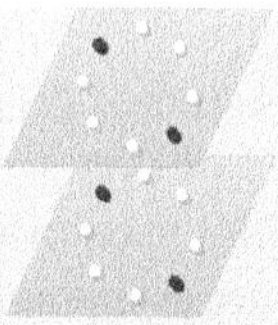	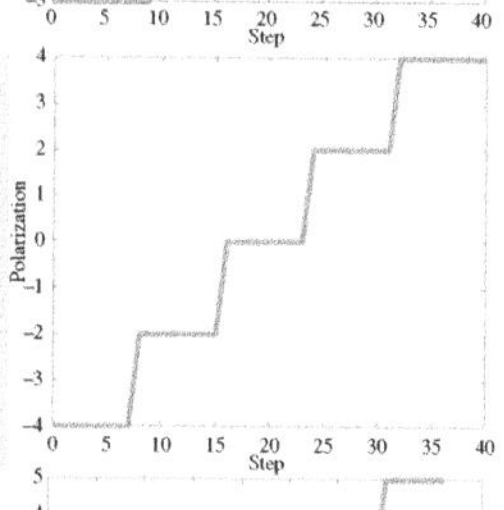
6	5	{−5, −3, −1,1,3,5}	1	2	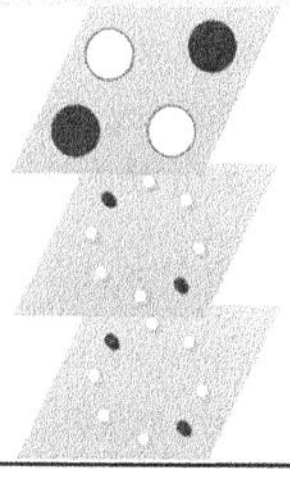	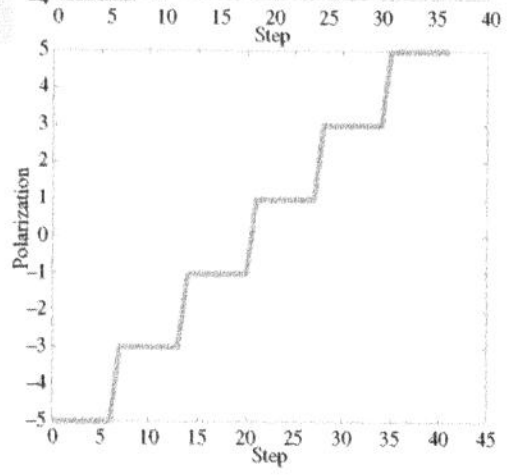

(continued)

Table 5.11 *(Continnued)*

7	6	{−6, −4, −2,0,2,4,6}	0	3	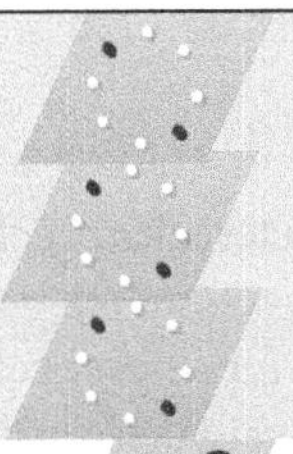	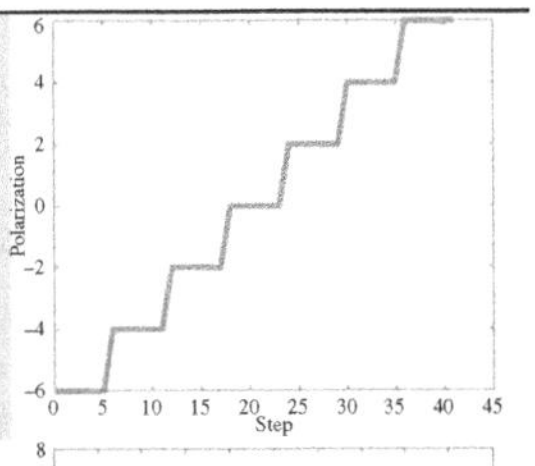
8	7	{−7, −5, −3, −1,1,3,5,7}	1	3	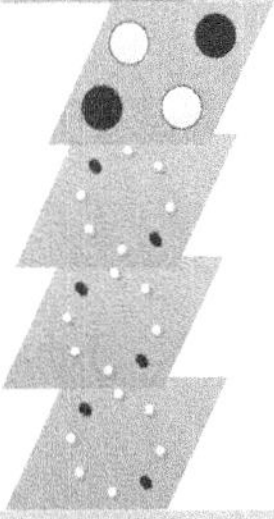	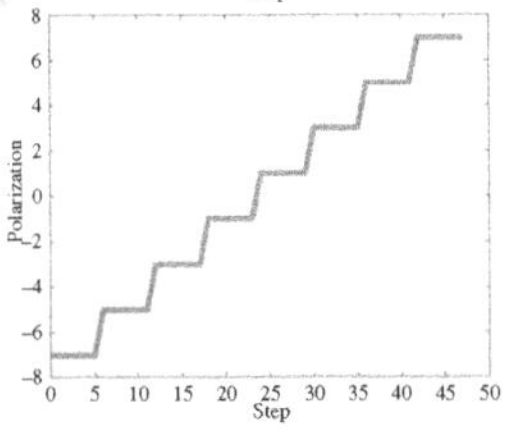
...	...	...	...	...	...	...

Source: Akbari-Hasanjani and Sabbaghi-Nadooshan [131]/with permission of John Wiley & Sons.

5.5 Fuzzy Logic Design Using MIN and MAX Functions

In this research, for the first time, a simple fuzzy function is simulated in QQCA using MIN and MAX gates. Each fuzzy system has three parts for fuzzification, linguistic rule base, and defuzzification. In the fuzzification and linguistic rule base parts, the input values are converted into electrical signals. The polarizations presented for QQCA can be used as numerical values in a quaternary fuzzy system after fuzzification. In fuzzy logic, the numerical values are between 0 and 1. According to the simulation results, four levels in quaternary logic can be considered between 0 and 1. Therefore, the proposed design is close to fuzzy logic.

Fuzzy rules can be implemented using MIN–MAX functions, and the output of these rules can be sent to the defuzzification part. SETs are used in the QQCA cell to apply input to the fuzzy rule part. Output SETs are also used in QQCA to receive output. A simple fuzzy system is shown in Figure 5.25.

According to Figure 5.25, the studied system has two inputs (In1 and In2) and one output (OUT). The input and output variables are very low, low, medium, and high. These variables are used in a pressure regulation system with the equivalent order of polarizations {−3: A, −1: E, +1: F, +3: D}. The fuzzy rules in the system are as follows:

If one of the inputs is A, then output = other input
If one of the inputs is D, then output = inverse of the other input
If both inputs are in E or F states, then output = E
If one of the inputs is in E and the other input is in F, then output = F

Table 5.12 shows the truth table of the system based on the stated fuzzy rules. This function is based on Eq. (5.33). Figure 5.26 shows the simulation results of the proposed fuzzy structure. The use of quantum cells in fuzzy logic reduces energy consumption and occupied area.

$$Y = ((\text{In1} \cdot \text{In2}') + (\text{In1}' \cdot \text{In2})) \tag{5.33}$$

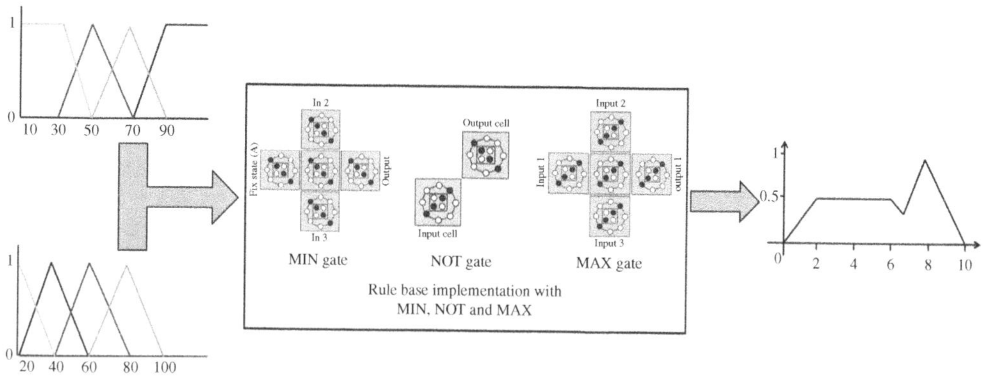

Figure 5.25 Block diagram of the studied fuzzy system.

Table 5.12 Truth table of the proposed fuzzy pressure control system.

In1	In2	Out	Out linguistic variables (defuzzification)
A	A	A	Very low
A	E	E	Low
A	F	F	Medium
A	D	D	High
E	A	E	Low
E	E	E	Low
E	F	F	Medium
E	D	F	Medium
F	A	F	Medium
F	E	F	Medium
F	F	E	Low
F	D	E	Low
D	A	D	High
D	E	F	Medium
D	F	E	Low
D	D	A	Very low

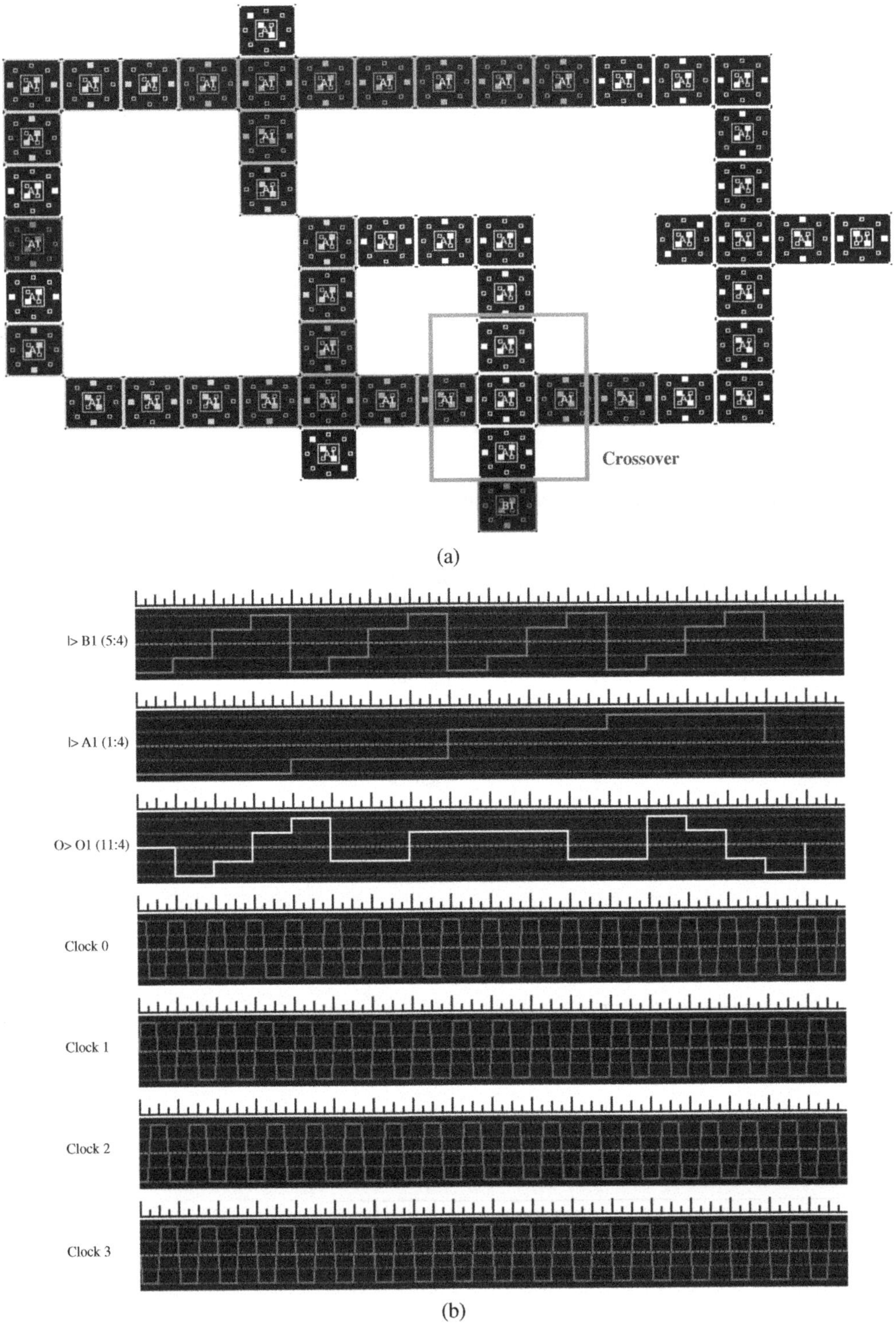

Figure 5.26 Fuzzy system rule base simulation: (a) the circuit representation and (b) the simulation results of the rule base.

5.6 Quantum Information

Quantum information (QI) is a field of study in quantum systems that processes and displays QI. The performance of QI is based on the principles of quantum mechanics, in which, unlike classical physics, information is stored in terms of qubits [136, 137]. In the representation of QI, |0> and |1> are equivalent to, respectively, zero and one in binary logic.

In QI, there are issues of special importance, such as the superposition principle in which information can have both |0> and |1> states because in quantum computations, the probability of information occurrence is used. Another issue that is particularly important in QI is particle entanglement. This phenomenon means that two or more qubits are related to each other in such a way that they cannot be represented by one qubit [138, 139].

The field of QI is divided into three branches: quantum computing, quantum cryptography, and quantum communication, which are briefly reviewed below.

5.6.1 Quantum Computation

In quantum computing, quantum mechanics is used to analyze and develop quantum systems. Quantum computers are used for quantum computations. One of the advantages of this computer system is the use of the superposition principle. Superposition principle allows quantum computers to process several inputs simultaneously. This increases the response speed of quantum computers, which can be called computers with parallel processing [140, 141].

Algorithms such as Shor's algorithm and Grover's algorithm, which are unsolvable for classical computers, can be easily solved by using quantum computers [138, 142]. However, the design and construction of quantum computers have many challenges, such as errors caused by environmental interactions and noise. In quantum computing, there are basic mathematical concepts, some of which are mentioned in the following.

- *Quantum State*:
  The quantum state of a qubit is represented as a linear combination of $|0\rangle$ and $|1\rangle$. Equation (5.34) represents the quantum state as a linear combination of $|0\rangle$ and $|1\rangle$.

$$|\psi\rangle = \alpha\,|\,0\rangle + \beta\,|\,1\rangle \tag{5.34}$$

  where α and β represent the probabilities of the qubit in the states $|0\rangle$ and $|1\rangle$, respectively. Also, the normalization condition is shown in Eq. (5.35).

$$|\alpha|^2 + |\beta|^2 = 1 \tag{5.35}$$

- *Quantum Gates*:
  Quantum gates, such as Hadamard gate, X gate, Z gate, and CNOT gate, are operators that are applied to qubits to perform specific operations [143, 144]. Figure 5.27 shows quantum gates.

- *Quantum Superposition*:
  Quantum superposition allows qubits to exist in a combination of multiple states. This means that a particle can exist in different states.
- *Quantum Measurement*:
  When a qubit is measured, it collapses into one of its eigenstates with a certain probability. Born's rule is used to calculate the probability of obtaining a specific measurement outcome. Equation (5.36) shows the probability of obtaining the eigenstates of a qubit.

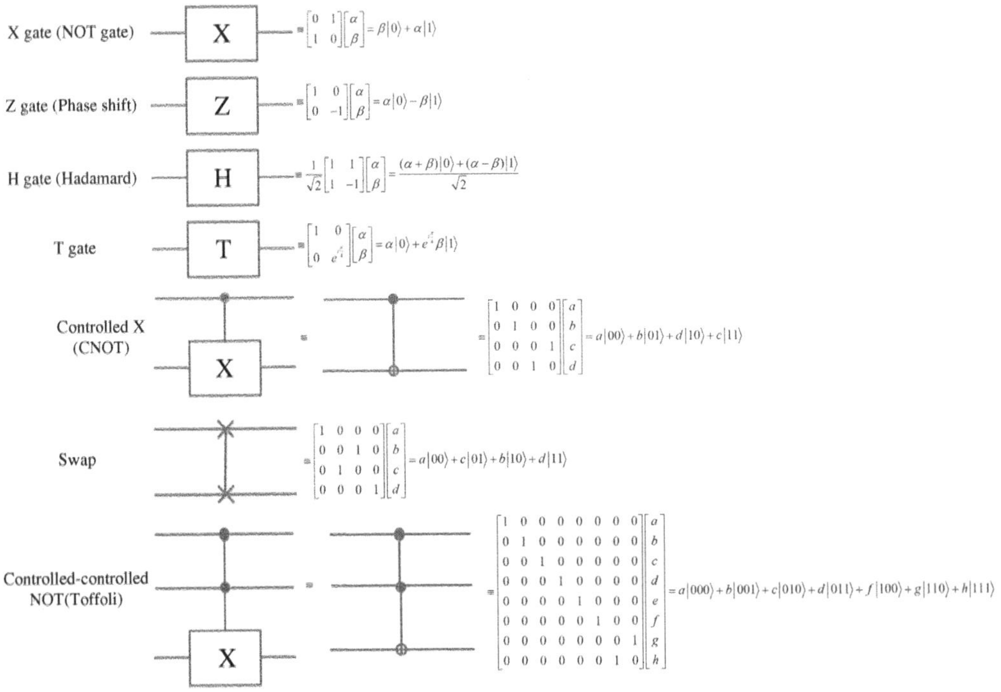

Figure 5.27 Quantum gates.

$$P(|0\rangle) = |\alpha|^2$$
$$P(|1\rangle) = |\beta|^2 \tag{5.36}$$

- *Quantum Entanglement:*
  Quantum entanglement shows the correlation between qubits even if they are physically separated. For example, a state in the entanglement of two qubits can be shown as follows:

$$|\psi\rangle = \alpha|00\rangle + \beta|11\rangle$$

5.6.2 Quantum Communication

Quantum communication uses the principles of quantum mechanics, such as superposition and entanglement, to create a secure connection between receiver and sender. In quantum communication, information is coded using qubits. One of the advantages of quantum communication is the ability to create secure communication using quantum encryption, which allows the receiver and sender to create a secret key that cannot be hacked. Another advantage of quantum communication is the creation of high-capacity communication in which by using entanglement in quantum systems, information can be sent to long distances regardless of the distance between the sender and receiver [145, 146].

5.6.3 Quantum Cryptography

Quantum encryption, also known as quantum key distribution (QKD), uses the basic properties of quantum physics to send information. In quantum cryptography or QKD, two parties

(Alice and Bob) can create secure cryptography, which is based on the principles of quantum mechanics, such as Heisenberg's uncertainty principle and the no-cloning theorem [147].

Quantum particles are used in the encryption process. The sender of information produces a series of quantum particles with quantum characteristics such as polarization or spin angle and sends them to the receiver [136]. The receiving party measures the received information and reads the information.

6

Effect of Polarization for Two Adjacent Cells and Power Consumption in Multi-Valued Cells

6.1 Effect of Two Adjacent Cells in Terms of Polarization

6.1.1 Effect of Two Adjacent Cells in TQCA

In a ternary quantum dot cellular automata (TQCA) cell with two particles, the effect of two neighboring cells is checked to confirm the accuracy of the calculations. In [130, 148–153], the effect of two neighboring binary cells is discussed, and Eq. (6.1) is used. A binary cell can be expressed as a single particle system with the quantum polarization $\{-1, 0, 1\}$.

$$P_j = \frac{\frac{E_k}{2\gamma}\overline{P}_j}{\sqrt{1+\left(\frac{E_k}{2\gamma}\overline{P}_j\right)^2}} \tag{6.1}$$

In TQCA cells, the calculation is based on a two-particle system with the polarization $\{2, 0, 0, -2\}$. Therefore, the performance of a two-particle system can be defined as follows [129]:

$$|\psi\rangle = (a\,|00\rangle + b\,|01\rangle\; + c\,|10\rangle\; + d\,|11\rangle\,) = |0\rangle \otimes (a\,|0\rangle + b\,|1\rangle\,) + |1\rangle \otimes (c\,|0\rangle + d\,|1\rangle\,) \tag{6.2}$$

The variables u and v are defined in Eq. (6.3) to account for the four variables or four degrees of freedom in Eq. (6.2) [129].

$$u = \sqrt{|a|^2 + |b|^2}$$
$$v = \sqrt{|c|^2 + |d|^2} \tag{6.3}$$

If the system is multiparticle, its polarization is defined as in Eq. (6.4). In this equation, n is the number of particles in the system [129].

$$P = n \cdot [|u|^2 - |v|^2] \tag{6.4}$$

Using Refs. [150, 153] and Eqs. (6.1) and (6.4), the effect of polarization of two neighboring cells in TQCA can be considered as follows [129]:

$$P_j = n \cdot \frac{\frac{E_k}{2\gamma}\overline{P}_j}{\sqrt{1+\left(\frac{E_k}{2\gamma}\overline{P}_j\right)^2}} \tag{6.5}$$

In Eq. (6.5), E_k is the kinetic energy, γ is the tunneling energy, $\overline{P_j}$ is the polarization of the input or drive cell, P_j is the polarization of the neighboring cell, and n is the number of particles. Figure 6.1 shows the effect of two adjacent cells on the tunneling energies of $0.1E_k$, $0.2E_k$, and $0.3E_k$.

Computing with Multi-Value Logic in Quantum Dot Cellular Automata, First Edition.
Reza Sabbaghi-Nadooshan, Reza Akbari-Hasanjani, Leila Dehbozorgi, Majid Haghparast, and Hamid Reza Akbari-Hasanjani.

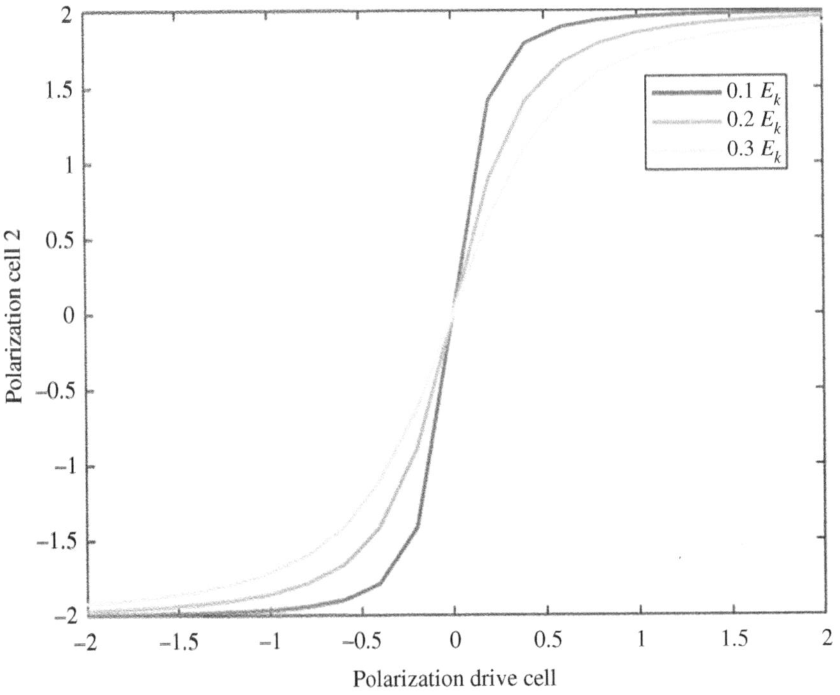

Figure 6.1 Effect of two neighboring TQCA cells at different kinetic energies. Source: Akbari-Hasanjani et al. [129]/with permission of Elsevier.

6.1.2 Effect of Two Quaternary Cells

This section will examine the effect of the two neighboring quaternary quantum dot cellular automata (QQCA) cells proposed in Chapter 3. A quaternary system is a three-particle system in which the proposed cell is presented as a two-layer structure to avoid the instability of the gates that are realized by this type of cell [130, 131]. In the first layer of the proposed quaternary cell, a ternary cell, which is a two-particle system, is used. In the second layer, a binary cell, which is a one-particle system, is used. The effect of two neighboring cells on binary and ternary cells is presented in Eq. (6.6) [130].

$$P_j = n \cdot \frac{\frac{E_k}{2\gamma}\overline{P}_j}{\sqrt{1+\left(\frac{E_k}{2\gamma}\overline{P}_j\right)^2}} \tag{6.6}$$

where n is 1 for the binary cell and 2 for the ternary cell, E_k is the kinetic energy, γ is the tunneling energy, P_j is the polarization of the neighboring cell, and $\overline{P}_j$ is the polarization of the drive cell.

According to the cell structure presented in [130], all simulations are performed assuming that the two layers are isolated. Accordingly, the effect of two neighboring quaternary cells can be determined as the algebraic sum of the electrostatic interactions with adjacent cells [130]. Equation (6.7) expresses the effect of two quaternary cells with $n_{Binary} = 1$ and $n_{Ternary} = 2$ [130]. Figures 6.2 and 6.3 show the interactions of two neighboring cells in the ternary and binary layers, plotted for $0.1E_k$, $0.2E_k$, and $0.3E_k$.

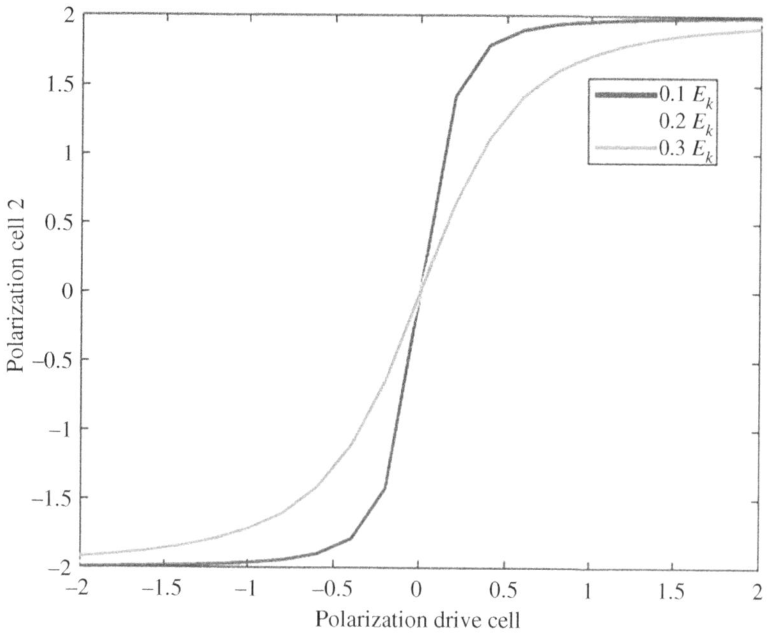

Figure 6.2 Interactions of two cells in the ternary layer. Source: Akbari-Hasanjani and Sabbaghi-Nadooshan [130]/with permission of John Wiley & Sons.

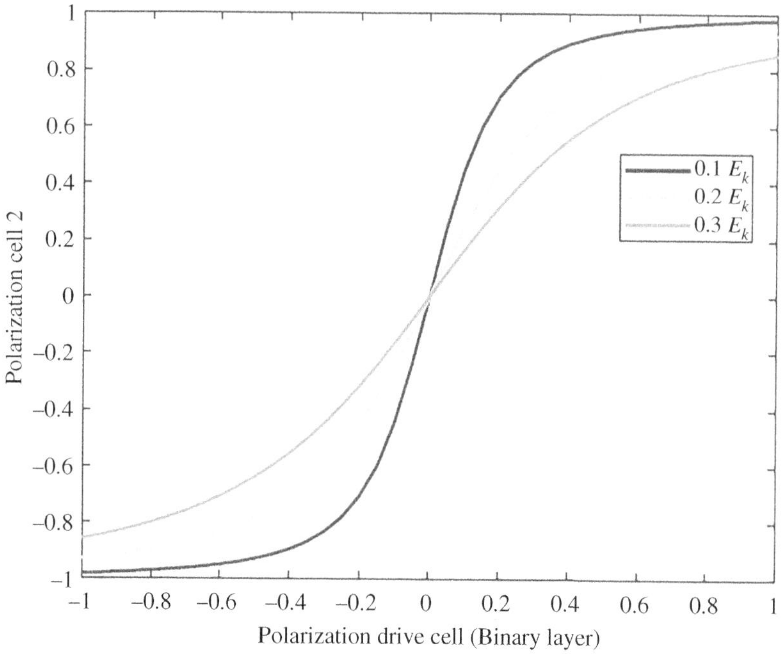

Figure 6.3 Interactions of two cells in the binary layer. Source: Akbari-Hasanjani and Sabbaghi-Nadooshan [130]/with permission of John Wiley & Sons.

$$P_{j_{\text{Quaternary}}} = n_{\text{ternary}} \cdot \frac{\frac{E_k}{2\gamma}\overline{P}_{j_{\text{Ternary}}}}{\sqrt{1+\left(\frac{E_k}{2\gamma}\overline{P}_{j_{\text{Ternary}}}\right)^2}} + n_{\text{Binary}} \cdot \frac{\frac{E_k}{2\gamma}\overline{P}_{j_{\text{Binary}}}}{\sqrt{1+\left(\frac{E_k}{2\gamma}\overline{P}_{j_{\text{Binary}}}\right)^2}} \tag{6.7}$$

6.1.3 Effect of Two QuQCA Cells

This section examines the interactions of two adjacent quinary quantum dot cellular automata (QuQCA) cells. In quantum analysis, a QuQCA cell is a four-particle system in which the proposed cell is presented as a two-layer structure to avoid the instability of the gates that are realized by this type of cell [131]. Each layer is a two-particle system in the form of a TQCA cell [131]. The interactions of two neighboring cells are calculated as presented in Eq. (6.5).

The polarization set in QuQCA is {−4, −2, 0, 2, 4}, which is obtained by combining the polarizations of ternary cells in the two layers as described in detail in Chapter 3. The simulations are performed assuming the isolation of the two layers. To calculate the interactions of two adjacent QuQCA cells, the contributions of two adjacent TQCA cells in the first layer and the two adjacent TQCA cells in the second layer are algebraically summed. Equation (6.8) expresses the impact of two adjacent QuQCA cells [131].

$$P_{j_{\text{PantaTernary}}} = n_{\text{ternary(layer1)}} \cdot \frac{\frac{E_k}{2\gamma}\overline{P}_{j_{\text{Ternary(layer1)}}}}{\sqrt{1+\left(\frac{E_k}{2\gamma}\overline{P}_{j_{\text{Ternary(layer1)}}}\right)^2}} + n_{\text{ternary(layer2)}} \cdot \frac{\frac{E_k}{2\gamma}\overline{P}_{j_{\text{Ternary(layer2)}}}}{\sqrt{1+\left(\frac{E_k}{2\gamma}\overline{P}_{j_{\text{Ternary(layer2)}}}\right)^2}} \tag{6.8}$$

In Eq. (6.8), $n_{\text{Ternary(layer1)}} = n_{\text{Ternary(layer2)}} = 2$, $\overline{P}_{j_{\text{Ternary(layer1)}}}$ is the drive cell polarization in the first layer, $\overline{P}_{j_{\text{Ternary(layer2)}}}$ is the drive cell polarization in the second layer, $P_{j_{\text{PantaTernary}}}$ is the polarization in the neighboring QuQCA cell, E_k is the kinetic energy, and γ is the tunneling energy. Figure 6.4

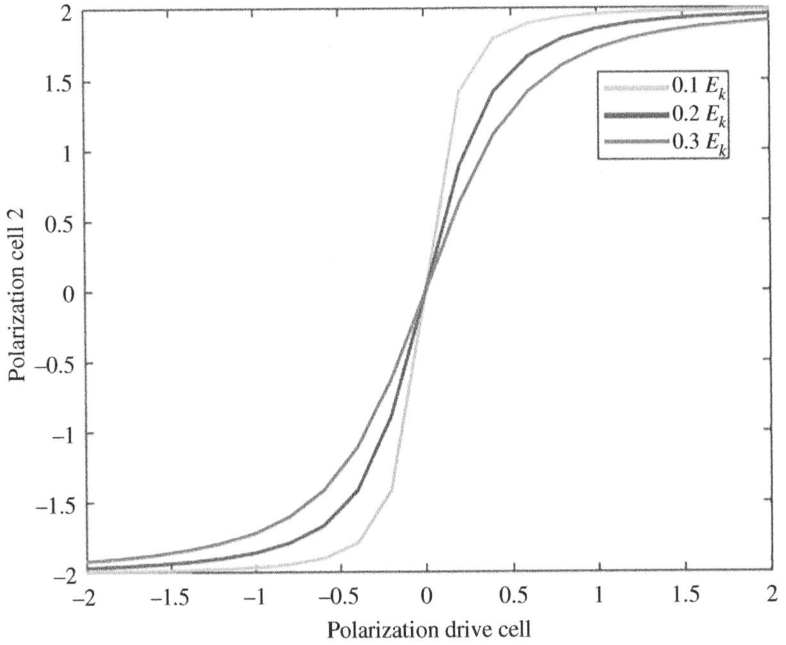

Figure 6.4 Interactions of two neighboring cells in the ternary layers of QuQCA cells. Source: Akbari-Hasanjani and Sabbaghi-Nadooshan [131]/with permission of John Wiley & Sons.

shows the interactions of two neighboring cells in the ternary layer for tunneling energies of $0.1E_k$, $0.2E_k$, and $0.3E_k$ [131].

6.2 Power Consumption in QCA Technology

6.2.1 Power Consumption in a TQCA Cell with Changes in Time and Polarization

In this section, the power consumption of a TQCA cell versus polarization and time will be examined. To evaluate the power consumption, the consumed power is measured over time while the polarization is varied. All calculations are presented in electron volts (eV), and simulations are performed at 298.15 °K [129–131]. Based on Eqs. (6.9) and (6.10), the amount of power consumed by each cell for changes in polarization is equal to [129–131, 154–156]:

$$P_{\text{diss}}(t) = \frac{\hbar}{2}\vec{\Gamma} \cdot \left(\frac{\mathrm{d}}{\mathrm{d}t}\lambda(t)\right) \tag{6.9}$$

$$P_{\text{diss}} = \frac{\mathrm{d}E}{\mathrm{d}t} = -\frac{1}{\tau}(E - E_{\text{ss}}) \tag{6.10}$$

where $\vec{\Gamma}$ is the value of the three-dimensional energy vector in the x, y, and z directions. Equation (6.11) gives the three-dimensional energy vector [88, 89]. In Eq. (6.10), E is the expected instantaneous energy, which is calculated by the TQCA Hamiltonian matrix. In Eq. (6.12), E_{ss} denotes the thermal equilibrium energy, which is obtained in Eq. (6.13) [129–131].

$$\vec{\Gamma}_i = \frac{Tr[\widehat{H}\delta_i]}{\hbar} \qquad i = x, y, z \tag{6.11}$$

where $\widehat{H}$ is the Hamiltonian matrix computed in Chapter 4, and δ_i is the matrix obtained after removing $\hbar/2$ from the S_x, S_y, and S_z matrices, which are also called quasi-Pauli matrices because they do not have all the characteristics of the Pauli matrices [129–131].

$$E = \frac{h}{2}\vec{\Gamma}\vec{\lambda} \tag{6.12}$$

$$E_{\text{ss}} = \frac{\hbar}{2}\vec{\Gamma}\overrightarrow{\lambda_{\text{ss}}} = -\left(\frac{\hbar|\vec{\Gamma}|}{2}\right)\tanh(\Delta) \tag{6.13}$$

where $\vec{\lambda}$ is the density-matrix form in coherence vector representation $\left(\vec{\lambda} = [\lambda_1 \;\; \lambda_2 \;\; \lambda_3]^{\mathrm{T}}\right)$, which is obtained as follows [123–125, 129–131]:

$$\vec{\lambda} = \langle \delta_i \rangle \qquad i = x, y, z \tag{6.14}$$

In Eq. (6.13), $\vec{\lambda}_{\text{ss}}$ is in the steady state with the coherence vector, which is obtained from the steady state density matrix in thermal equilibrium as given in Eq. (6.14). In Eqs. (6.13) and (6.15), the temperature ratio is used [129–131, 154, 155].

$$\vec{\lambda}_{\text{ss}} = -\frac{\vec{\Gamma}}{|\vec{\Gamma}|}\tanh(\Delta) \tag{6.15}$$

$$\Delta = \frac{\hbar|\vec{\Gamma}|}{2K_{\text{b}}T} \tag{6.16}$$

where K_{b} is the Boltzmann constant and T is temperature in Kelvin.

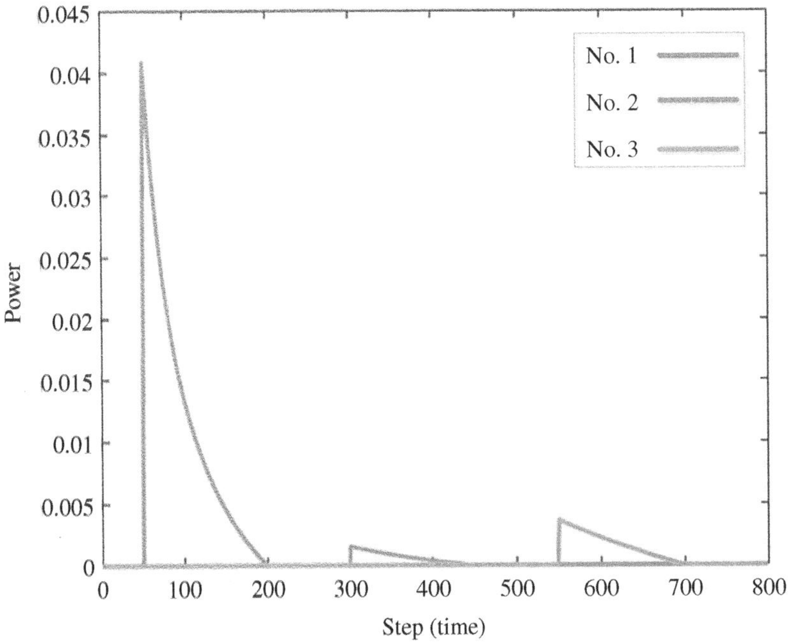

Figure 6.5 Power consumption in TQCA as per polarization set {−2, 0, 2}.

The maximum power consumption is obtained from Eq. (6.17). By inserting the values of E and E_{ss} into this equation, the maximum power consumption in a cell is given by [129–131]

$$P_{diss} = -\frac{1}{\tau}(E - E_{ss}) = -\frac{\hbar}{2\tau}\vec{\Gamma}(\vec{\lambda} - \overrightarrow{\lambda_{ss}}) \tag{6.17}$$

where τ is the time interval during which the cell is in the relax phase. Here, temporal changes in polarization are evaluated over a wide range to better understand power consumption. Figure 6.5 shows the power consumption for polarization {−2, 0, 0, 2}. In this figure, colors No. 1 to No. 3 represent power consumption for polarizations −2, 0, and 2, which occur at 50, 300, and 550, respectively.

6.2.2 Power Consumption in QQCA Cell with Changes in Time and Polarization

This section will briefly review the power consumption of the proposed QQCA quantum cell. A discussion on power consumption in QQCA based on quantum computation and polarization was first presented in Chapter 3.

6.2.2.1 Power Consumption in QQCA Using Quantum Computation

The previous section evaluated the power consumed in a TQCA cell. In this section, Eqs. (6.9) to (6.17) are used to calculate power consumption in a QQCA cell based on the polarization set {−3, −1, 1, 3} and the corresponding Hamiltonian matrix [130]. According to Eqs. (6.9) to (6.17), Figure 6.6 shows the amount of power consumed per quantum computation in QQCA. In this figure, the colors No. 1 to No. 4 show the amounts of power in polarizations −3, −1, 1, and 3, respectively [130].

6.2.2.2 Power Consumption of QQCA Cell

In this book, the calculations for the QQCA system are based on a three-particle system with the polarization values {−3, −1, 1, 3}. To avoid instability of the gates that are based on quaternary cells,

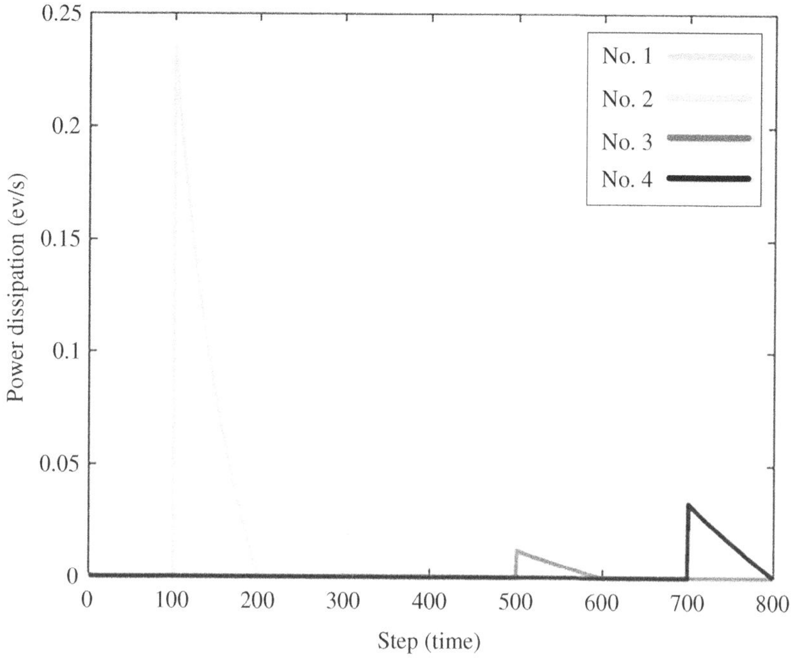

Figure 6.6 Energy consumption in QQCA using quantum computation. Source: Akbari-Hasanjani and Sabbaghi-Nadooshan [130]/with permission of John Wiley & Sons.

the three particles are placed on two layers, one containing a binary quantum dot cellular automata (bQCA) cell, which is a single-particle system, and the other one containing a TQCA cell, which is a two-particle system [130]. In the proposed cell, QQCA polarization is obtained from the algebraic sum of the polarizations of the first (TQCA) and second (bQCA) layer, as thoroughly discussed in the previous sections.

In calculating the power of the QQCA cell, we must first calculate the power consumption in each layer using Eqs. (6.9) to (6.17). Then, the power of the two layers can be algebraically summed to obtain the total power consumption of the cell [130]. Figure 6.7 shows the power consumption of the proposed QQCA cell in different polarizations. In this figure, the colors No. 1 to No. 4 show the amounts of power in polarizations −3, −1, 1, and 3, respectively [130].

6.2.3 Power Consumption in QuQCA Cell with Changes in Time and Polarization

In this section, the power consumption in QuQCA will be evaluated based on the quantum calculations of the quinary system and the proposed QuQCA cell. Calculation results are presented in electron volts (eV), and simulations are performed at room temperature.

6.2.3.1 Power Consumption in QuQCA Using Quantum Computation

To calculate power consumption as shown in Figure 6.8, Eqs. (6.9) to (6.17) are used, and the power consumption in a QuQCA cell is based on the polarizations {−4, −2, 0, 2, 4} and the corresponding Hamiltonian matrix, which was calculated in Chapter 5 [131].

Figure 6.8 shows the amount of power consumption of the QuQCA cell per quantum computation in all polarizations. In this figure, the colors No. 1 to No. 5 curves show the amounts of power in polarizations −4, −2, 0, 2, and 4, respectively [130].

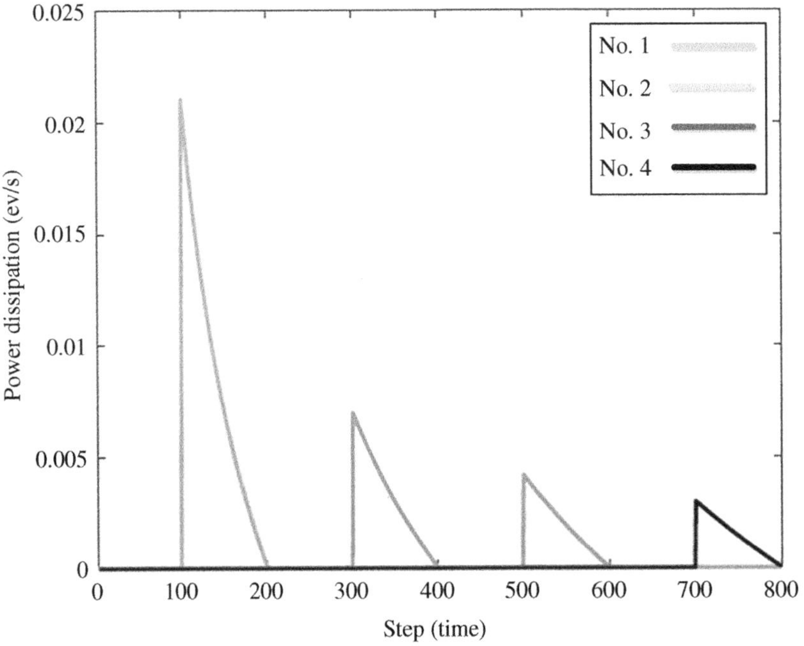

Figure 6.7 Power consumption in the proposed QQCA cell. Source: Akbari-Hasanjani and Sabbaghi-Nadooshan [130]/with permission of John Wiley & Sons.

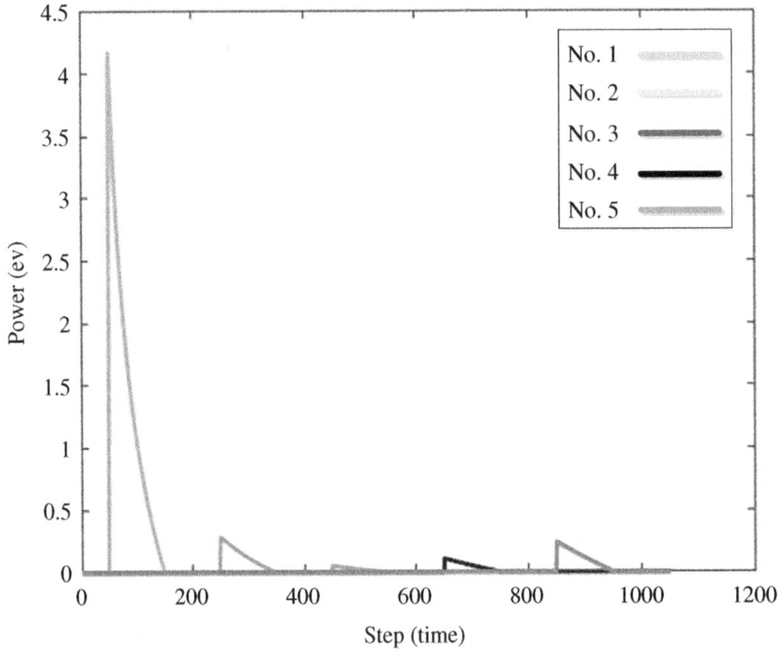

Figure 6.8 Power consumption using QuQCA quantum calculations. Source: Akbari-Hasanjani and Sabbaghi-Nadooshan [131]/Adapted from John Wiley & Sons.

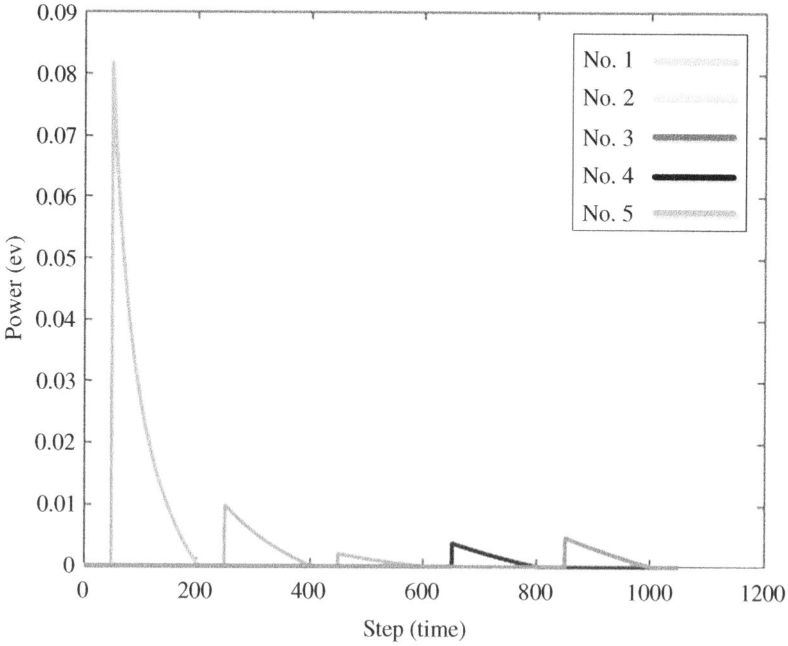

Figure 6.9 Power consumption in the proposed QuQCA cell. Source: Akbari-Hasanjani and Sabbaghi-Nadooshan [131]/with permission of John Wiley & Sons.

6.2.3.2 Power Consumption in QuQCA Cell

The QuQCA cell is considered a four-particle system with the polarization values 4, −4, −2, 0, 2, 4. To avoid instability of the gates, these three particles are placed on two layers, each of which contains a TQCA cell, which is a two-particle system [131].

To calculate the power consumption of the proposed QuQCA cell, first, the contribution of each layer in a given polarization is calculated from Eqs. (6.9) to (6.17). Then, the total power consumption of the QuQCA cell is equal to the algebraic sum of the amounts of power consumed by the layers. Figure 6.9 shows the power consumption of the proposed QuQCA cell, plotted for different polarizations [131]. In this figure, the colors No. 1 to No. 5 curves show the amounts of power in polarizations −4, −2, 0, 2, and 4, respectively.

7

Structure of Basic Gates Using the Proposed Cells in Multi-Value QCA

This chapter uses MATLAB to evaluate the basic gates based on the proposed cells in quantum dot cellular automata (QCA) technology. The results of multi-valued gates are presented below.

7.1 Structure of Basic Gates Using TQCA

The quantum polarization of the ternary quantum dot cellular automata (TQCA) cell was calculated in the previous section using the Hartree–Fock approximation. In this book, we design each gate in one clock to correct the error in the AND and OR gates expressed in Ref. [102]. In our proposed structure, the distance of the circle on which the potential wells are located from the cell wall is reduced to 1.5 nm, whereas in the previous work, this distance was 2.5 nm.

7.1.1 NOT Gate

A NOT gate generally functions based on Eq. (7.1). In this equation, x is the input value applied to the gate input, and “1” is the maximum input value.

$$\text{NOT}\ (x) = 1 - x \tag{7.1}$$

As stated in Chapter 3, the polarization in the presence of two particles in a system assumes four states {−2, 0, 0, 2} or {A, C, D, B}, which are equivalent to {0, 1/2, 1/2, 1}, and in Eq. (5.1), we put the equivalent values. The simulation of NOT gate is performed as in standard binary quantum dot cellular automata (bQCA) in one-phase clock. Figure 7.1 shows the NOT gate using TQCA cells. Table 7.1 shows the NOT gate results, and Table 7.2 shows the Coulomb energy values for different outputs. In Table 5.2, the final value is selected to be the minimum Coulomb energy in different output states.

7.1.2 Simulation of Two Adjacent Cells

This section investigates the interactions between two adjacent cells, one of which is input and the other is output. Two-cell simulation can be performed to evaluate the basis of the QCA quantum wire structure. Figure 7.2 shows the structure of the simulated two cells. Table 7.3 shows the output of this structure.

Computing with Multi-Value Logic in Quantum Dot Cellular Automata, First Edition.
Reza Sabbaghi-Nadooshan, Reza Akbari-Hasanjani, Leila Dehbozorgi, Majid Haghparast, and Hamid Reza Akbari-Hasanjani.

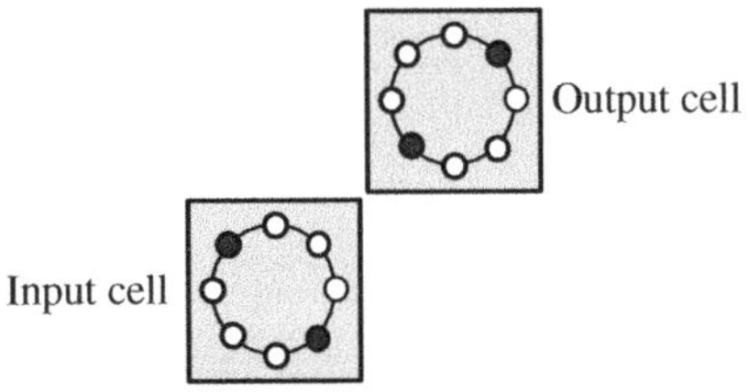

Figure 7.1 Ternary NOT gate.

Table 7.1 Output of ternary NOT gate.

Input		Output	
Symbol	**Polarization**	**Symbol**	**Polarization**
A	−2	B	2
C	0	C	0 (90°)
D	0	D	0 (0°)
B	2	A	−2

Table 7.2 Outputs of the ternary inverter gate for all polarization states.

	Output cell in state A	**Output cell in state C**	**Output cell in state D**	**Output cell in state B**	**Final result**
Input cell in state A	$E_t(J) = 0.2943*10^{-20}$	$E_t(J) = 0.2395*10^{-20}$	$E_t(J) = 0.2395*10^{-20}$	$E_t(J) = 0*10^{-20}$	$E_t(J) = 0*10^{-20}$
Input cell in state C	$E_t(J) = 0.2395*10^{-20}$	$E_t(J) = 0.1174*10^{-20}$	$E_t(J) = 0.5194*10^{-20}$	$E_t(J) = 0.4982*10^{-20}$	$E_t(J) = 0.1174*10^{-20}$
Input cell in state D	$E_t(J) = 0.2395*10^{-20}$	$E_t(J) = 0.5194*10^{-20}$	$E_t(J) = 0.1174*10^{-20}$	$E_t(J) = 0.4982*10^{-20}$	$E_t(J) = 0.1174*10^{-20}$
Input cell in state B	$E_t(J) = 0*10^{-20}$	$E_t(J) = 0.4982*10^{-20}$	$E_t(J) = 0.4982*10^{-20}$	$E_t(J) = 0.8029*10^{-20}$	$E_t(J) = 0*10^{-20}$

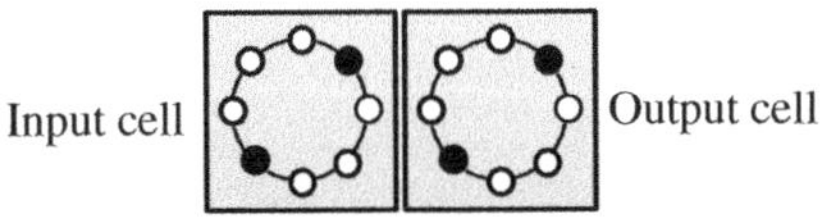

Figure 7.2 Two adjacent TQCA cells (wire).

Input A creates polarization A in the output of the TQCA wire, and input B produces polarization B in the output. Moreover, input C creates polarization D in the output, and input D produces polarization C in the output. To produce polarization C in the output from input C, an odd number of TQCA cells has to be used in the wire. Table 7.4 shows the Coulomb energy values of outputs in MATLAB for two adjacent cells. In this table, the final value is selected to be the minimum Coulomb energy in various output states.

Table 7.3 Interactions of two adjacent ternary cells (TQCA quantum wire).

Input		Output	
Symbol	Polarization	Symbol	Polarization
A	−2	A	−2
C	0	D	0 (0°)
D	0	C	0 (90°)
B	2	B	2

Table 7.4 Outputs of the interactions of two adjacent ternary cells for all output states.

	Output cell in state A	Output cell in state C	Output cell in state D	Output cell in state B	Final result
Input cell in state A	$E_t(J) = 0.0196^* 10^{-19}$	$E_t(J) = 0.0679^* 10^{-19}$	$E_t(J) = 0.1469^* 10^{-19}$	$E_t(J) = 0.1552^* 10^{-19}$	$E_t(J) = 0.0196^* 10^{-19}$
Input cell in state C	$E_t(J) = 0.6787^* 10^{-20}$	$E_t(J) = 0.7493^* 10^{-20}$	$E_t(J) = 0^* 10^{-20}$	$E_t(J) = 0.6787^* 10^{-20}$	$E_t(J) = 0^* 10^{-20}$
Input cell in state D	$E_t(J) = 0.1469^* 10^{-19}$	$E_t(J) = 0^*10^{-19}$	$E_t(J) = 0.2872^* 10^{-19}$	$E_t(J) = 0.1469^* 10^{-19}$	$E_t(J) = 0^*10^{-19}$
Input cell in state B	$E_t(J) = 0.1552^* 10^{-19}$	$E_t(J) = 0.0679^* 10^{-19}$	$E_t(J) = 0.1469^* 10^{-19}$	$E_t(J) = 0.0196^* 10^{-19}$	$E_t(J) = 0.0196^* 10^{-19}$

7.1.3 Implementation of AND and OR Functions

This section presents three-level AND and OR functions using QCA technology. Moreover, it has attempted to implement the QCA standard as AND and OR gates. This book aims to eliminate errors in AND and OR gates. Reference [102] reports the occurrence of an error in the gates presented in Refs. [93, 102, 104, 106, 108]. This error is eliminated by changing the clock and presenting other models to obtain AND and OR functions.

Using the provided structures and changing one of the dimensions of the structures, we intend to fix the error of AND and OR gates and obtain an output like in bQCA using one-phase clock. In this model, we consider the distance from the potential wells to the cell wall to be 1.5 nm. In Figure 7.3, the values of states {A, C, D, B} are applied to the majority gate inputs, and the output results are given in Table 7.5.

of the ternary majority gate are provided for AND and OR gates [102]. Note that a logic value cannot be assigned to state D because it has no function for input and output. Also, for one of the inputs in polarization state −2 or A, the output function has an AND gate response. If one of the inputs is in polarization state 2 or B, we have the OR gate response at the output. Bold letters are the errors that are fixed in the AND and OR gates. Equation (7.2) shows the AND and OR gate functions.

$$
\begin{aligned}
Y &= \text{AND}\ (X_1, X_2) = \min\ (X_1, X_2) \\
Y &= \text{OR}\ (X_1, X_2) = \max\ (X_1, X_2)
\end{aligned}
\tag{7.2}
$$

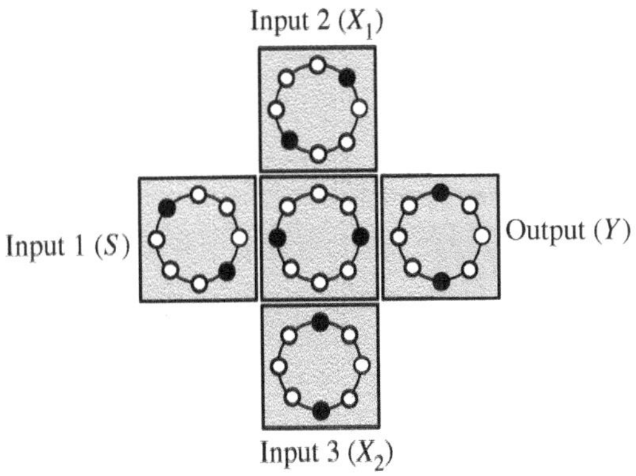

Figure 7.3 Majority ternary gate.

Table 7.5 Majority ternary gate output.

Y	X_2	X_1	S	Y	X_2	X_1	S	Y	X_2	X_1	S	Y	X_2	X_1	S
A	A	A	A	A	A	A	B	A	A	A	C	C	A	A	D
A	C	A	A	C	C	A	B	C	C	A	C	C	C	A	D
A	D	A	A	D	D	A	B	C	D	A	C	D	D	A	D
A	B	A	A	B	B	A	B	C	B	A	C	D	B	A	D
A	A	C	A	C	A	C	B	C	A	C	C	C	A	C	D
C	C	C	A	C	C	C	B	C	C	C	C	C	C	C	D
C	D	C	A	C	D	C	B	C	D	C	C	D	D	C	D
C	B	C	A	B	B	C	B	C	B	C	C	B	B	C	D
A	A	D	A	D	A	D	B	C	A	D	C	D	A	D	D
C	C	D	A	C	C	D	B	C	C	D	C	D	C	D	D
D	D	D	A	D	D	D	B	D	D	D	C	D	D	D	D
D	B	D	A	B	B	D	B	C	B	D	C	D	B	D	D
A	A	B	A	B	A	B	B	C	A	B	C	D	A	B	D
C	C	B	A	B	C	B	B	C	C	B	C	C	C	B	D
D	D	B	A	B	D	B	B	C	D	B	C	D	D	B	D
B	B	B	A	B	B	B	B	B	B	B	C	C	B	B	D
$Y = \text{AND}(X_1, X_2)$								$Y = \text{OR}(X_1, X_2)$							

As shown in Table 7.5, by eliminating state D, the errors of the AND and OR gates are eliminated. To eliminate these errors in Ref. [102], two-phase clock is used, while here we use only one-phase clock. One of the advantages of using one-phase clock is the reduction of circuit delay. Figure 7.4 shows the AND and OR gate structures and truth table separately, while the D state is removed in Table 7.6.

Table B.1 shows the external Coulomb energy values for each output of the AND gate in MATLAB. In Table B.1, the final value is selected to be the minimum Coulomb energy in different output states.

Table B.2 shows the Coulomb energy values for each output of the OR gate in MATLAB. In Table B.2, the final value is selected to be the minimum Coulomb energy in different output states.

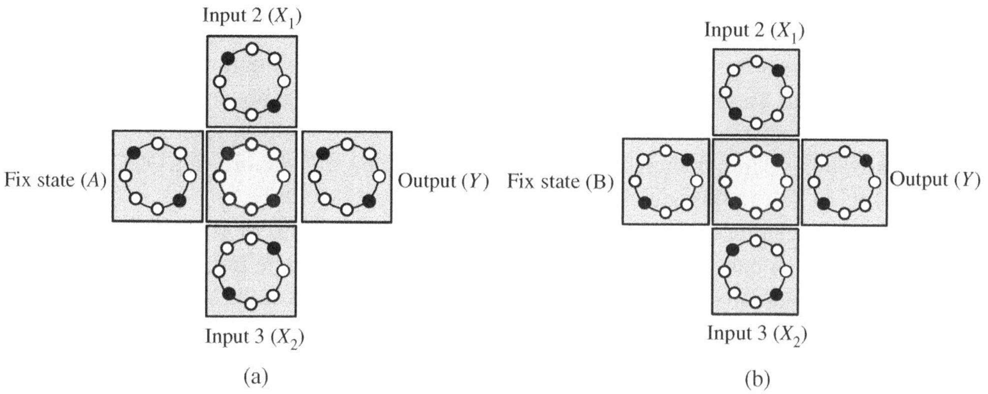

Figure 7.4 (a) AND gate structure with a fixed input in state A and (b) OR gate structure with a fixed input in state B.

Table 7.6 Results of AND and OR gates with three values.

Y	X_2	X_1	S	Y	X_2	X_1	S
A	A	A	A	A	A	A	B
A	C	A	A	C	C	A	B
A	B	A	A	**B**	**B**	**A**	B
A	A	C	A	C	A	C	B
C	C	C	A	C	C	C	B
C	B	C	A	B	B	C	B
A	**A**	**B**	A	B	A	B	B
C	C	B	A	B	C	B	B
B	B	B	A	B	B	B	B
AND gate response				OR gate response			

7.2 Structure of Basic QQCA Gates

In the quantum computation of QQCA, the polarization calculated from the Hartree–Fock approximation is in the form $\{-3, -1, 1, 3\}$. Placing the three particles of a QQCA cell in a single layer causes instability when the quantum numbers of the two particles are the same. This in turn prevents the proper function of the cell and correct response of the logic gates.

As stated in Chapter 5, the proposed QQCA cell has two layers. In the first layer, a TQCA cell is used, and in the second layer, a bQCA cell is used. The algebraic sum of the polarizations of these two layers is the polarization of the QQCA cell. In this part, the goal is to design QQCA basic gates like in bQCA using one-phase clock.

In a QQCA cell, the radius of the circle on which the potential wells are located is 2.5 nm, and the distance from the wells to the cell wall is 1.5 nm. In the first layer, the wells are at an angular distance of 45°, and in the second layer, the angle is 90° (on the mentioned circle).

7.2.1 QQCA-Based NOT Gate

Equation (5.1) expresses the NOT gate relation, in which x is the gate input and "1" denotes the maximum input value. To set the QQCA polarization, the polarizations are converted into values between 0 and 1, i.e., polarization 3 is converted into 1, polarization 1 into 2/3, polarization −1 into 1/3, and polarization −3 into 0; these values are then replaced by x. Therefore, in a QQCA cell, the polarization set {−3, −1, 1, 3} is equivalent to {A, B, C, D} or {0, 1/3, 2/3, 1}.

The NOT gate simulation using the proposed QQCA cell is performed in MATLAB in a similar way to bQCA based on one-phase clock. Figure 7.5 shows the NOT gate in QQCA, plotted with the symbolic representation of QQCA cells. Table 7.7 shows the output of the NOT gate states, and Table 7.8 shows the Coulomb energy values for each output state. Table 7.8 presents the final values, which are selected to be the minimum external electrostatic energy in different output states.

7.2.2 Simulation of Two Adjacent Cells

This section investigates the effects of two adjacent QQCA cells on each other. The simulation of the two adjacent cells is performed on the basis of the QQCA wire structure. Figure 7.6 shows the structure of the two QQCA cells drawn using the symbolic QQCA cell model. Table 7.9 shows the output truth table of the adjacent cells, and Table 7.10 shows the output states resulting from the interaction of the two cells.

In Table 7.9, for each input, the output is equal to the input. The advantage of using the proposed method is that, similar to a ternary quantum wire, there is no need to care about the number of cells to be even or odd in designing the quantum wire. In Table 7.10, the values of Coulomb energy

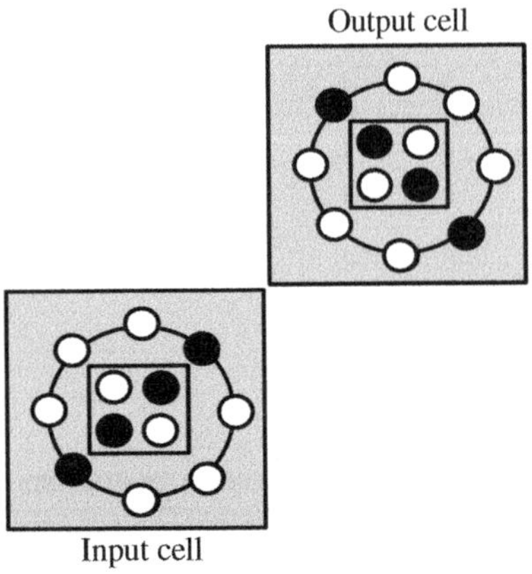

Figure 7.5 QQCA-based NOT gate.

Table 7.7 Output of the QQCA-based NOT gate.

Input		Output	
Symbol	**Polarization**	**Symbol**	**Polarization**
A	−3	D	3
B	−1	C	1
C	1	B	−1
D	3	A	−3

Table 7.8 Results of the QQCA-based NOT gate in different states and corresponding final values.

	Output cell in A state	Output cell in B state	Output cell in C state	Output cell in D state	Final result
Input cell in A state	$E_{t_{bQCA}} = 0.2943*10^{-20}$	$E_{t_{bQCA}} = 0*10^{-20}$	$E_{t_{bQCA}} = 0.2943*10^{-20}$	$E_{t_{bQCA}} = 0*10^{-20}$	$E_{t_{bQCA}} = 0*10^{-20}$
	$E_{t_{TQCA}} = 0.2943*10^{-20}$	$E_{t_{TQCA}} = 0.2943*10^{-20}$	$E_{t_{TQCA}} = 0*10^{-20}$	$E_{t_{TQCA}} = 0*10^{-20}$	$E_{t_{TQCA}} = 0*10^{-20}$
	$E_{t_{QQCA}} = 5.886*10^{-20}$	$E_{t_{QQCA}} = 0.2943*10^{-20}$	$E_{t_{QQCA}} = 0.5004*10^{-20}$	$E_{t_{QQCA}} = 0*10^{-20}$	$E_{t_{QQCA}} = 0*10^{-20}$
Input cell in B state	$E_{t_{bQCA}} = 0*10^{-19}$	$E_{t_{bQCA}} = 0.8029*10^{-20}$	$E_{t_{bQCA}} = 0*10^{-20}$	$E_{t_{bQCA}} = 0.8029*10^{-20}$	$E_{t_{bQCA}} = 0*10^{-20}$
	$E_{t_{TQCA}} = 0.2943*10^{-20}$	$E_{t_{TQCA}} = 0.2943*10^{-20}$	$E_{t_{TQCA}} = 0*10^{-20}$	$E_{t_{TQCA}} = 0*10^{-20}$	$E_{t_{TQCA}} = 0*10^{-20}$
	$E_{t_{QQCA}} = 0.2943*10^{-20}$	$E_{t_{QQCA}} = 1.0972*10^{-20}$	$E_{t_{QQCA}} = 0*10^{-20}$	$E_{t_{QQCA}} = 0.8029*10^{-20}$	$E_{t_{QQCA}} = 0*10^{-20}$
Input cell in C state	$E_{t_{bQCA}} = 0.2943*10^{-20}$	$E_{t_{bQCA}} = 0*10^{-20}$	$E_{t_{bQCA}} = 0.2943*10^{-20}$	$E_{t_{bQCA}} = 0*10^{-20}$	$E_{t_{bQCA}} = 0*10^{-20}$
	$E_{t_{TQCA}} = 0*10^{-20}$	$E_{t_{TQCA}} = 0*10^{-20}$	$E_{t_{TQCA}} = 0.8029*10^{-20}$	$E_{t_{TQCA}} = 0.8029*10^{-20}$	$E_{t_{TQCA}} = 0*10^{-20}$
	$E_{t_{QQCA}} = 0.2943*10^{-20}$	$E_{t_{QQCA}} = 0*10^{-20}$	$E_{t_{QQCA}} = 1.0972*10^{-20}$	$E_{t_{QQCA}} = 0.8029*10^{-20}$	$E_{t_{QQCA}} = 0*10^{-20}$
Input cell in D state	$E_{t_{bQCA}} = 0*10^{-20}$	$E_{t_{bQCA}} = 0.8029*10^{-20}$	$E_{t_{bQCA}} = 0*10^{-20}$	$E_{t_{bQCA}} = 0.8029*10^{-20}$	$E_{t_{bQCA}} = 0*10^{-20}$
	$E_{t_{TQCA}} = 0*10^{-20}$	$E_{t_{TQCA}} = 0*10^{-20}$	$E_{t_{TQCA}} = 0.8029*10^{-20}$	$E_{t_{TQCA}} = 0.8029*10^{-20}$	$E_{t_{TQCA}} = 0*10^{-20}$
	$E_{t_{QQCA}} = 0*10^{-20}$	$E_{t_{QQCA}} = 0.8029*10^{-20}$	$E_{t_{QQCA}} = 0.8029*10^{-20}$	$E_{t_{QQCA}} = 1.6058*10^{-20}$	$E_{t_{QQCA}} = 0*10^{-20}$

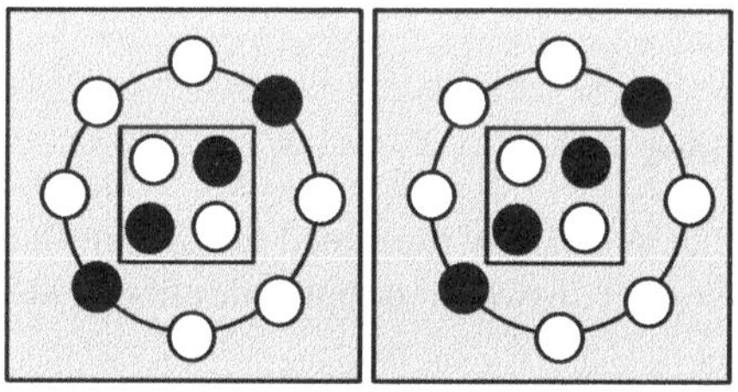

Figure 7.6 Two QQCA cells (QQCA-based wire).

Table 7.9 Effect of two adjacent four-valued cells (QQCA-based wire).

Input		Output	
Symbol	**Polarization**	**Symbol**	**Polarization**
A	−3	A	−3
B	−1	B	−1
C	1	C	1
D	3	D	3

Table 7.10 Output states of the interactions of two quaternary cells and the final results.

	Output cell in A state	Output cell in B state	Output cell in C state	Output cell in D state	Final result
Input cell in A state	$E_{tbQCA}=0.196*10^{-20}$	$E_{tbQCA}=1.552*10^{-20}$	$E_{tbQCA}=0.196*10^{-20}$	$E_{tbQCA}=1.552*10^{-20}$	$E_{tbQCA}=0.196*10^{-20}$
	$E_{tTQCA}=0.196*10^{-20}$	$E_{tTQCA}=0.196*10^{-20}$	$E_{tTQCA}=1.552*10^{-20}$	$E_{tTQCA}=1.552*10^{-20}$	$E_{tTQCA}=0.196*10^{-20}$
	$E_{tQQCA}=0.392*10^{-20}$	$E_{tQQCA}=1.748*10^{-20}$	$E_{tQQCA}=1.748*10^{-20}$	$E_{tQQCA}=3.104*10^{-20}$	$E_{tQQCA}=0.392*10^{-20}$
Input cell in B state	$E_{tbQCA}=1.552*10^{-20}$	$E_{tbQCA}=0.196*10^{-20}$	$E_{tbQCA}=1.552*10^{-20}$	$E_{tbQCA}=0.196*10^{-20}$	$E_{tbQCA}=0.196*10^{-20}$
	$E_{tTQCA}=0.196*10^{-20}$	$E_{tTQCA}=0.196*10^{-20}$	$E_{tTQCA}=1.552*10^{-20}$	$E_{tTQCA}=1.552*10^{-20}$	$E_{tTQCA}=0.196*10^{-20}$
	$E_{tQQCA}=1.748*10^{-20}$	$E_{tQQCA}=0.392*10^{-20}$	$E_{tQQCA}=3.104*10^{-20}$	$E_{tQQCA}=1.748*10^{-20}$	$E_{tQQCA}=0.392*10^{-20}$
Input cell in C state	$E_{tbQCA}=0.196*10^{-20}$	$E_{tbQCA}=1.552*10^{-20}$	$E_{tbQCA}=0.196*10^{-20}$	$E_{tbQCA}=1.552*10^{-20}$	$E_{tbQCA}=0.196*10^{-20}$
	$E_{tTQCA}=1.552*10^{-20}$	$E_{tTQCA}=1.552*10^{-20}$	$E_{tTQCA}=0.196*10^{-20}$	$E_{tTQCA}=0.196*10^{-20}$	$E_{tTQCA}=0.196*10^{-20}$
	$E_{tQQCA}=1.748*10^{-20}$	$E_{tQQCA}=3.104*10^{-20}$	$E_{tQQCA}=0.392*10^{-20}$	$E_{tQQCA}=1.748*10^{-20}$	$E_{tQQCA}=0.392*10^{-20}$
Input cell in D state	$E_{tbQCA}=1.552*10^{-20}$	$E_{tbQCA}=0.196*10^{-20}$	$E_{tbQCA}=1.552*10^{-20}$	$E_{tbQCA}=0.196*10^{-20}$	$E_{tbQCA}=0.196*10^{-20}$
	$E_{tTQCA}=1.552*10^{-20}$	$E_{tTQCA}=1.552*10^{-20}$	$E_{tTQCA}=0.196*10^{-20}$	$E_{tTQCA}=0.196*10^{-20}$	$E_{tTQCA}=0.196*10^{-20}$
	$E_{tQQCA}=3.104*10^{-20}$	$E_{tQQCA}=1.748*10^{-20}$	$E_{tQQCA}=1.748*10^{-20}$	$E_{tQQCA}=0.392*10^{-20}$	$E_{tQQCA}=0.392*10^{-20}$

are presented for different outputs. The minimum Coulomb energy is selected for a fixed input to obtain the final value in the output.

7.2.3 Implementation of AND and OR Functions

In this section, the AND and OR functions in QQCA cells are examined to confirm the accuracy of the proposed cell. This study tries to give the correct answer to design gates like in bQCA using one-phase clock.

As mentioned, the dimensions of the cells are 8 nm×8 nm. The wells are located on a circle with a radius of 2.5 nm and a distance of 1.5 nm from the cell wall. Also, in the first layer, a ternary cell is used, and each well is located at an angular distance of 45° from the other well, and the first well is at 0°. In the second layer, a bQCA cell is used, and the angular distance between the wells is 90°, and the distance from the wells to the cell wall is 1.5 nm. First, the states {A, B, C, D} are applied to the inputs of the majority gate, as shown in Figure 7.7. Figure 7.7a shows a standard majority gate, and Figure 7.7b shows a diagonal majority gate. The responses of the two proposed gates with states {A, B, C, D} are shown in Table 7.11. Both proposed gates are the same in terms of the output response and differ only in the amount of electrostatic energy.

Before examining the AND and OR gates using QQCA cells, the truth table of these gates is obtained [157–159]. Table 7.12 shows the truth table in which the quaternary logic levels are expressed in the form of {0, 1, 2, 3} or its equivalent polarization {−3, −1, 1, 3}.

In Table 7.12, for input polarizations 1 and −1 (B and C), the output of the AND gate is equal to polarization −3 or A; for input polarizations 1 and −1 (B and C), the output of the OR gate is equal to 3 or D. Considering these values, the definition of the MIN function as the AND function and

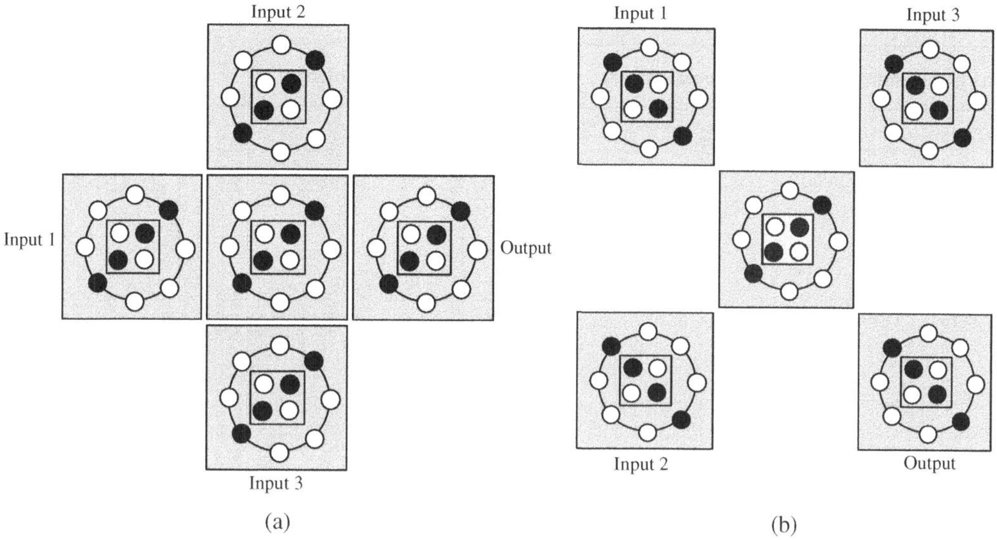

Figure 7.7 QQCA majority gates: (a) standard majority gate and (b) diagonal majority gate.

Table 7.11 QQCA majority gate output.

In1	In2	In3	Out	In1	In2	In3	Out	In1	In2	In3	Out	In1	In2	In3	Out
A	A	A	A	B	A	A	A	C	A	A	A	D	A	A	A
A	A	B	A	B	A	B	B	C	A	B	A	D	A	B	B
A	A	C	A	B	A	C	A	C	A	C	C	D	A	C	C
A	A	D	A	B	A	D	B	C	A	D	C	D	A	D	D
A	B	A	A	B	B	A	B	C	B	A	A	D	B	A	B
A	B	B	B	B	B	B	B	C	B	B	B	D	B	B	B
A	**B**	**C**	**A**	B	B	C	B	C	B	C	C	**D**	**B**	**C**	**D**
A	B	D	B	B	B	D	B	C	B	D	D	D	B	D	D
A	C	A	A	B	C	A	A	C	C	A	C	D	C	A	C
A	**C**	**B**	**A**	B	C	B	B	C	C	B	C	**D**	**C**	**B**	**D**
A	C	C	C	B	C	C	C	C	C	C	C	D	C	C	C
A	C	D	C	B	C	D	D	C	C	D	C	D	C	D	D
A	D	A	A	B	D	A	B	C	D	A	C	D	D	A	D
A	D	B	B	B	D	B	B	C	D	B	D	D	D	B	D
A	D	C	C	B	D	C	D	C	D	C	C	D	D	C	D
A	D	D	D	B	D	D	D	C	D	D	B	D	D	D	D

the MAX function as the OR function is no longer correct. Therefore, the AND and OR gates based on the proposed QQCA cell should have the results presented in Table 7.12.

According to Table 7.11, which presents the majority gate output, if one of the inputs is in state A or polarization 3, the majority gate becomes an AND gate. If one of the inputs is in state D or polarization 3, the majority gate acts as an OR gate. In Figure 7.8, the structure of the AND and OR

Table 7.12 Validity of QQCA-based AND and OR gates.

In1 (polarization)	In2 (polarization)	AND Gate (polarization)	OR Gate (polarization)
0 (−3)	0 (−3)	0 (−3)	0 (−3)
0 (−3)	1 (−1)	0 (−3)	1 (−1)
0 (−3)	2 (1)	0 (−3)	2 (1)
0 (−3)	3 (3)	0 (−3)	3 (3)
1 (−1)	1 (−1)	1 (−1)	1 (−1)
1 (−1)	2 (1)	0 (−3)	3 (3)
1 (−1)	3 (3)	1 (−1)	3 (3)
2 (1)	2 (1)	2 (1)	2 (1)
2 (1)	3 (1)	2 (1)	3 (3)
3 (3)	3 (3)	3 (3)	3 (3)

Source: Adapted from Jahangir et al. [157].

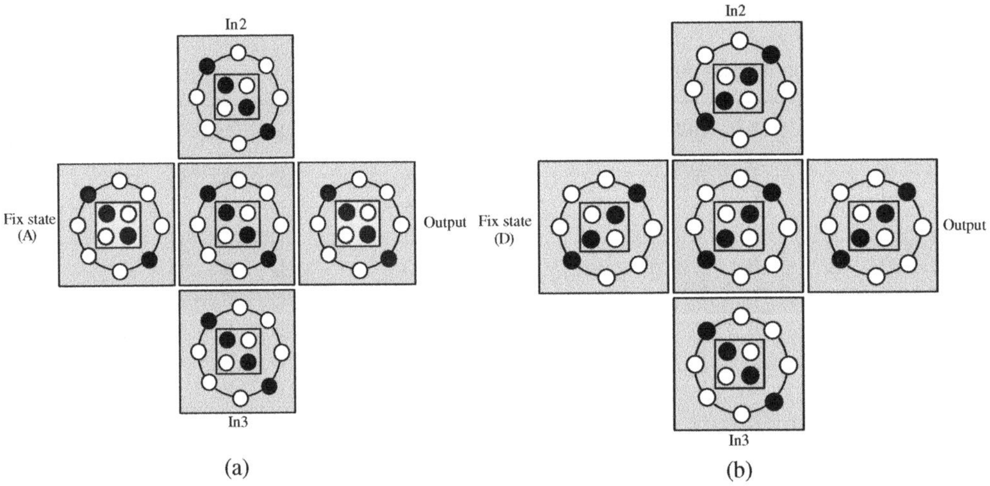

Figure 7.8 AND and OR gates with two inputs and one fixed input (a) AND gate structure with one fixed input in state A and (b) OR gate structure with one fixed input in state D.

gates with two inputs and one fixed input is shown, and in Table 7.13, the truth table of the AND and OR gates is presented.

In the following, the external electrostatic energies of the AND and OR gates are evaluated to confirm the accuracy of the obtained results. Table B.3 gives the MATLAB analysis results for the amount of external electrostatic energy of the AND gate. In this table, the final value or output is selected to be equal to the minimum amount of electrostatic energy for all polarizations of the output cell.

Table B.4 shows the amount of electrostatic energy of the OR gate, simulated in MATLAB as done for the AND gate. The output value is selected to be equal to the minimum electrostatic energy for all output polarizations.

Table 7.13 Quaternary AND and OR gate truth table.

In1	In2	In3	Out	In1	In2	In3	Out
A	A	A	A	D	A	A	A
A	A	B	A	D	A	B	B
A	A	C	A	D	A	C	C
A	A	D	A	D	A	D	D
A	B	A	A	D	B	A	B
A	B	B	B	D	B	B	B
A	B	C	A	D	B	C	D
A	B	D	B	D	B	D	D
A	C	A	A	D	C	A	C
A	C	B	A	D	C	B	D
A	C	C	C	D	C	C	C
A	C	D	C	D	C	D	D
A	D	A	A	D	D	A	D
A	D	B	B	D	D	B	D
A	D	C	C	D	D	C	D
A	D	D	D	D	D	D	D
AND gate				OR gate			

7.3 Structure of Basic Gates Using QuQCA

In the proposed QuQCA cell, the polarizations were calculated as {−4, −2, 0, 2, 4} using quantum measurements. In this part of the book, the AND, OR, and NOT gates consisting of two adjacent cells, which are the basis of QuQCA-based wire, are evaluated. The proposed QuQCA cell is presented in two ternary layers, which are at a distance of 25 nm from each other to achieve isolation. In each layer of the QuQCA cell, the radius of the circle on which the potential wells are located is 2.5 nm, and the distance from the wells to the cell wall is 1.5 nm. The angular distance of the wells from each layer is 45°, and the first well is at 0°.

7.3.1 QuQCA-Based NOT Gate

Equation (7.1) presents the NOT gate relation, in which x is the NOT gate input in QuQCA, and the value given in this equation is 35 times the maximum input value. To use Eq. (7.1), the polarization value must be between 0 and 1. Since the QuQCA polarizations range from −4 to 0, polarization 2 is converted into 1/4, polarization 0 into 1/2, polarization 2 into 3/4, and polarization 4 into 1. Then, the converted values can be replace x.

To simulate QuQCA-based NOT gates, MATLAB 2018b is used, and all gates are simulated in one clock according to the QCA standard. Figure 7.9 shows the NOT gate using the QuQCA symbolic model. Table 7.14 shows the output of the NOT gate, and Table 7.15 shows the amounts of Coulomb energy for all output states. The final value in Table 7.15 is selected to be the minimum external electrostatic energy at the output.

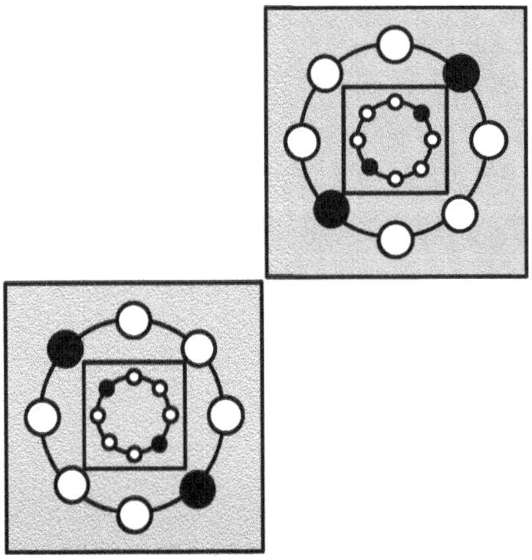

Figure 7.9 Representation of a NOT gate using the QuQCA-based symbolic model.

Table 7.14 NOT gate states in QuQCA.

Input		Output	
Symbol	**Polarization**	**Symbol**	**Polarization**
A	−4	E	4
B	−2	D	2
C	0	C	0
D	2	B	−2
E	4	A	−4

7.3.2 Simulation of Two Adjacent Cells in QuQCA

This section examines two adjacent cells, which make a basic quantum wire. Figure 7.10 shows the structure of two adjacent QuQCA cells using the symbolic QuQCA model. Table 7.16 shows the QuQCA-based wire truth table. Table 7.17 presents the external electrostatic energies of two adjacent QuQCA cells for all output states and the final value, which is the minimum external electrostatic energy in the output cell.

According to Table 7.17, to design a quantum wire, the number of cells must be odd because each layer has a polarization of 0 (90°) in the ternary cell. In a QuQCA-based wire, if the input is 0 (90°), the output is 0 (0°) with even cells, and for the output to be 0 (90°), the number of cells must be odd. Table 7.17 shows that the final value is selected based on the minimum external electrostatic energy in the output state.

7.3.3 Implementation of AND and OR Functions

This section investigates the accuracy of the proposed QuQCA cell. The simulated AND and OR gates are presented in the same simulation clock as in the QCA standard.

Table 7.15 QuQCA NOT gate for all states and determining the final value (×1e-20).

Output input	A	B	C	D	E	Final result
A	$E_{t_{TQCA1}} = 0.2943$	$E_{t_{TQCA1}} = 0.2943$	$E_{t_{TQCA1}} = 0.2395$	$E_{t_{TQCA1}} = 0$	$E_{t_{TQCA1}} = 0$	$E_{t_{PQCA}} = 0$
	$E_{t_{TQCA2}} = 0.2943$	$E_{t_{TQCA2}} = 0.2395$	$E_{t_{TQCA2}} = 0.2395$	$E_{t_{TQCA2}} = 0.2395$	$E_{t_{TQCA2}} = 0$	Out state = E state
	$E_{t_{PQCA}} = 0.5886$	$E_{t_{PQCA}} = 0.5338$	$E_{t_{PQCA}} = 0.479$	$E_{t_{PQCA}} = 0.2395$	$E_{t_{PQCA}} = 0$	
B	$E_{t_{TQCA1}} = 0.2943$	$E_{t_{TQCA1}} = 0.2943$	$E_{t_{TQCA1}} = 0.2395$	$E_{t_{TQCA1}} = 0$	$E_{t_{TQCA1}} = 0$	$E_{t_{PQCA}} = 0.1174$
	$E_{t_{TQCA2}} = 0.2395$	$E_{t_{TQCA2}} = 0.1174$	$E_{t_{TQCA2}} = 0.1174$	$E_{t_{TQCA2}} = 0.1174$	$E_{t_{TQCA2}} = 0.4982$	Out state = D state
	$E_{t_{PQCA}} = 0.5338$	$E_{t_{PQCA}} = 0.4117$	$E_{t_{PQCA}} = 0.3569$	$E_{t_{PQCA}} = 0.1174$	$E_{t_{PQCA}} = 0.4982$	
C	$E_{t_{TQCA1}} = 0.2395$	$E_{t_{TQCA1}} = 0.2395$	$E_{t_{TQCA1}} = 0.1174$	$E_{t_{TQCA1}} = 0.5194$	$E_{t_{TQCA1}} = 0.5194$	$E_{t_{PQCA}} = 0.2348$
	$E_{t_{TQCA2}} = 0.2395$	$E_{t_{TQCA2}} = 0.1174$	$E_{t_{TQCA2}} = 0.1174$	$E_{t_{TQCA2}} = 0.1174$	$E_{t_{TQCA2}} = 0.5194$	Out state = C state
	$E_{t_{PQCA}} = 0.479$	$E_{t_{PQCA}} = 0.3569$	$E_{t_{PQCA}} = 0.2348$	$E_{t_{PQCA}} = 0.6368$	$E_{t_{PQCA}} = 1.0388$	
D	$E_{t_{TQCA1}} = 0$	$E_{t_{TQCA1}} = 0$	$E_{t_{TQCA1}} = 0.4982$	$E_{t_{TQCA1}} = 0.4982$	$E_{t_{TQCA1}} = 0.4982$	$E_{t_{PQCA}} = 0.1174$
	$E_{t_{TQCA2}} = 0.2395$	$E_{t_{TQCA2}} = 0.1174$	$E_{t_{TQCA2}} = 0.1174$	$E_{t_{TQCA2}} = 0.1174$	$E_{t_{TQCA2}} = 0.5194$	Out state = B state
	$E_{t_{PQCA}} = 0.2395$	$E_{t_{PQCA}} = 0.1174$	$E_{t_{PQCA}} = 0.6156$	$E_{t_{PQCA}} = 0.6156$	$E_{t_{PQCA}} = 1.0176$	
E	$E_{t_{TQCA1}} = 0$	$E_{t_{TQCA1}} = 0$	$E_{t_{TQCA1}} = 0.4982$	$E_{t_{TQCA1}} = 0.8029$	$E_{t_{TQCA1}} = 0.8029$	$E_{t_{PQCA}} = 0$
	$E_{t_{TQCA2}} = 0$	$E_{t_{TQCA2}} = 0.4982$	$E_{t_{TQCA2}} = 0.4982$	$E_{t_{TQCA2}} = 0.4982$	$E_{t_{TQCA2}} = 0.8029$	Out state = A state
	$E_{t_{PQCA}} = 0$	$E_{t_{PQCA}} = 0.4982$	$E_{t_{PQCA}} = 0.9964$	$E_{t_{PQCA}} = 1.3011$	$E_{t_{PQCA}} = 1.6058$	

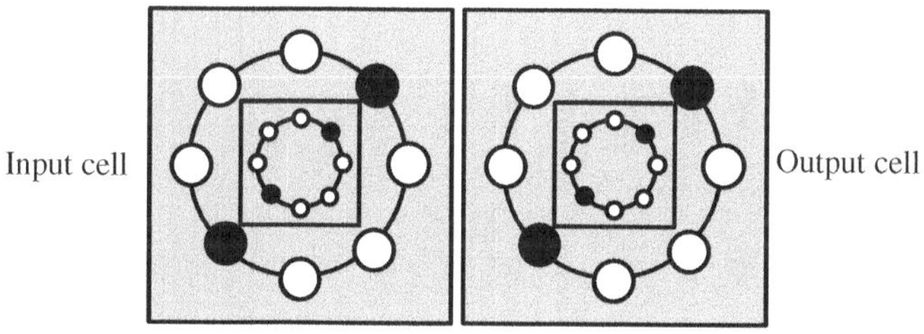

Figure 7.10 Two neighboring cells using the QuQCA symbolic model.

Table 7.16 QuQCA-based wire truth table.

Input		Output	
Symbol	**Polarization**	**Symbol**	**Polarization**
A	−4	A	−4
B	−2	B	−2
C	0	C	0
D	2	D	2
E	4	E	4

Table 7.17 Output states of the interactions of two QuQCA cells and final results (× 1e-20).

Out In	A	B	C	D	E	Intercellular effect	Final result
A	$E_{t_{TQCA1}}=0.196$	$E_{t_{TQCA1}}=0.196$	$E_{t_{TQCA1}}=0.679$	$E_{t_{TQCA1}}=1.552$	$E_{t_{TQCA1}}=1.552$		$E_{t_{PQCA}}=0.392$
	$E_{t_{TQCA2}}=0.196$	$E_{t_{TQCA2}}=0.679$	$E_{t_{TQCA2}}=0.679$	$E_{t_{TQCA2}}=0.679$	$E_{t_{TQCA2}}=1.552$	-----------	
	$E_{t_{PQCA}}=0.392$	$E_{t_{PQCA}}=0.875$	$E_{t_{PQCA}}=1.358$	$E_{t_{PQCA}}=2.231$	$E_{t_{PQCA}}=3.104$		Out state = A state
B	$E_{t_{TQCA1}}=0.196$	$E_{t_{TQCA1}}=0.196$	$E_{t_{TQCA1}}=0.679$	$E_{t_{TQCA1}}=1.552$	$E_{t_{TQCA1}}=1.552$	$E_{t_{TQCA1}}=0.196$	$E_{t_{PQCA}}=0.196$
	$E_{t_{TQCA2}}=0.6787$	$E_{t_{TQCA2}}=0.7493$	$E_{t_{TQCA2}}=0.7493$	$E_{t_{TQCA2}}=0.7493$	$E_{t_{TQCA2}}=0.6787$	$E_{t_{TQCA2}}=0$	
	$E_{t_{PQCA}}=0.8747$	$E_{t_{PQCA}}=0.9453$	$E_{t_{PQCA}}=1.4283$	$E_{t_{PQCA}}=2.3013$	$E_{t_{PQCA}}=2.2307$	$E_{t_{PQCA}}=0.196$	Out state = Intercellular
C	$E_{t_{TQCA1}}=0.6787$	$E_{t_{TQCA1}}=0.6787$	$E_{t_{TQCA1}}=0.7493$	$E_{t_{TQCA1}}=0.6787$	$E_{t_{TQCA1}}=0.6787$	$E_{t_{TQCA1}}=0$	$E_{t_{PQCA}}=0$
	$E_{t_{TQCA2}}=0.6787$	$E_{t_{TQCA2}}=0.7493$	$E_{t_{TQCA2}}=0.7493$	$E_{t_{TQCA2}}=0.7493$	$E_{t_{TQCA2}}=0.6787$	$E_{t_{TQCA2}}=0$	
	$E_{t_{PQCA}}=1.3574$	$E_{t_{PQCA}}=1.428$	$E_{t_{PQCA}}=1.4986$	$E_{t_{PQCA}}=1.428$	$E_{t_{PQCA}}=1.3574$	$E_{t_{PQCA}}=0$	Out state = Intercellular
D	$E_{t_{TQCA1}}=1.552$	$E_{t_{TQCA1}}=1.552$	$E_{t_{TQCA1}}=0.679$	$E_{t_{TQCA1}}=0.196$	$E_{t_{TQCA1}}=0.196$	$E_{t_{TQCA1}}=0.196$	$E_{t_{PQCA}}=0.196$
	$E_{t_{TQCA2}}=0.6787$	$E_{t_{TQCA2}}=0.7493$	$E_{t_{TQCA2}}=0.7493$	$E_{t_{TQCA2}}=0.7493$	$E_{t_{TQCA2}}=0.6787$	$E_{t_{TQCA2}}=0$	
	$E_{t_{PQCA}}=2.2307$	$E_{t_{PQCA}}=2.3013$	$E_{t_{PQCA}}=1.4283$	$E_{t_{PQCA}}=0.9453$	$E_{t_{PQCA}}=0.8747$	$E_{t_{PQCA}}=0.196$	Out state = Intercellular
E	$E_{t_{TQCA1}}=1.552$	$E_{t_{TQCA1}}=1.552$	$E_{t_{TQCA1}}=0.679$	$E_{t_{TQCA1}}=0.196$	$E_{t_{TQCA1}}=0.196$		$E_{t_{PQCA}}=0.392$
	$E_{t_{TQCA2}}=1.552$	$E_{t_{TQCA2}}=0.679$	$E_{t_{TQCA2}}=0.679$	$E_{t_{TQCA2}}=0.679$	$E_{t_{TQCA2}}=0.196$	--------	
	$E_{t_{PQCA}}=3.104$	$E_{t_{PQCA}}=2.231$	$E_{t_{PQCA}}=1.358$	$E_{t_{PQCA}}=0.875$	$E_{t_{PQCA}}=0.392$		Out state = E state

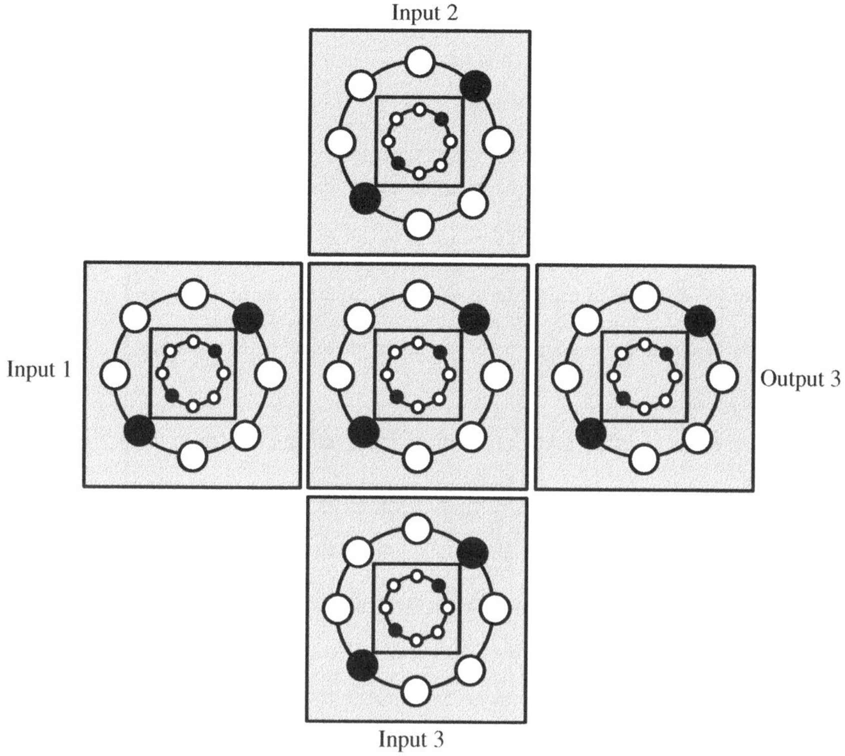

Figure 7.11 QuQCA majority gate.

The dimensions of the proposed cell are 8 nm×8 nm, and the quantum dots are located on a circle with a radius of 2.5 nm at an angular distance of 45° from each other, and the distance from the wells to the cell wall is 1.5 nm. The proposed cell has two layers, and there is a ternary cell in each layer. The polarization set calculated in QuQCA is {−4, −2, 0, 2, 4}, which can be expressed as {A, B, C, D, E}.

Figure 7.11 shows the majority gate with QuQCA cells. The states {A, B, C, D, E} are applied to the input of the majority gate. Table 7.18 shows the majority gate simulation results of the states {A, B, C, D, E}, which are simulated in MATLAB.

According to Table 7.18 and Eq. (7.3), in the majority gate, if one of the inputs is in polarization −4 or state A, the output function has the response of an AND gate, while if one of the inputs is in polarization +4 or state E, the output function has the OR gate response.

$$\begin{aligned} Y &= \text{AND}\ (X_1, X_2) = \min\ (X_1, X_0) \\ Y &= \text{OR}\ (X_1, X_2) = \max\ (X_1, X_0) \end{aligned} \tag{7.3}$$

Figure 7.12 shows the AND and OR gates with one of the inputs fixed in states A and E. Table 7.19 shows the AND and OR gate truth tables.

The external electrostatic energies of the AND and OR gates are evaluated below.

Table B.5 shows the values of Coulomb energy in the AND gate for a fixed input and different outputs. The final value is equal to the minimum Coulomb energy in each state. Table B.6 shows the values of Coulomb energy in the OR gate for a fixed input and different outputs. The final value is equal to the minimum Coulomb energy in each state.

Table 7.18 Majority gate truth table in QuQCA.

In1	In2	In3	Out	In1	In2	In3	Out	In1	In2	In3	Out	In1	In2	In3	Out	In1	In2	In3	Out
A	A	A	A	B	A	A	A	C	A	A	A	D	A	A	A	E	A	A	A
A	A	B	A	B	A	B	B	C	A	B	A	D	A	B	B	E	A	B	B
A	A	C	A	B	A	C	B	C	A	C	C	D	A	C	C	E	A	C	C
A	A	D	A	B	A	D	B	C	A	D	C	D	A	D	D	E	A	D	D
A	A	E	A	B	A	E	B	C	A	E	C	D	A	E	D	E	A	E	E
A	B	A	A	B	B	A	B	C	B	A	B	D	B	A	B	E	B	A	B
A	B	B	B	B	B	B	B	C	B	B	B	D	B	B	B	E	B	B	B
A	B	C	B	B	B	C	B	C	B	C	C	D	B	C	C	E	B	C	C
A	B	D	B	B	B	D	B	C	B	D	C	D	B	D	D	E	B	D	D
A	B	E	B	B	B	E	B	C	B	E	C	D	B	E	D	E	B	E	E
A	C	A	A	B	C	A	B	C	C	A	C	D	C	A	C	E	C	A	C
A	C	B	B	B	C	B	B	C	C	B	C	D	C	B	C	E	C	B	C
A	C	C	C	B	C	C	C	C	C	C	C	D	C	C	C	E	C	C	C
A	C	D	C	B	C	D	C	C	C	D	C	D	C	D	D	E	C	D	D
A	C	E	C	B	C	E	C	C	C	E	C	D	C	E	D	E	C	E	E
A	D	A	A	B	D	A	B	C	D	A	C	D	D	A	D	E	D	A	D
A	D	B	B	B	D	B	B	C	D	B	C	D	D	B	D	E	D	B	D
A	D	C	C	B	D	C	C	C	D	C	C	D	D	C	D	E	D	C	D
A	D	D	D	B	D	D	D	C	D	D	D	D	D	D	D	E	D	D	D
A	D	E	D	B	D	E	D	C	D	E	D	D	D	E	D	E	D	E	E
A	E	A	A	B	E	A	B	C	E	A	C	D	E	A	D	E	E	A	E
A	E	B	B	B	E	B	B	C	E	B	C	D	E	B	D	E	E	B	E
A	E	C	C	B	E	C	C	C	E	C	C	D	E	C	D	E	E	C	E
A	E	D	D	B	E	D	D	C	E	D	D	D	E	D	D	E	E	D	E
A	E	E	E	B	E	E	E	C	E	E	E	D	E	E	E	E	E	E	E

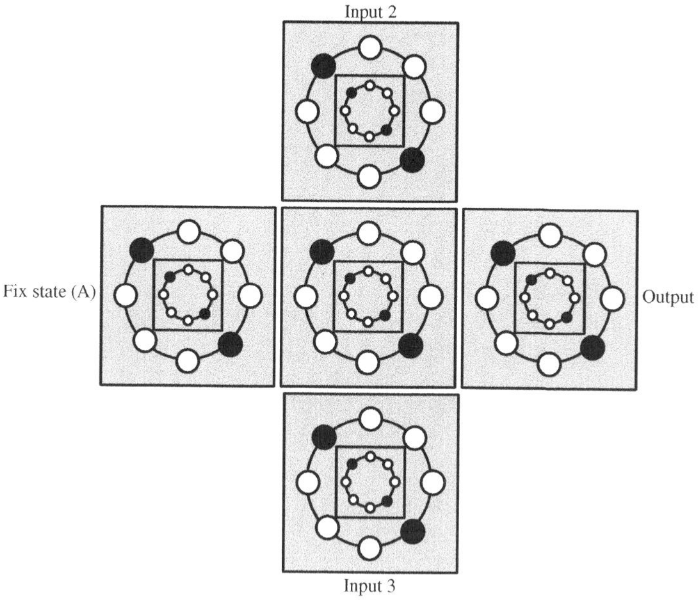

(a)

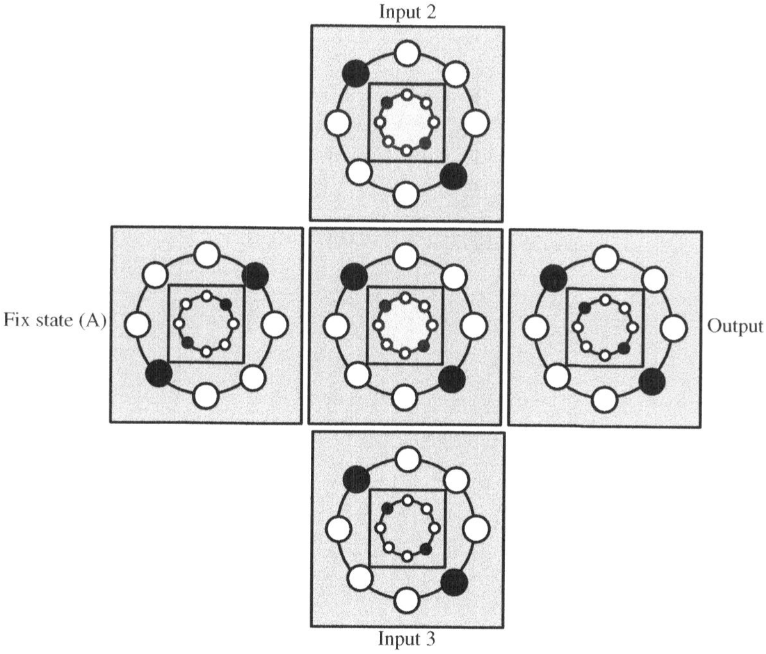

(b)

Figure 7.12 QQCA-based AND and OR gate structures: (a) AND gate structure with a fixed input in state A and (b) OR gate structure with a fixed input in state E.

Table 7.19 Truth tables of the quinary AND and OR gates.

In1	In2	In3	Out	In1	In2	In3	Out
A	A	A	A	E	A	A	A
A	A	B	A	E	A	B	B
A	A	C	A	E	A	C	C
A	A	D	A	E	A	D	D
A	A	E	A	E	A	E	E
A	B	A	A	E	B	A	B
A	B	B	B	E	B	B	B
A	B	C	B	E	B	C	C
A	B	D	B	E	B	D	D
A	B	E	B	E	B	E	E
A	C	A	A	E	C	A	C
A	C	B	B	E	C	B	C
A	C	C	C	E	C	C	C
A	C	D	C	E	C	D	D
A	C	E	C	E	C	E	E
A	D	A	A	E	D	A	D
A	D	B	B	E	D	B	D
A	D	C	C	E	D	C	D
A	D	D	D	E	D	D	D
A	D	E	D	E	D	E	E
A	E	A	A	E	E	A	E
A	E	B	B	E	E	B	E
A	E	C	C	E	E	C	E
A	E	D	D	E	E	D	E
A	E	E	E	E	E	E	E
AND gate				OR gate			

8

Implementation of Ternary and Quaternary Basic Gates Using QQCASim and TQCASim

In this chapter, the results of the proposed ternary and quaternary gates, which are based on ternary quantum dot cellular automata (TQCA) and quaternary quantum dot cellular automata (QQCA) cells, respectively, are confirmed. These gates were simulated in Chapter 7 using MATLAB. The ternary and quaternary gates are simulated in this chapter using specialized software, TQCASim and QQCASim [129, 130].

Also, fabrication of quantum dot cellular automata (QCA) cells requires lower temperatures to reduce thermal noise and improve device performance. However, Lent could make QCA cells. According to the article by Lent and coworkers on QCA construction, a binary quantum dot cellular automata (bQCA) cell can be created by connecting two metal points in a series to two other metal points through a tunneling barrier and a capacitor [160]. The Lent method for TQCA involves connecting a cell from four metal points in a series, arranged in a semicircle, and connected to four other metal points through a tunneling dam and capacitor. The movement of electrons in the cells is facilitated by Coulomb repulsion [160].

The Lent method uses standard electron beam lithography with two layers of resist on a Si or Si/SiO_2 substrate. Oxygen and aluminum deposition are used to create oxide electrodes, ensuring that the resistance of the connections is large enough to prevent quantum load fluctuations. QCA quantum dots are made using nanometer-sized aluminum islands, and the position of electrons in the output cell is detected using a single-electron transistor [160].

In order to construct a TQCA cell of the provided size (8 nm × 8 nm), manufacturing technology with sizes below 10 nm is required. Electron beam lithography with polymethmetacrylate (PMMA) resist has been reported to achieve dimensions of 5–7 nm [161]. Plasmonic lithography is another technology that can be used for sizes below 9 nm [162]. Extreme ultra-violet (EUV) interference lithography has been used to create a 6 nm pattern using different masks for miniaturization [163]. The field emission scanning probe lithography (FE-SPL) method, which utilizes low-energy electrons, is another approach for making sub-10 nm devices [164]. The minimum diameter of the quantum dot made is 1 nm, using the In2Te3 semiconductor [165]. These methods are suitable for creating sizes below 10 nm, which aligns with the proposed cell size and allows for its use in the design of three-value logic circuits.

Computing with Multi-Value Logic in Quantum Dot Cellular Automata, First Edition.
Reza Sabbaghi-Nadooshan, Reza Akbari-Hasanjani, Leila Dehbozorgi, Majid Haghparast, and Hamid Reza Akbari-Hasanjani.

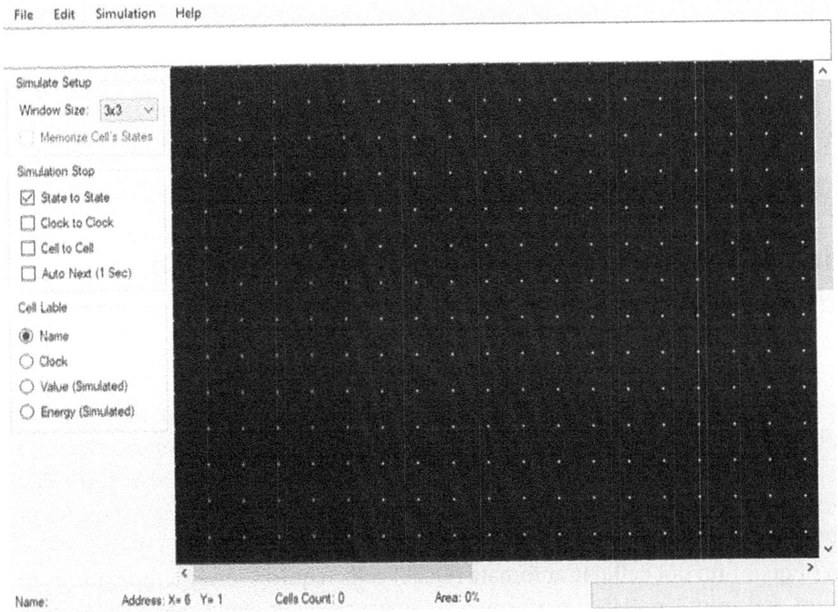

Figure 8.1 Environment of TQCASim software version 1.0.11.1. Source: Akbari-Hasanjani et al. [129]/with permission of Elsevier.

8.1 TQCA Simulator

In this section, the designs of the ternary gates, which were presented in Chapter 7, are simulated using the new TQCASim software [129]. Based on quantum measurements, the polarization set of a TQCA cell is given by $\{-2, 0, 0, 2\}$. In the literature, the polarization set $\{-1, 0, 0, 1\}$ has been used, but these values are wrong in terms of quantum measurement; the findings and reasons presented here confirm the accuracy of this claim.

To design ternary gates and eliminate errors, the dimensions of the TQCA cell were changed, and the basic gates were simulated and analyzed using MATLAB in Chapter 7. Now, to confirm the calculations of MATLAB, the TQCASim software is upgraded, and the quantum calculations and cell dimensions presented in this study are applied to it.

The TQCA basic gates in this book are simulated by the TQCASim software version 1.0.11.1 [129]. Figure 8.1 shows the environment of the TQCASim software version 1.0.11.1 [129].

In TQCASim, TQCA cells are classified as input, intermediate, normal, output, and fixed cells. Figure 8.2 shows the TQCA cell options.

TQCASim version 1.0.11.1 is used to evaluate the following basic gates.

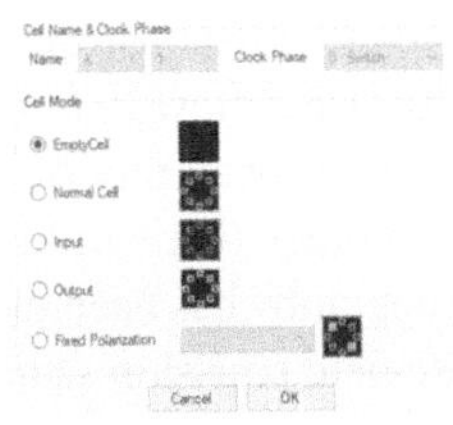

Figure 8.2 TQCA cell options. Source: Akbari-Hasanjani et al. [129]/with permission of Elsevier.

8.1.1 Majority Gate

In a majority gate, the output equals the maximum input states. Each majority gate consists of five TQCA cells [129]. Using TQCASim software, the majority gate is simulated in one-phase clock. Figure 8.3 shows the majority gate designed with TQCASim software, the results of the majority

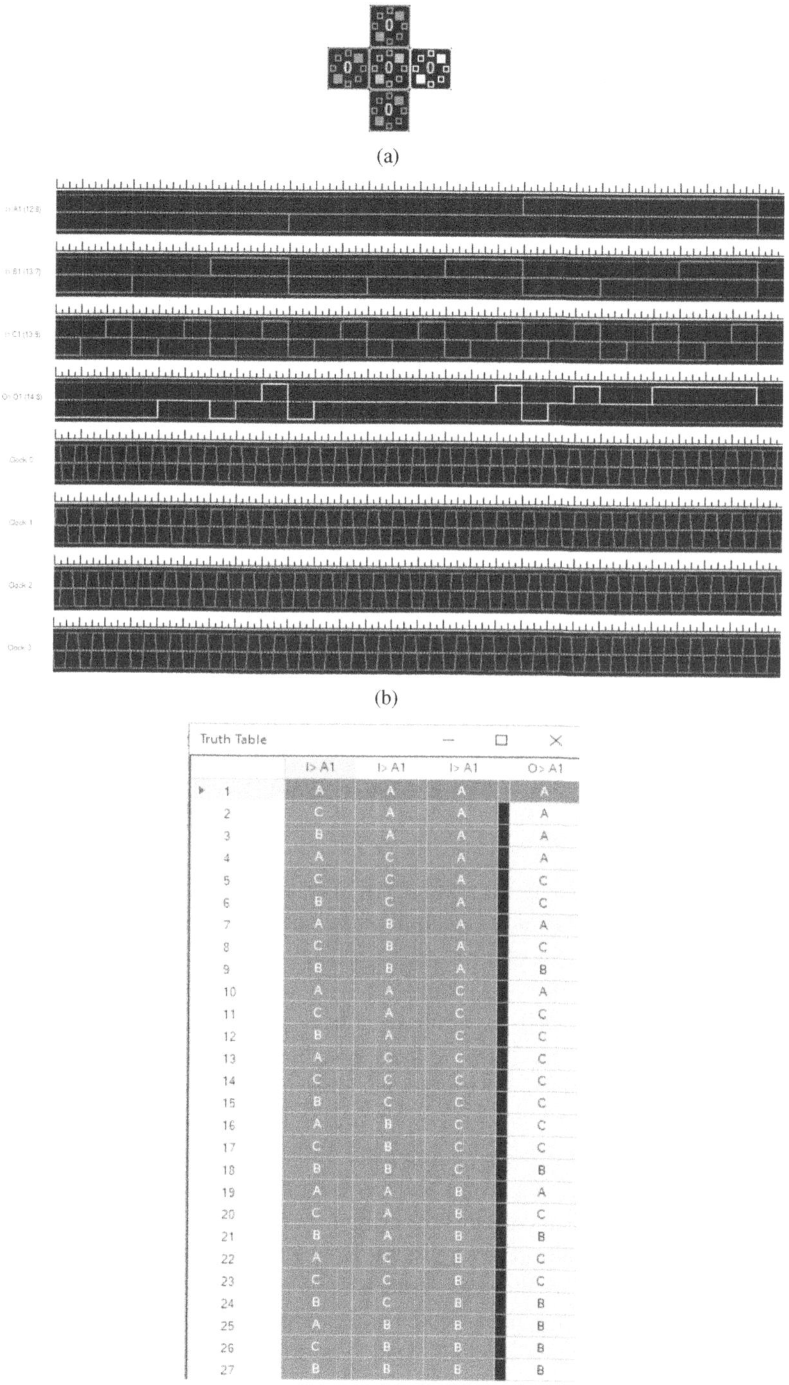

	I> A1	I> A1	I> A1	O> A1
1	A	A	A	A
2	C	A	A	A
3	B	A	A	A
4	A	C	A	A
5	C	C	A	C
6	B	C	A	C
7	A	B	A	A
8	C	B	A	C
9	B	B	A	B
10	A	A	C	A
11	C	A	C	C
12	B	A	C	C
13	A	C	C	C
14	C	C	C	C
15	B	C	C	C
16	A	B	C	C
17	C	B	C	C
18	B	B	C	B
19	A	A	B	A
20	C	A	B	C
21	B	A	B	B
22	A	C	B	C
23	C	C	B	C
24	B	C	B	B
25	A	B	B	B
26	C	B	B	B
27	B	B	B	B

Figure 8.3 Simulation of a three-value majority gate: (a) majority gate designed with TQCASim, (b) majority gate waveform, and (c) the truth table obtained from the software in TQCA. Source: Akbari-Hasanjani et al. [129]/with permission of Elsevier.

gate waveform, and the truth table obtained from the software [129]. In Figure 8.3, the software writes the phase clock of each cell inside the cell.

In the calculation of polarization set, its equivalent, i.e., {A, C, D, B}, is used. In this set, A is equivalent to polarization −3, C is 0 (90°), D is 0 (0°), and B is 3. According to the truth table and waveform of the majority gate simulated with TQCASim, the results are equal to those of the majority gate in Table 7.5, which shows the accuracy of the calculations.

8.1.2 AND Gate

If one of the inputs in the majority gate is kept fixed in polarization −3 or A, the majority gate functions as an AND gate, and the output is equal to the MIN selection in the input [129]. Figure 8.4 shows the AND gate designed by the TQCASim software, the AND gate waveform simulation results, and the AND gate truth table obtained from the software.

Comparing the results of AND gate simulation in TQCASim with those in MATLAB shows that the results of both simulations are the same.

8.1.3 OR Gate

If in the majority gate, one of the inputs is kept fixed in polarization 3 or B, the majority gate response is equivalent to the OR gate response, and the output is equal to MAX selection in the input [129].

Figure 8.5 shows the OR gate designed by TQCASim, the OR gate waveform, and the OR gate truth table obtained by TQCASim [129].

Comparing the results of the OR gate simulation in TQCASim with those in MATLAB shows that both simulations are the same.

8.1.4 Ternary Wire

The TQCA-based wire is made by putting together several TQCA cells. In TQCA technology, if the number of cells is odd, the applied input is exactly received at the output. However, if the number of cells is even, for an input of 0 (90°) or C, the output is 0 (0°) or D. Other polarizations are received at the output the same as at the input [129].

In this section, the interaction of two cells is simulated first, and the simulation results of TQCASim are compared with those of MATLAB. Figure 8.6 shows the two cells simulated with TQCASim, the waveform of two cells, and the truth table of the two cells simulated in TQCASim [129].

Next, wires with odd and even numbers of cells are simulated in TQCASim. In Figure 8.7, a wire with odd number of cells and the obtained results are shown. According to Figure 8.7b and c, the information applied to the input is exactly received at the output [129].

Figure 8.8 shows a wire with an even number of cells and its results. According to Figure 8.8b and c, the input and the received output are the same. According to the results of Figure 8.8b and c, in a wire with an even number of cells, the input 0 (90°) or C is converted into 0 (0°) or D in the output. In the output, if the intracellular state D occurs, TQCASim software warns the designer by turning the background color of the corresponding cell in the truth table into red [129].

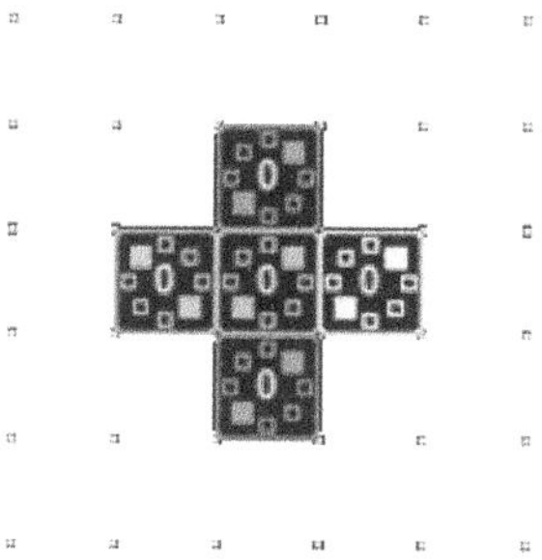

(a)

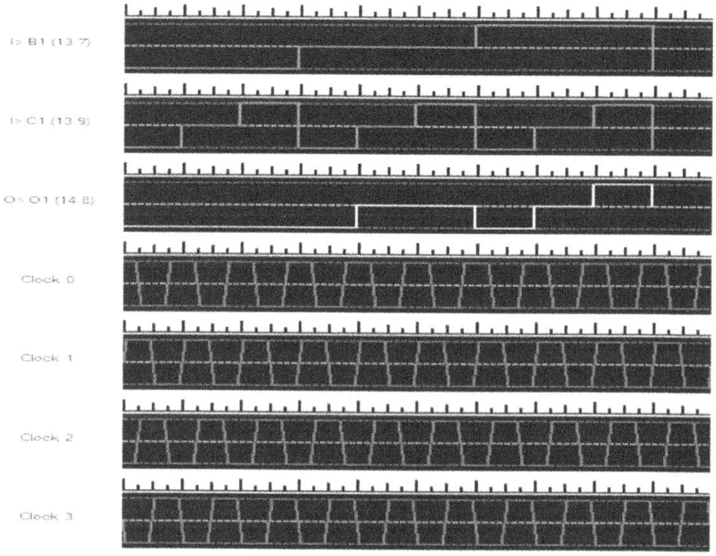

(b)

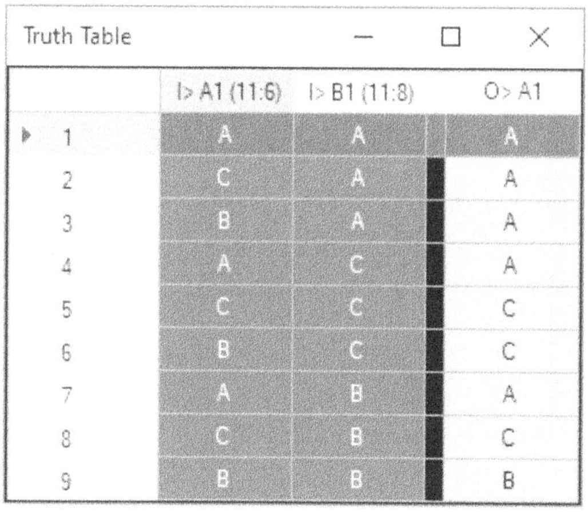

Truth Table

	I> A1 (11:6)	I> B1 (11:8)	O> A1
1	A	A	A
2	C	A	A
3	B	A	A
4	A	C	A
5	C	C	C
6	B	C	C
7	A	B	A
8	C	B	C
9	B	B	B

(c)

Figure 8.4 Simulation of ternary AND gates; (a) AND gates designed with TQCASim software, (b) AND gate waveform obtained by TQCASim, and (c) AND gate truth table obtained by TQCASim.
Source: Akbari-Hasanjani et al. [129]/with permission of Elsevier.

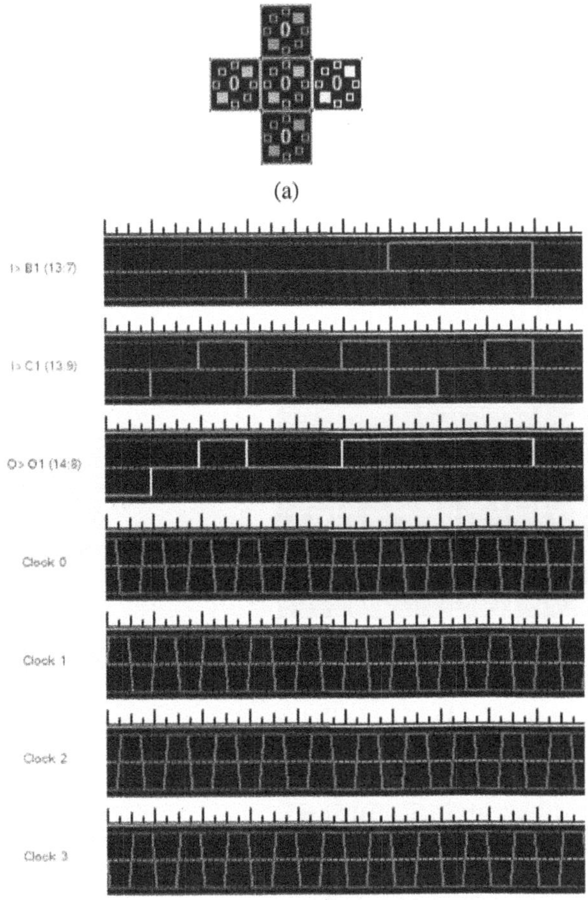

(a)

(b)

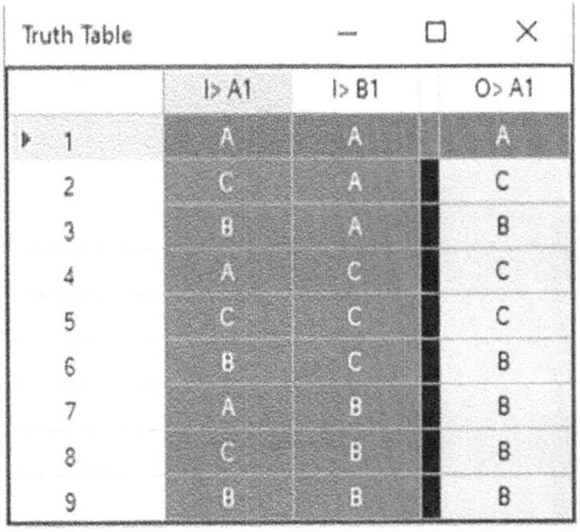

Truth Table

	I> A1	I> B1	O> A1
1	A	A	A
2	C	A	C
3	B	A	B
4	A	C	C
5	C	C	C
6	B	C	B
7	A	B	B
8	C	B	B
9	B	B	B

(c)

Figure 8.5 Simulation of the ternary OR gate: (a) OR gate designed by TQCASim, (b) OR gate waveform using TQCASim, and (c) OR gate truth table obtained from TQCASim. Source: Akbari-Hasanjani et al. [129]/with permission of Elsevier.

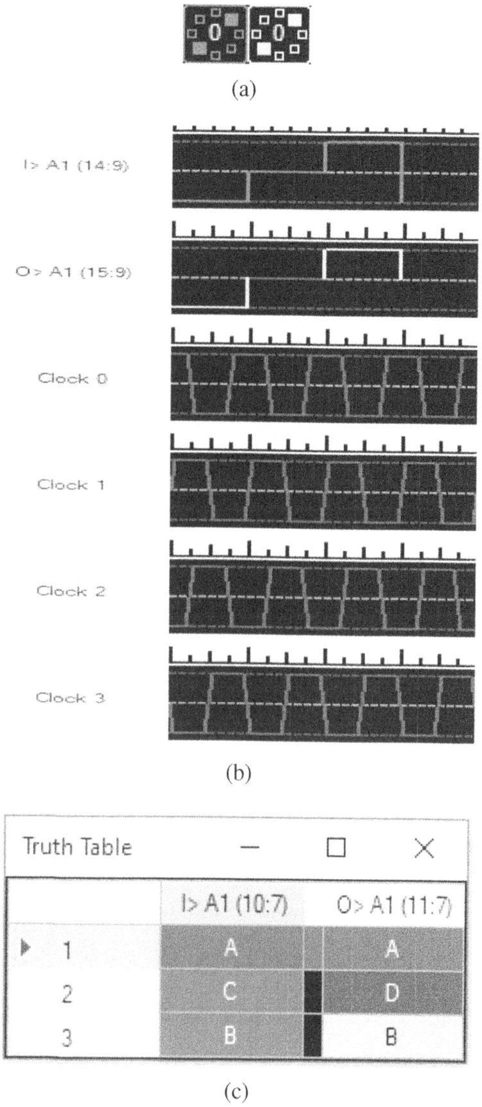

Figure 8.6 Simulation of two TQCA cells: (a) the two cells simulated in TQCASim, (b) two-cell interaction waveform using TQCASim, and (c) the truth table of two cells simulated in TQCASim. Source: Akbari-Hasanjani et al. [129]/with permission of Elsevier.

8.2 QQCA Simulator for QQCA Circuit Simulation

In this section, QQCASim software is reviewed using the provided calculations. The QQCASim software can implement the AND and OR functions of the quaternary inverse function [130]. With the introduction of QQCASim, quaternary circuits based on QCA technology can be simulated.

The polarization set in the quaternary system is calculated as $\{-3, -1, 1, 3\}$ [130]. The combination of polarizations in bQCA and TQCA is used to construct the polarization in QQCA. The reason for using both bQCA and TQCA is that the quaternary system is a three-particle system. If three particles are placed in one layer, quaternary logic gates become unstable [129].

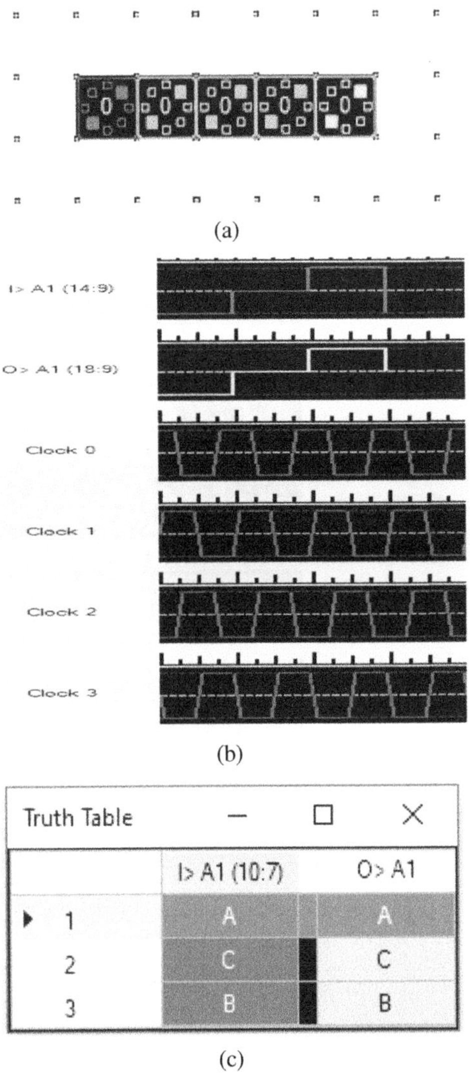

Figure 8.7 Simulation of a ternary wire with an odd number of cells: (a) odd number of cells in a wire simulated in TQCASim, (b) interaction waveform of the wire in TQCASim, and (c) truth table of the wire in TQCASim. Source: Akbari-Hasanjani et al. [129]/with permission of Elsevier.

The dimensions are considered to be 8 nm × 8 nm. According to Ref. [160], size reduction in QCA technology improves performance, so we examine these dimensions for QCA metal technology [130]. MATLAB and QQCASim results indicate performance and energy improvement in QCA cells.

Quantum gate simulations based on QQCA are performed by QQCASim software version 1.0.0.1. Figure 8.9 shows the simulation environment of QQCASim version 1.0.0.1 [123]. The proposed symbolic model is used in the software [130].

In QQCASim, cells are classified into input, middle, output, and fixed cells. Figure 8.10 shows the QQCA cell modes used in QQCASim software version 1.0.01 [123].

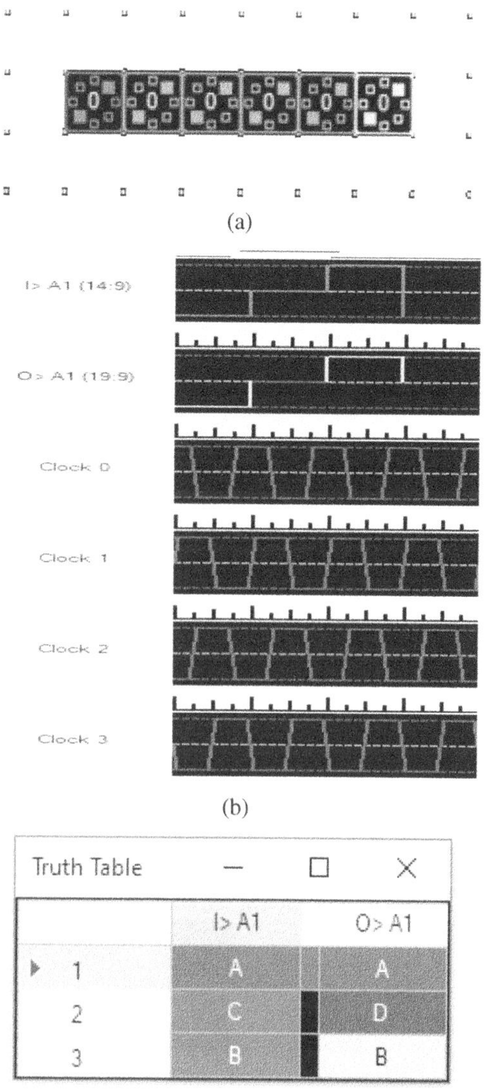

Figure 8.8 Simulation of a ternary wire with an even number of cells: (a) ternary wire with an even number of TQCA cells, (b) interaction waveform of the wire in TQCASim, and (c) truth table of the wire simulated in TQCASim. Source: Akbari-Hasanjani et al. [129]/with permission of Elsevier.

Next, the quaternary AND, OR, and NOT basic gates and quaternary wire are examined using QQCASim.

8.2.1 Majority Gate

The majority gate structure in the proposed QQCA model is similar to the majority gate structures in bQCA and TQCA because these two structures are used in the design [130]. In a majority gate, the output response equals the state with the most repetition in the input. According to the QCA standard, a majority gate is designed in one-phase clock. Figure 8.11 shows the main results and the results provided by QQCASim software.

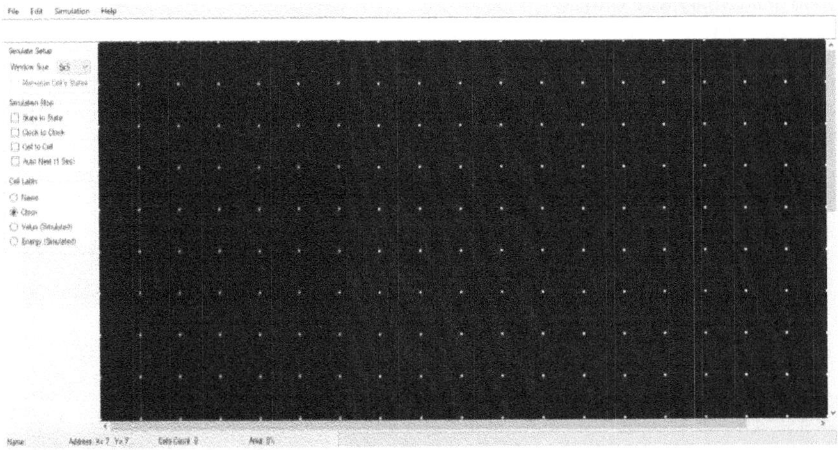

Figure 8.9 Environment of QQCASim software version 1.0.0.1. Source: Mianroodi et al. [123]/with permission of Journal Management System/Scientia Iranica.

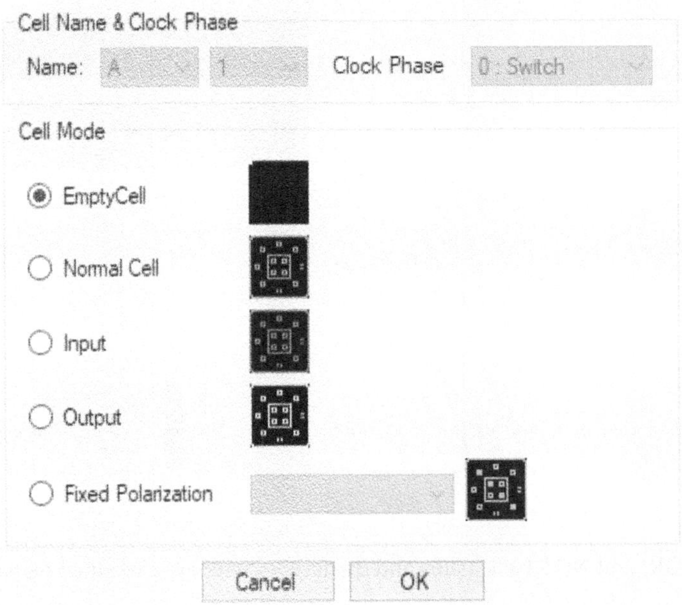

Figure 8.10 Specifications of QQCA cells. Source: Mianroodi et al. [123]/with permission of Journal Management System/Scientia Iranica.

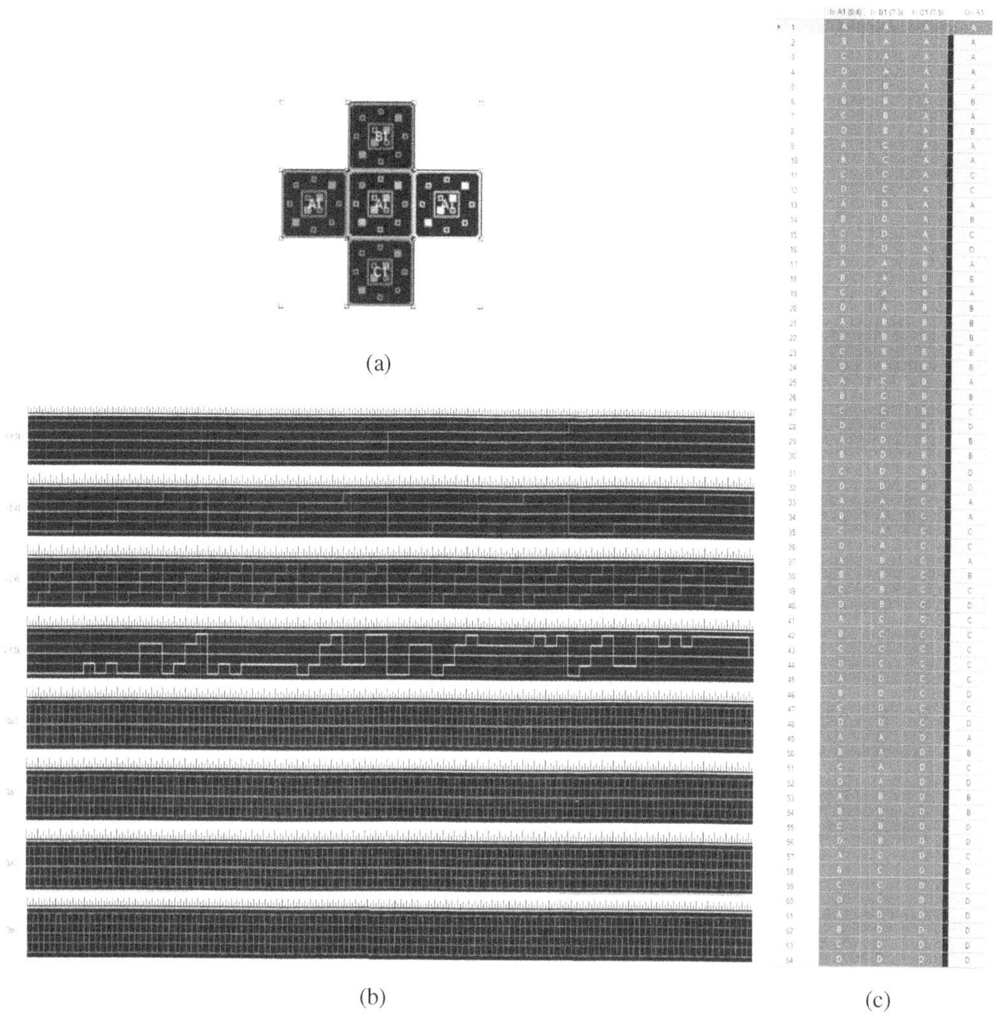

Figure 8.11 Simulation of a quaternary majority gate: (a) majority gate designed with QQCASim, (b) the majority gate waveform, and (c) truth table obtained from the software. Source: Akbari-Hasanjani et al. [133]IEEE/CC-BY 4.0.

8.2.2 AND Gate

If one of the QQCA majority gate inputs is fixed in state A, the majority gate response is the same as the AND gate response [130]. Figure 8.12 shows the AND gate and the simulation results of the QQCASim software.

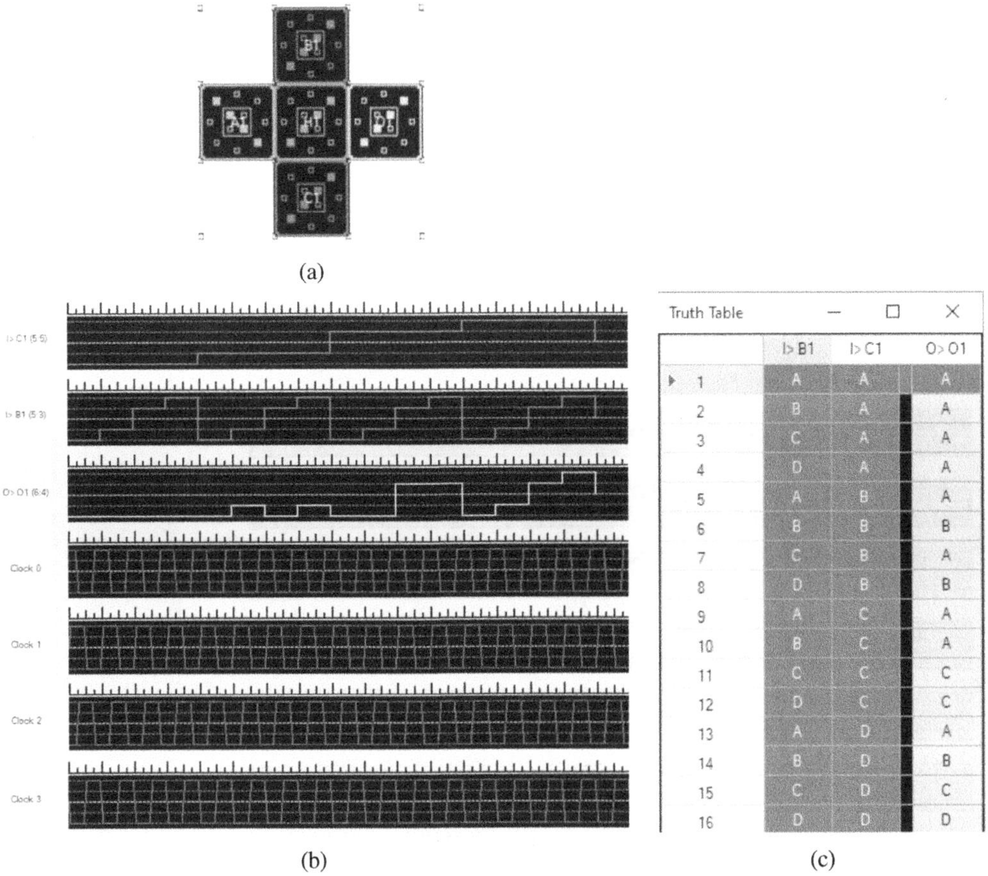

	I> B1	I> C1	O> O1
1	A	A	A
2	B	A	A
3	C	A	A
4	D	A	A
5	A	B	A
6	B	B	B
7	C	B	A
8	D	B	B
9	A	C	A
10	B	C	A
11	C	C	C
12	D	C	C
13	A	D	A
14	B	D	B
15	C	D	C
16	D	D	D

Figure 8.12 Simulation of the quaternary AND gate: (a) AND gate designed using QQCASim, (b) AND gate waveform in QQCASim, and (c) AND gate truth table obtained by QQCASim. Source: Akbari-Hasanjani and Sabbaghi-Nadooshan [130]/with permission of John Wiley & Sons.

8.2.3 OR Gate

If one of the majority gate inputs is fixed in state D, the majority gate response is the same as the OR gate response [130]. Figure 8.13 shows quaternary OR gate with the simulation results of QQCASim.

8.2.4 NOT Gate

A NOT gate in QQCA consists of two diagonal cells. In the NOT gate, the inverted input appears at the output. For example, for the input polarization −3, the polarization +3 can be obtained at

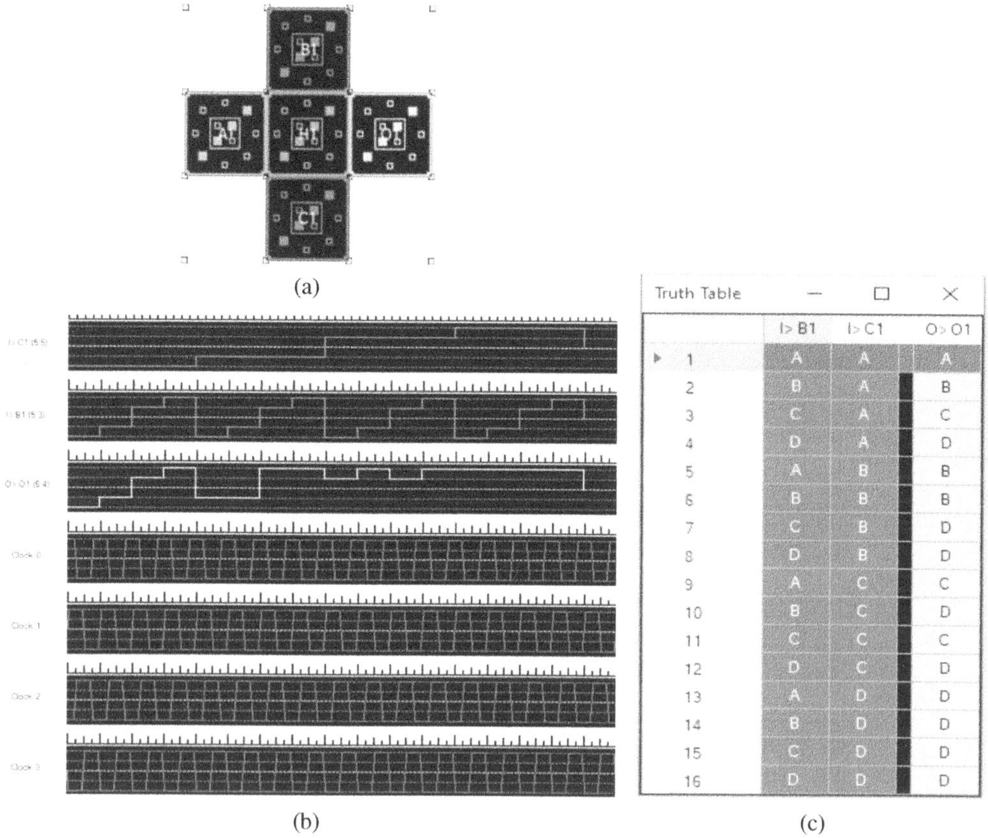

	I> B1	I> C1	O> O1
1	A	A	A
2	B	A	B
3	C	A	C
4	D	A	D
5	A	B	B
6	B	B	B
7	C	B	D
8	D	B	D
9	A	C	C
10	B	C	D
11	C	C	C
12	D	C	D
13	A	D	D
14	B	D	D
15	C	D	D
16	D	D	D

Figure 8.13 Simulation of the quaternary OR gate: (a) OR gate designed in QQCASim, (b) OR gate waveform using QQCASim, and (c) OR gate truth table obtained using QQCASim. Source: Akbari-Hasanjani and Sabbaghi-Nadooshan [130]/with permission of John Wiley & Sons.

the output. Figure 8.14 shows the NOT gate model and the QQCASim simulation results using the proposed cell.

8.2.5 Quaternary Quantum Wire

The QQCA quantum wire is formed by the juxtaposition of QQCA cells [130]. In QQCA cells, the total polarization of TQCA and bQCA is used to select each state, so data transfer is not affected by the number of cells (being even or odd) [129]. Figure 8.15 shows the QQCA wire simulation and its results.

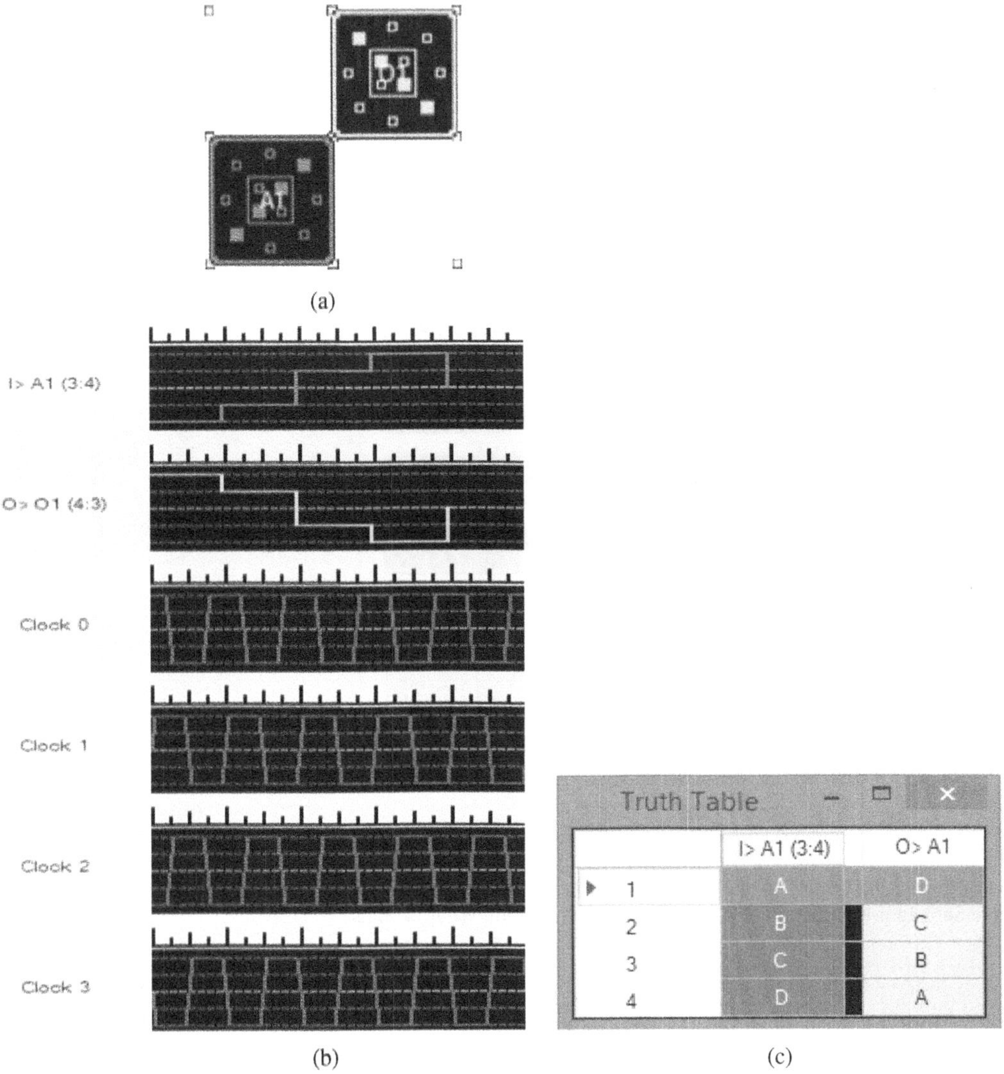

Figure 8.14 Quaternary NOT gate simulation: (a) NOT gate in QQCASim, (b) NOT gate output waveform, and (c) NOT gate truth table simulation using QQCASim. Source: Adapted from Akbari-Hasanjani and Sabbaghi-Nadooshan [130].

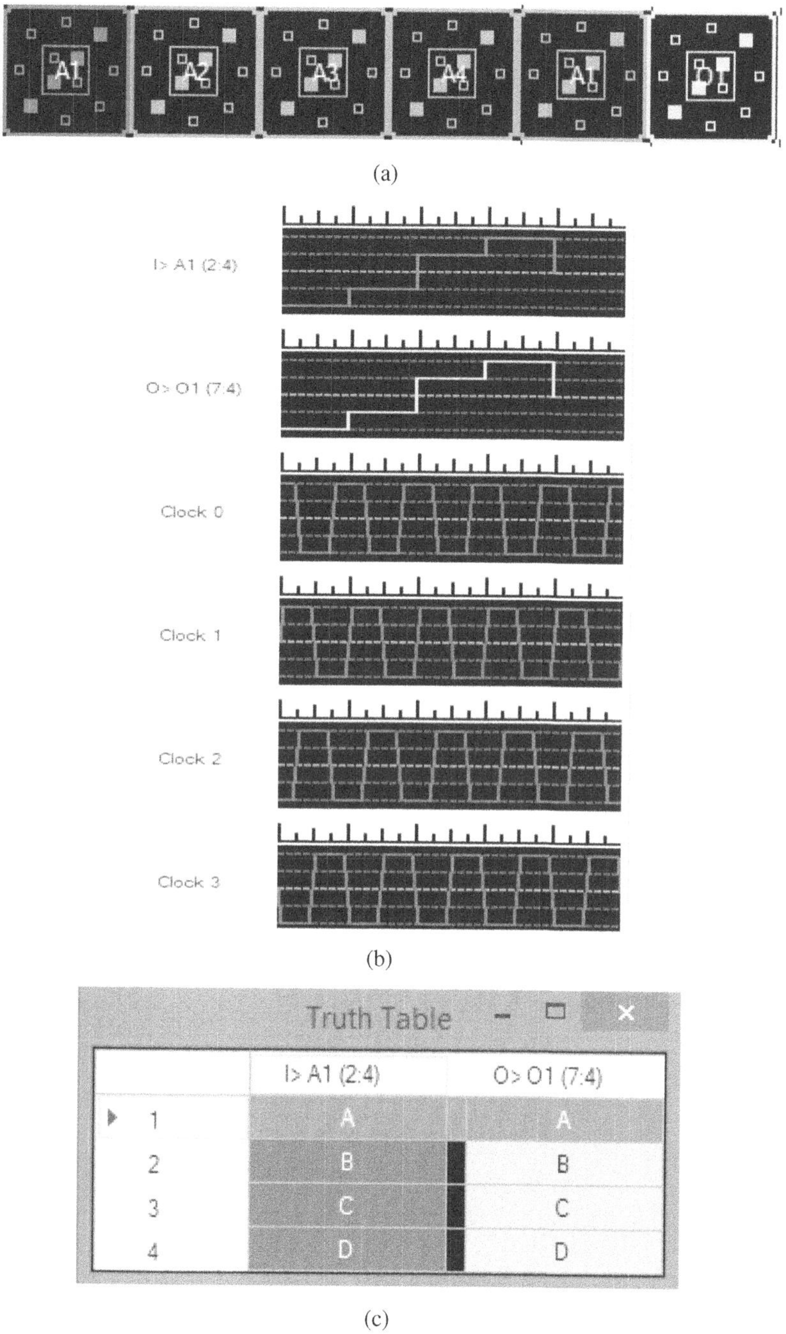

Figure 8.15 A quaternary quantum wire: (a) quantum wire, (b) quantum wire output waveform, and (c) quantum wire truth table, all obtained using QQCASim. Source: Adapted from Refs. [129, 130].

9

A Review of Literature on Memory Structures

9.1 Introduction

It is expected that the number of devices connected to the Internet exceed one hundred billion worldwide by 2030. In this ever-increasing trend, tens of billions of computing systems, from personal computers to smartphones and many other devices, will be connected to the Internet. This gigantic growth in computing devices causes data explosion, so millions of zettabytes of memory are needed to support this large scale of calculations. Consequently, in the not-so-distant future, memories will play an essential and sensitive role in computing systems. In the remainder of this book, one of the most basic components of circuits, namely memory, is analyzed [166].

Over many years, the ever-increasing growth of transistors in all technologies has been accompanied by the integration density and efficiency optimization of devices. A study conducted on the production of static RAM (SRAM) between 2003 and 2017 showed that SRAM bit-cell area had decreased by a factor of 0.5 in each production process [166]. This scale reduction has been due to various innovations in circuit lithography, such as using leakage reduction techniques and high coefficient gate dielectric with metal-gate technology.

For an effective nanoscale SRAM design, a reliable architecture is required. The manufacturing process becomes optimal if the effects of design limitations, stress, and approximation are taken into account. For more efficiency and a higher density of bits, design innovation and optimization are required.

The design and manufacturing techniques use the minimum supply voltage to achieve high performance. Furthermore, some techniques are employed to improve write/read operations and reduce leakage under ideal operating conditions of the circuits.

In modern SRAM designs, the reliability parameter is of particular importance. Hot carrier injection (HCI), erratic bits, noise, time-dependent dielectric breakdown limit the operating voltage range and lifetime of SRAM arrays. Usually, some architecture-dependent techniques are used along with other techniques, including parity protection, reconfigurable caches, error correction codes, and redundancy.

The power supply reduces the sensitivity of the circuit parameters against process variations. The RAM bit-cells that are designed in nanodimensions with minimum transistor dimensions are vulnerable to process variations. Process variations include line-edge roughness (LER), random dopant fluctuations, and some other sources of variations. These variations cause a mismatch between the threshold voltages of adjacent transistors in a memory cell. Moreover, process variations, which lead to read disturbance, hold failure, access time failure, and write failure, limit the minimum operating voltage (VMIN). Nanoscale technologies, especially quantum dot

Computing with Multi-Value Logic in Quantum Dot Cellular Automata, First Edition.
Reza Sabbaghi-Nadooshan, Reza Akbari-Hasanjani, Leila Dehbozorgi, Majid Haghparast, and Hamid Reza Akbari-Hasanjani.

cellular automata (QCA), lead to a significant reduction in dimensions, ultra-low energy consumption, and high switching speed, as a potentially strong alternative to complementary metal–oxide–semiconductor (CMOS) technology. The design of optimal memory units and SRAM cells is very important in digital systems. Therefore, the design of optimal SRAM cells in QCA technology is a very critical issue in this technology that has to be addressed.

Generally, there are two types of common memory designs including line-based memories and loop-based memories. In the next chapters of this book, some new binary and ternary designs for the loop-based processing in memory (PIM) approach are presented. A loop-based memory has feedback with four clock zones, while a line-based memory stores the previous values using the QCA line.

In this chapter, flip-flop, RAM, and PIM structures are introduced based on binary QCA basic gates.

9.2 A Review of Basic Memory Structures

9.2.1 D Flip-Flop in Binary QCA

Flip-flops are one of the basic structures for storing information. D flip-flop (DFF) is one of the most widely used flip-flops. A DFF has two inputs including a D input and a clock input. The next output does not depend on the previous output, but it is equal to the D input, as represented by Eq. (9.1) [167].

$$Q(t+1) = D \tag{9.1}$$

Many models have been proposed for DFFs in binary QCA [167, 168]. By using a three-input majority gate and connecting a feedback loop to memory, a relatively optimal DFF can be realized. Flip-flops have been used in many applications, such as memory cell designs and frequency divider circuits.

9.2.2 RAM Cell in Binary QCA

A RAM cell consists of three inputs and one output. Two of three inputs in a RAM cell are control inputs. One of these inputs controls the read/write operation, and the other input controls select operation. A write operation is performed when the read/write input is "1." When the read/write input is "0," a read operation is performed. Whenever the second control input, i.e., select, becomes "1," the memory cell is activated. If the select input becomes "0," the memory cell is disabled and no information is written or read.

9.2.3 PIM and Akers Logic Array

A PIM structure is an integrated unit that can store information and perform arithmetic operations. The PIM architecture has become increasingly popular due to its complex applications, such as big data analysis, machine learning, soft computing, and other emerging computing paradigms. Unlike the conventional von Neumann architecture, which requires two separate units to process and store data, the PIM architecture performs both tasks in a single integrated unit, as illustrated in Figure 9.1a. Therefore, PIM is much more optimal in terms of speed and power consumption. Various methods have been presented for implementing the PIM architecture [169, 170].

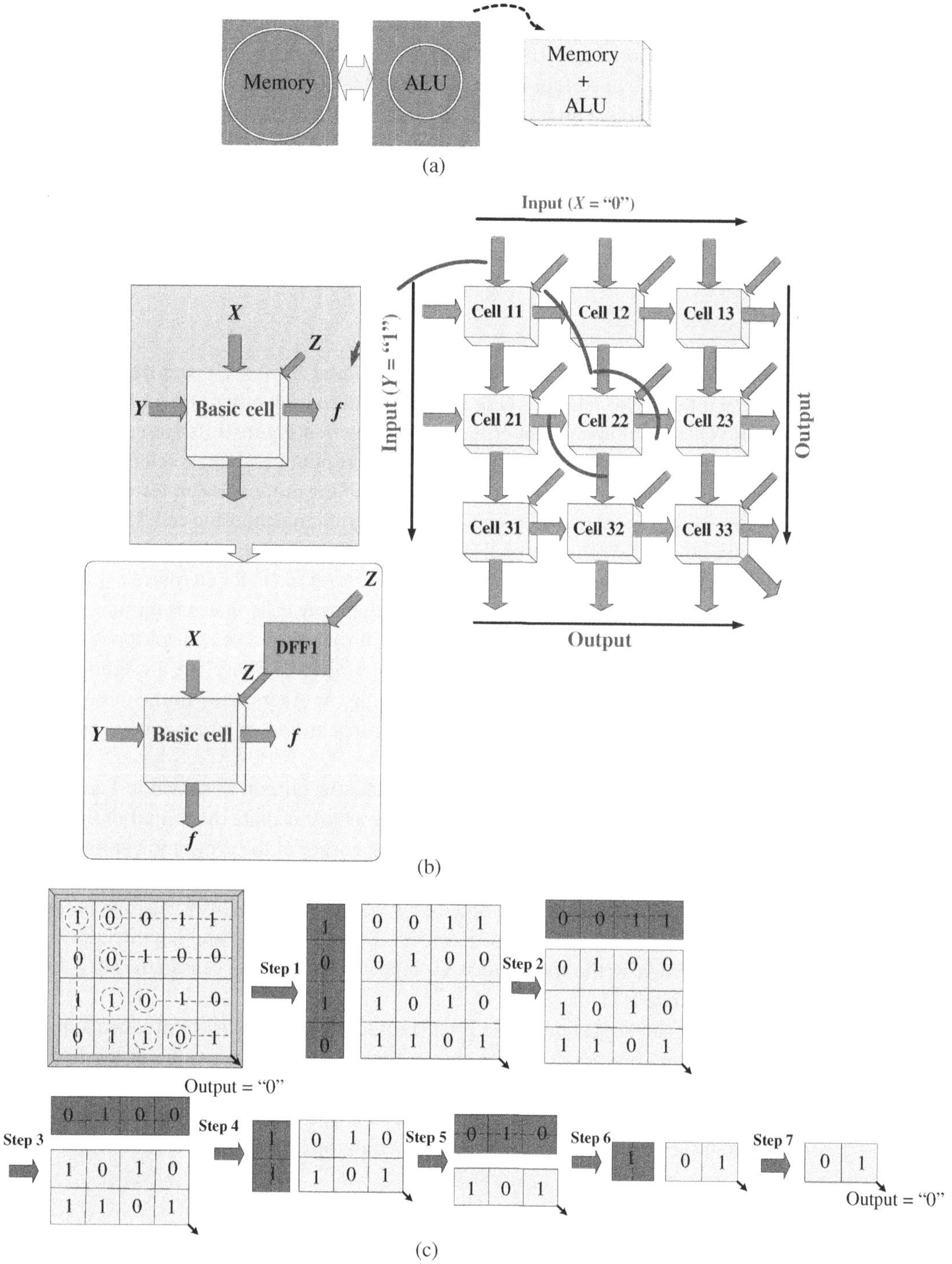

Figure 9.1 (a) Comparison of von Neumann and PIM structures, (b) representation of Akers array and a logic cell with three inputs X, Y, and Z and two identical outputs, and (c) stripping process in Akers array to obtain output. Source: Dehbozorgi et al. [171]/with permission of Springer Nature.

Akers logic array was proposed by the famous mathematician Sheldon B. Akers in 1972. Akers rectangular logic array can perform computations in memory. Figure 9.1 shows a 3×3 Akers logic array, which is a two-dimensional array consisting of identical logic cells connected in a rectangular grid [104]. Akers has provided four alternative relationships, each of which can be used to design PIM computing structures. Equations (9.2) to (9.5) show the four alternative relationships. To design a PIM structure, one of these relationships must be logically realized.

$$F1 = Y\,(X \text{ AND } Z) \tag{9.2}$$

$$F2 = \text{Majority}\,(X, Y, Z) = XY + XZ + YZ \tag{9.3}$$

$$F3 = X\overline{Z} + YZ \tag{9.4}$$

$$F4 = X + YZ \tag{9.5}$$

The basic cell of Akers logic array has three inputs X, Y, and Z ($X \le Y$), and the output F is produced from these inputs. The cells in Akers array are next to each other horizontally and vertically. Each output is the input to a neighboring cell, as shown in Figure 9.1b. Here, X and Y are considered inputs, and Z is considered a control input. The output of each logic cell in the array is transferred to two neighboring logic cells. Accordingly, one of the outputs is connected to the bottom cell, and another output is connected to the input of the right neighboring cell. The same data transfer mechanism is applied to other vertical and horizontal neighbors in the array. As shown in Figure 9.1b, input X is set to "0" for all columns, and input Y is set to "1" for all rows ($X \le Y$) [104].

The Boolean functions are implemented by arranging the array cells in a certain manner. The final output function is received at the bottom-right cell in the array. The control input Z stores the data in the cell, and the whole array performs Boolean functions. Therefore, an Akers array is considered a basic structure for PIM. An advantage of using the Akers array in the design of the PIM architecture is the possibility to design regular structures for use in larger circuits optimally and more easily.

A stripping process is used for the Akers array to calculate the output, as shown in Figure 9.1c. Each Akers cell functions as a majority gate. A common way to calculate the output of the Akers array is to start calculating from the topmost cell at the left corner of the array and continue until reaching the bottom-right corner of the array. As shown, the initial input to the upper left cell is "1," while this cell receives "1" from the left. Since each cell acts as a majority gate, the output of this cell is "1." Therefore, the bottom cell also receives "1" from both top and left directions, and the output is "1" again. Similarly, the above process is repeated for all lower cells in the left column, and the output of all of them is "1."

In fact, the left column can be separated as a layer from the original matrix to produce a smaller matrix. Following the same process, it can be shown that when the input value of the first cell on the upper left corner of the array is "0," the upper row of the matrix is stripped. Therefore, to calculate the output of the Akers array, columns and rows are stripped according to the previous considerations so that the final value reaches the last cell in the bottom-right corner of the array and the output is determined.

9.3 Review of Literature on Memory Design

Memory is a very important structure in circuit design. There are generally two types of QCA-based memory designs, i.e., loop-based and line-based memory structures. A loop-based memory structure uses a feedback loop with four clock zones [172]. A line-based memory stores previous values using a line of QCA cells [172]. Molecular QCA structures can achieve a data processing speed beyond 1 THz, which theoretically can increase up to 25 THz. In the following, some studies on memory designs are reviewed.

In Ref. [173], a hybrid architecture is proposed in which for writing information, a serial memory architecture is used to reduce the occupied area, while for reading information, a parallel memory is used to increase the reading speed. Because six gates are used in this structure, this design is not optimal in terms of cell count, delay, and occupied area. Moreover, only one control input is implemented in this memory to control the read and write functions.

A new DFF structure is designed in [174]. Since four gates are used in this design, it shows better performance in terms of cell count and speed compared to some similar structures. Furthermore, similar to previous studies, only one control line is used in this circuit.

In Ref. [175], two improved models are proposed for RAM structures. Compared to the first model, in the second model, cell count, occupied area, and delay are reduced, so the performance of the structure is improved. An advantage of this approach compared to previous studies is that there are two control inputs besides the primary input in both models. These control inputs are used to control write operations and select memory cells. However, the design of the first model is not optimal in terms of number of cells and gates.

In Ref. [176], a 2:1 multiplexer is used. In this structure, the wires are not crossed. This structure has read/write, SET/RESET, and select control pins/inputs. Although by adding these pins, the complexity and occupied area increase, there is more control over the memory cell.

A RAM cell is designed in [177] using one five-input majority gate and three three-input majority gates. The proposed structure includes a SET/RESET control. The addition of control inputs increases the number of QCA cells in the RAM structure. In the following, the studies on PIM in binary QCA are reviewed.

A PIM-based XOR gate is implemented in [178]. In this structure, four basic PIM cells in the Akers array are used. A total of 184 cells are used in this design with an occupied area of 0.23 μm^2. It should be noted that in this structure, only some multiplexers are used to realize the Akers array, and no part is dedicated to storing information. At first, this structure seems to perform well in terms of QCA cell count and occupied area, but the drawback of the design is the omission of the memory unit and only retaining the processing unit in the PIM-based XOR gate.

In Ref. [179], NAND and NOR gates based on the Akers array and a binary QCA structure based on basic PIM cells are designed. In the design of the basic cell, the flip-flop structure is used. The cell count of the NAND and NOR gates is equal to 147, and the occupied areas are 0.31 μm^2 and 0.34 μm^2, respectively. The computation and storing in the above structure are performed in a single unit. However, a large number of cells are used in this structure since the basic cell must be designed to be used as a switch in Akers array for realizing logic gates. This structure can be redesigned with fewer binary QCA cells.

In Ref. [180], an Akers structure is used to design an arithmetic circuit. First, a basic cell is presented for designing a circuit to implement the relationship $F = PQ + P'R'$. The von Neumann version of the above circuit is also designed and compared with the Akers structure. According to this comparison, the PIM-based structure is superior to non-PIM structures in terms of occupied area and delay by 23.94% and 33.33%, respectively.

Reference [181] presents the only ternary memory model in the literature. A memory cell in ternary QCA is designed in this reference. A pipeline mechanism with an adiabatic technique is used for designing this memory model. Accordingly, the pipeline uses a clock pulse to create the delay required to store information in the memory. This structure consists of four consecutive clock zones. The data stored in the memory are always circulating in the memory loop. However, when a new instruction is applied to the write input, the previous data is replaced by the new data. Moreover, the read instruction does not affect the data circulation in the memory loop. This structure represents a ternary memory cell, which includes an inverter and two majority gates. In this memory cell, a control input is dedicated to write/read operations. The performance of this design in terms of the number of gates and QCA cells is relatively acceptable.

In many structures, cell count is increased to improve fault tolerance [182], while the complexity and occupied area also increase. Figure 9.2 shows different fault-tolerant three-input majority gates. In Figure 9.2a, fault tolerance against the omission of two cells is 16.6% and against the omission of one cell is equal to 55.6%. In Figure 9.2b, the fault tolerance against the omission of two cells is 31.4% and against the omission of one cell is equal to 60%. In Figure 9.2c, fault tolerance against the omission of two cells is 37.5%, and against the omission of one cell is equal to 62.5%. Therefore, the fault rate decreases with the increase in cell count.

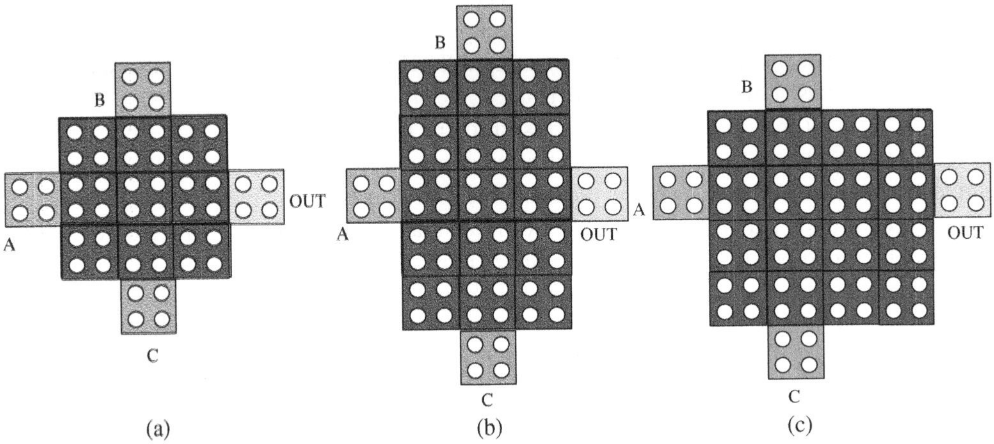

Figure 9.2 Three different fault-tolerant three-input majority gates. Source: Adapted from Taheri et al. [182].

10

Proposed PIM Cells and Their Fault Analysis in Binary QCA

10.1 Introduction

Memory is one of the basic structures in various circuits. By optimizing the memory, larger circuit structures with higher performance can be realized. In this section, binary quantum dot cellular automata (QCA) technology for memory designs is discussed. These designs can be implemented in both binary and ternary structures. In this book, the loop-based memory cell design is presented. Due to the greater capabilities of processing in memory (PIM) structures, they are analyzed here. In the following sections, some PIM structures are proposed based on binary QCA. To verify the operation of the basic cells proposed for PIM, they are implemented in some logic gates in the next chapters. In the remainder of this chapter, the effect of manufacturing defects on memory and PIM structures is analyzed for the first time. For this purpose, the effects of cell omission and extra-cell deposition defects in the basic structures of the proposed memory cells are investigated.

The proposed designs are simulated in binary QCA using QCADesigner V2.0.3 software. In all binary simulations, default values are set in QCADesigner. A QCA cell is a square with sides of 18 nm, including four 5-nm-diameter quantum dots. Bistable parameters are given in Table 10.1. In this chapter, binary structures are designed and analyzed.

To analyze and compare the binary and ternary QCA structures, various parameters are defined as follows:

1. *Complexity:* This parameter represents the number of QCA cells used in a circuit. Fewer cells in circuit design mean a better optimization opportunity.
2. *Occupied Area*: This parameter denotes the area occupied by all QCA cells in a circuit and is expressed in square micrometers.
3. *Delay:* Delay is a measure of the circuit speed and is equal to the number of clock pulses from the input to the output of the circuit.
4. *Number of Circuit Layers*: This parameter is very important in the design of QCA circuits.
5. *Cost:* The cost of multilayer crossing is three times that of coplanar crossing [182]. Accordingly, cell count increases in multilayer structures, so the delay also increases in these structures.
6. *Number of Majority Gates*: This parameter is important in designing QCA-based circuits because majority gates receive three inputs and produce one output. This causes significant energy loss.
7. *Number of Crossings:* This parameter is a measure of the difficulty of circuit manufacturing.

Computing with Multi-Value Logic in Quantum Dot Cellular Automata, First Edition.
Reza Sabbaghi-Nadooshan, Reza Akbari-Hasanjani, Leila Dehbozorgi, Majid Haghparast, and Hamid Reza Akbari-Hasanjani.

Table 10.1 Parameters used in QCADesigner.

Parameter	Value
Cell size	18 nm
Number of samples	12,800
Convergence tolerance	0.001000
Radius of effect	65 nm
Relative permittivity	12.900000
Clock low	$3.8e^{-0.23}$
Clock high	$9.8e^{-0.22j}$
Clock shift	0
Clock amplitude factor	2
Layer separation	11.5
Maximum iterations per sample	100

There is no specific technique for designing and optimizing QCA circuits. However, some basic considerations can be made in all of these circuits to design more optimal structures compared to conventional technologies. These considerations lead to novel and more optimal QCA-based structures by using more optimal and fewer gates, as well as designing coplanar circuits, reducing the distance between different parts of the circuits, and avoiding wire crossing as much as possible. The criteria for more optimal circuits are a smaller occupied area, lower complexity, less delay and energy consumption, and better fault tolerance.

In the following, two cost functions are defined based on the aforementioned descriptions [182]:

- *Area–delay Cost Function*: This function is the product of the occupied area and circuit delay, according to Eq. (10.1).

$$\text{Cost}\, A-T = A \times T \tag{10.1}$$

where A is the occupied area and T is the circuit delay. For the sake of brevity, the occupied area is expressed in square micrometers and delay in picoseconds. Generally, the cost function is used to evaluate and compare complementary metal–oxide–semiconductor (CMOS) structures. If the value of one or both parameters of delay and occupied area increases, the cost function also increases. However, when the value of one or both of these parameters decreases, the amount of cost function is reduced.

- *Area–Complexity Cost Function*: This function denotes the ratio of the occupied area to the cell count of the circuit, as given in Eq. (10.2). In this equation, A and C represent the occupied area and cell count, respectively.

$$\text{Cost}\, A/C = A/C \tag{10.2}$$

If the cell count increases in a certain occupied area, the circuit density increases and the cost function decreases. Additionally, if the occupied area of a certain number of cells is reduced, the cost function decreases. The overall cost function is obtained from the product of the area–delay cost function and area–complexity cost function.

10.2 Binary QCA-Based Structures

10.2.1 First Model of Binary PIM

The first binary PIM model is called BPIM1. As mentioned in Section 9.2.3, the Akers array is used to design basic PIM cells. To realize BPIM1, Eq. (9.4) is used. Therefore, a multiplexer structure is required to create Akers array switches, and a data flip-flop (DFF) structure is used to store information in the design of the basic PIM cell. In the BPIM1 structure, the DFF is placed at the beginning of the circuit and before the multiplexer. As a result, BPIM1 can store the input information.

According to Figure 10.1, if "0" is applied to the control input Z, the signal X is propagated to the output, while if "1" is applied to Z, the signal Y is propagated to the output. Moreover, the input information of Z is always stored in the flip-flop. Accordingly, both information processing and storing operations are performed in this structure, so it can be used when the input information must be stored in the circuit [183].

The BPIM1 structure contains 59 cells with an occupied area of 0.11 μm^2. The delay of this structure to the main output is 0.5 clock cycles and to the flip-flop output is 1.25 clock cycles. By inserting the design parameters into Eqs. (10.1) and (10.2), the cost functions can be calculated as follows:

$$\text{Cost}_{\text{A-T}} = A \times T = 0.11 \times 0.5 = 0.055$$

$$\text{Cost}_{\text{A/C}} = A/C = 0.11/59 = 0.001864 = 1.864 \times 10^{-3}$$

The structure of BPIM1 has relatively optimal performance in terms of speed (its delay is only 0.5 clock cycles) since the PIM architecture consists of two parts including memory and switches.

10.2.2 Second Model of Binary PIM

In the second binary QCA-based PIM structure (BPIM2), the flip-flop is placed at the end of the circuit and stores the output information. The effect of relocating the flip-flop part in these two structures on the performance of the PIM design is analyzed. According to Figure 10.2, when the control input Z is equal to "1," the input Y is transferred to the main output, and when Z is equal to "0," X is transferred to the main output. Since the main output is connected to the input of the flip-flop, the output information is stored and maintained at the same time. This structure can be used in circuits, whose output information needs to be stored after processing. The proposed structure includes 46 binary QCA cells with an occupied area of 0.07 μm^2 and a delay of 0.75 clock cycles to the main output and 1.25 clock cycles to the flip-flop output. By inserting the values of the design parameters into Eqs. (10.1) and (10.2), the cost functions can be calculated as follows:

$$\text{Cost}_{\text{A-T}} = A \times T = 0.08 \times 0.5 = 0.04$$

$$\text{Cost}_{\text{A/C}} = A/C = 0.07/46 = 0.00152 = 1.52 \times 10^{-3}$$

The BPIM2 structure performs optimally in terms of speed (its delay is only 0.75 clock cycles). The values of the cost functions are acceptable because the PIM structure consists of two parts, i.e., memory and switches. Moreover, due to the more optimal parameters and cost functions of BPIM2, it exhibits a better performance than BPIM1.

10.2.3 Third Structure of PIM

The third proposed PIM model, BPIM3, whose basic cell is based on Eq. (9.4), is shown in Figure 10.3. In the design of this structure, no majority gate is directly used in the multiplexer

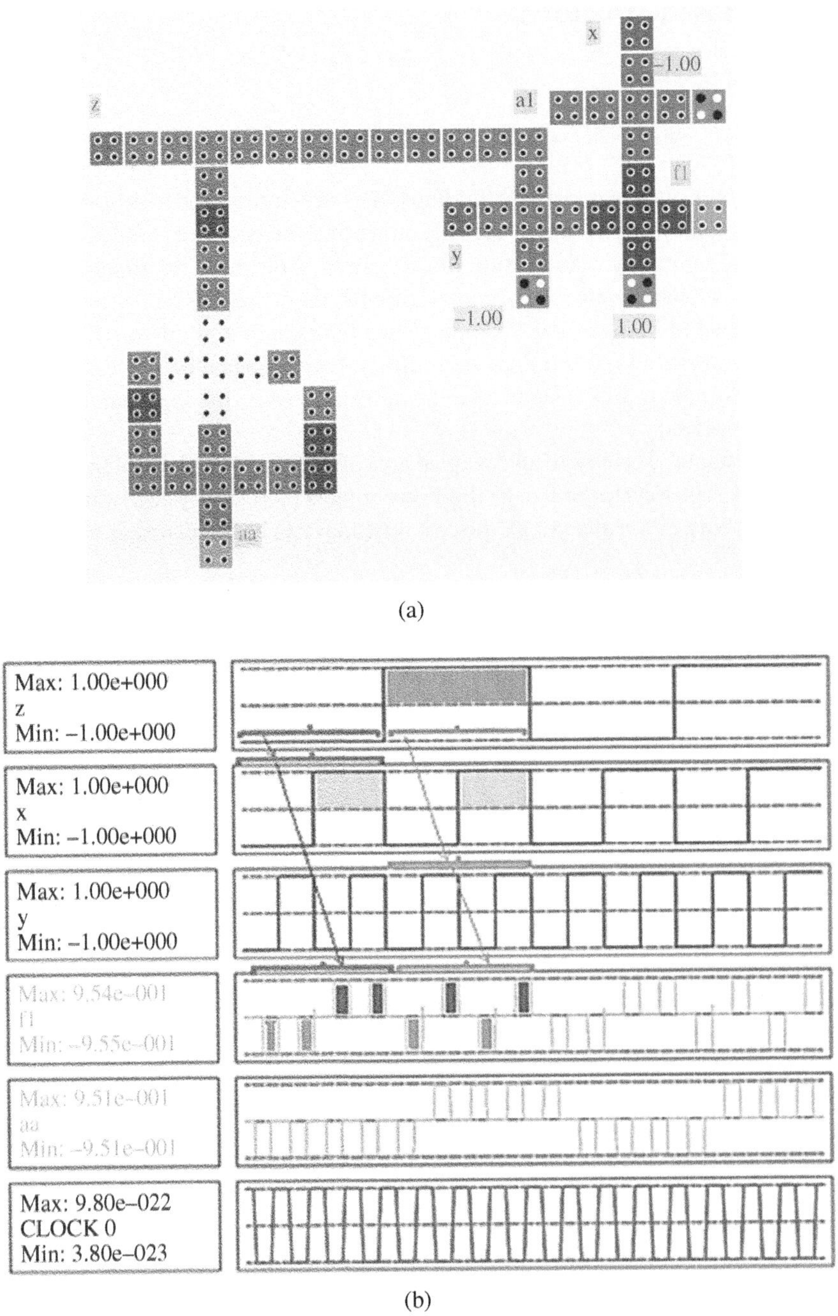

Figure 10.1 BPIM1: (a) implementation in binary QCA and (b) simulation results. Source: Dehbozorgi et al. [183]/with permission of Springer Nature.

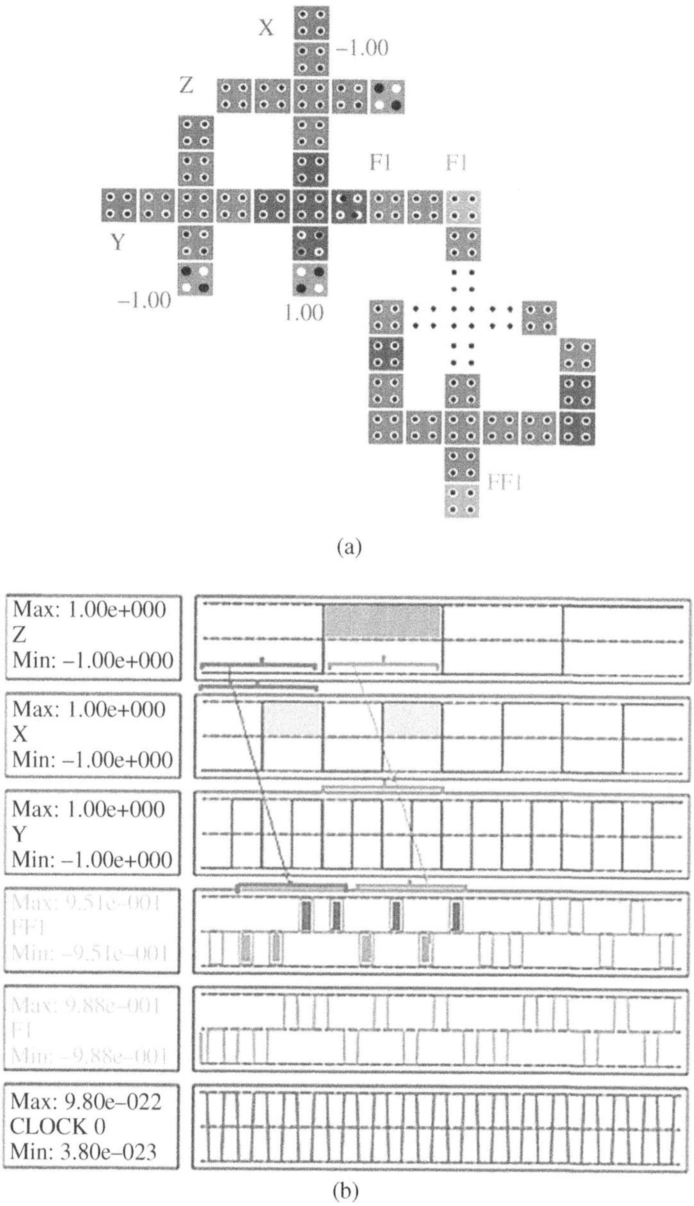

Figure 10.2 BPIM2: (a) implementation in binary QCA and (b) simulation results. Source: Dehbozorgi et al. [183]/with permission of Springer Nature.

part, while a multiplexer is used to achieve better performance in terms of cell count and occupied area. As shown in Figure 10.3, the fixed input "0" is applied to X, and the fixed input "1" is applied to Y. Moreover, in the design of this model, besides Z, another input is used to control the write operation, or in other words, it acts as a clock pulse. Table 10.2 summarizes the operating principle of the basic PIM cell in binary QCA [171].

The output of DFF is connected to the select input/line of the multiplexer and controls its operation. If the clock pulse (write) is "1," the input is written into the memory loop and then transferred

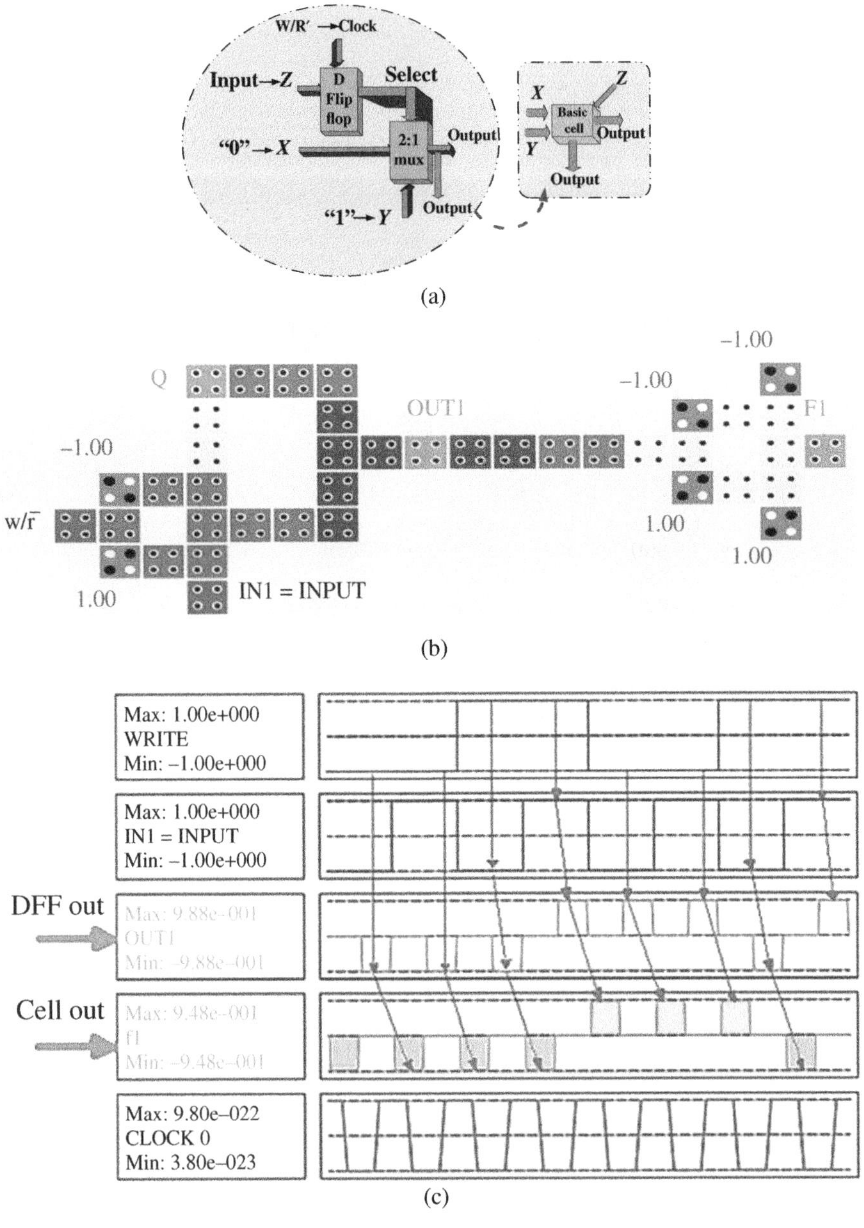

Figure 10.3 BPIM3: (a) block diagram of the proposed model, (b) implementation in binary QCA, and (c) simulation results. Source: Dehbozorgi et al. [171]/with permission of Springer Nature.

to the 2:1 multiplexer, which acts as the select line of the multiplexer. If the select input is "1," the input Y, which is "1," is transferred to the output; otherwise, the input X or "0" is transferred. The BPIM3 cell in Figure 10.3 has 40 cells with an occupied area of 0.05 μm^2 and a delay of 1 clock cycle. Equations (10.1) and (10.2) can be used to calculate cost functions as follows:

$$\text{Cost}_{\text{A-T}} = A \times T = 0.05 \times 1 = 0.05$$

$$\text{Cost}_{\text{A/C}} = A/C = 0.05/40 = 0.00125 = 1.25 \times 10^{-3}$$

Table 10.2 Outputs of different inputs applied to BPIM3188.

Mode	Write/$\overline{\text{Read}}$	Input	Memory loop	F1 = Output
Read	0	0	Unchanged	Stored value
Read	0	1	Unchanged	Stored value
Write	1	0	0	0
Write	1	1	1	1

Source: Dehbozorgi et al. [171]/with permission of Springer Nature.

Due to the more optimal parameters and cost functions of BPIM3, its performance in terms of cost functions and QCA cell count is better than BPIM1 and BPIM2. Therefore, optimizing cell count and occupied area is more effective in improving the design parameters.

10.2.4 Fourth Model of PIM

The fourth proposed PIM model, BPIM4, is the most optimal binary QCA-based PIM, which consists of an AND gate, a simple memory loop, and a multiplexer with an optimal design. In this structure, the smallest possible number of cells is used. According to the Akers array architecture, fixed values "0" and "1" are applied to X and Y inputs, and another control input is dedicated to controlling the write operation (clock pulse), as shown in Figure 10.4. The BPIM4 structure includes 33 cells with an occupied area of 0.04 μm^2 and a delay of 1.5 clock cycles. Using Eqs. (10.1) and (10.2), cost functions can be calculated as follows:

$$\text{Cost}_{\text{A-T}} = A \times T = 0.04 \times 1.5 = 0.06$$

$$\text{Cost}_{\text{A/C}} = A/C = 0.04/33 = 0.001212 = 1.212 \times 10^{-3}$$

Based on these calculations, the cell count and occupied area of BPIM4 are smaller than those of the previous models.

Until now, in this book, the parameters, including complexity, delay, and occupied area, as well as area–delay and area–complexity cost functions, of the proposed structures were examined and compared. In the next section of this chapter, we analyze the effect of faults.

10.3 Fault Analysis

Defects may occur in any technology during the manufacturing of various circuits. The QCA technology is no exception. Since QCA circuits are typically on the nanoscale, the possibility of faults increases during manufacturing (chemical and deposition processes). However, faults that occur in practice during manufacturing and can also be simulated are investigated. In this book, the defects of cell omission and deposition of an extra cell are investigated. Fault tolerance of each circuit against cell omission and extra-cell deposition defects is obtained by dividing the number of cells that produce a correct output by the total number of cells (except for input, output, or fixed cells) as given in Eq. (10.3) [182]:

$$\text{Fault tolerance} = \frac{\text{Number of correct output patterns}}{\text{Number of detective patterns}} \times 100 \qquad (10.3)$$

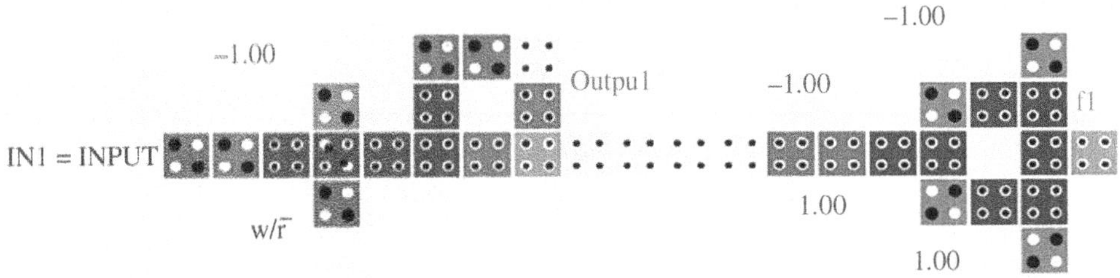

(a)

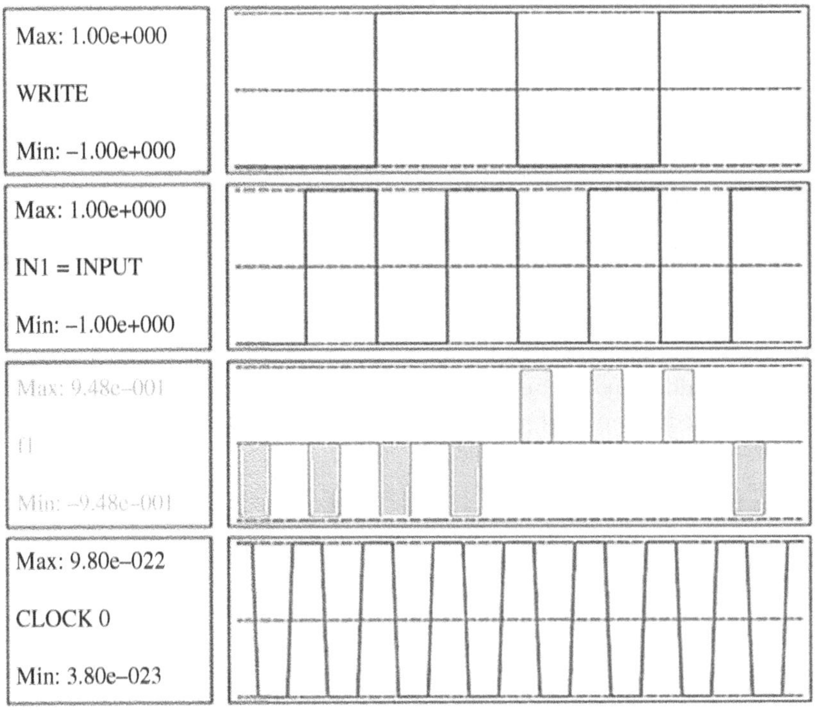

(b)

Figure 10.4 BPIM4: (a) block diagram, (b) implementation in binary QCA, and (c) simulation results. Source: Dehbozorgi et al. [171]/with permission of Springer Nature.

10.3.1 Fault Tolerance Analysis Against Cell Omission and Extra-Cell Deposition Defects in Proposed Models

In this subsection, we examine the fault tolerance of the presented models against defects of cell omission and deposition of an extra cell. It is assumed that each time, only a single cell is deposited or omitted during the manufacturing process. The circuit outputs of all input states are examined to determine whether they are correct or not. For this purpose, first, the cells of each structure are numbered, and then a cell is deposited or omitted. Next, the output waveforms for all input states are obtained using the corresponding simulator. Finally, the outputs of all input states are examined to determine whether they are correct or not.

In many studies on fault evaluation for different gates and circuits in binary QCA, it has been mentioned that "by increasing the cost of cells and occupied area in conventional gates, fault tolerance against manufacturing defects is improved" [182]. In these studies, the extra-cell deposition technique is used to realize more fault-tolerant gates. In other words, by depositing an extra cell next to the cells whose omission usually leads to a fault, a more fault-tolerant design is obtained. However, this technique increases the complexity and occupied area of the circuit. This subsection proposes a method to increase fault tolerance. This method redesigns the binary QCA-based circuit structure in ternary QCA without increasing cell count and occupied area. Thanks to this method, the new circuit is more fault-tolerant against manufacturing defects, without increasing occupied area.

Although cell count increases in this case, the use of ternary QCA cells, which have smaller dimensions, prevents a larger overall occupied area. Based on this method and by transforming the structure from binary to ternary, the cost functions, especially the area–complexity cost function, Eq. (10.2), can be reduced. In this way, for a given occupied area, cell count increases. All comparisons in binary QCA are made based on cell dimensions of $18 \times 18\,\text{nm}^2$ and a distance of 2 nm between two neighboring cells. In ternary QCA, cell dimensions are $8 \times 8\,\text{nm}^2$, and the distance between two neighboring cells is 0.5 nm. In this chapter, the faults of the proposed binary structures are investigated, and in the following chapters, the effect of faults on the proposed ternary structures is analyzed.

10.3.1.1 Fault Evaluation of BPIM1

In binary QCA-based PIM structures, which were proposed in Section 2.3, the memory unit is next to the arithmetic logic unit. The cells of the BPIM1 model are numbered in Figure 10.5.

To evaluate the cell omission defect, individual cells are omitted one by one, and the output waveforms are compared with the correct fault-free waveform. If a cell omission changes the output waveform for all input states, this cell omission affects the output; otherwise, it is ineffective. According to Table 10.3, out of 51 evaluated cells, the omission of 12 cells does not affect the output and leads to correct output.

By inserting the aforementioned values in Eq. (10.3), the fault tolerance against cell omission defect is equal to 23.52%.

$$\text{(Cell omission) fault tolerance of BPIM1} = \frac{12}{51} \times 100 = 23.52\%$$

Next, cells A1–A10 are added in the empty positions of the structure, and the effect of adding each cell on the output is evaluated; the results are shown in Table 10.4. The addition of 4 out of 10 cells does not affect the output. By inserting these numbers into Eq. (10.3), the fault tolerance against extra-cell deposition is 40%.

$$\text{(Extra cell) fault tolerance of BPIM1} = \frac{4}{10} \times 100 = 40\%$$

According to calculations, the fault tolerance against single-cell omission is relatively optimal for BPIM1. In this structure, five three-input majority gates and two NOT gates are used to design the multiplexer. This has led to more empty positions and lower fault tolerance against the extra-cell deposition defect.

10.3.1.2 Fault Evaluation of BPIM2

The numbering of BPIM2 cells is shown in Figure 10.6. By omitting individual cells, the output waveforms are obtained for all input states, as shown in Table 10.5. A total of 38 cells are evaluated, and correct outputs are obtained for 22 cells. According to Eq. (10.3), fault tolerance against cell omission is 57.89%.

$$\text{(Cell omission) fault tolerance of BPIM2} = \frac{22}{38} \times 100 = 57.89\%$$

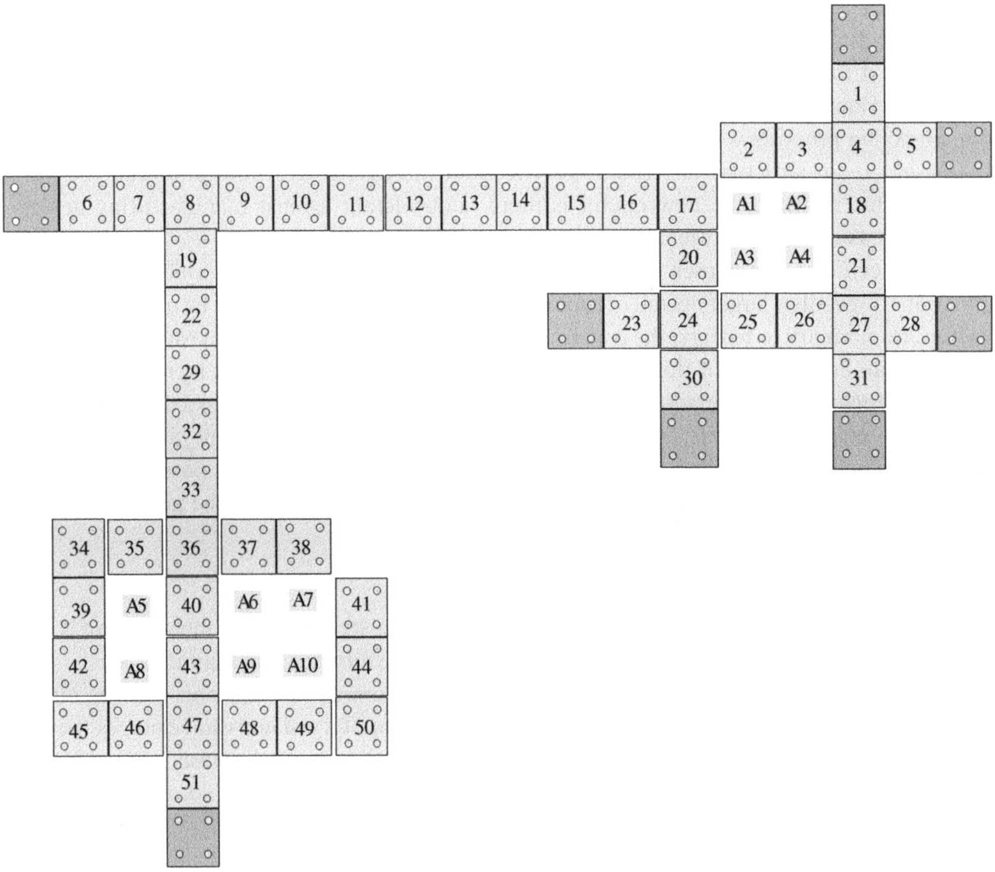

Figure 10.5 Cell numbering in BPIM1. Source: Dehbozorgi et al. [183]/with permission of Springer Nature.

Table 10.3 Cell omission fault evaluation in BPIM1.

Impact on output	Cell number	Impact on output	Cell number	Impact on output	Cell number	Impact on output	Cell number	Impact on output	Cell number
Yes	45	No	34	Yes	23	Yes	12	Yes	1
Yes	46	No	35	Yes	24	Yes	13	Yes	2
Yes	47	Yes	36	No	25	Yes	14	Yes	3
No	48	No	37	No	26	Yes	15	Yes	4
No	49	No	38	Yes	27	Yes	16	Yes	5
No	50	Yes	39	Yes	28	Yes	17	Yes	6

Source: Dehbozorgi et al. [183]/with permission of Springer Nature.

Table 10.4 Extra-cell deposition fault evaluation in BPIM1187.

Impact on output	Cell number	Impact on output	Cell number
No	6	Yes	1
Yes	7	Yes	2
No	8	Yes	3
No	9	Yes	4
No	10	Yes	5

Source: Dehbozorgi et al. [183]/with permission of Springer Nature.

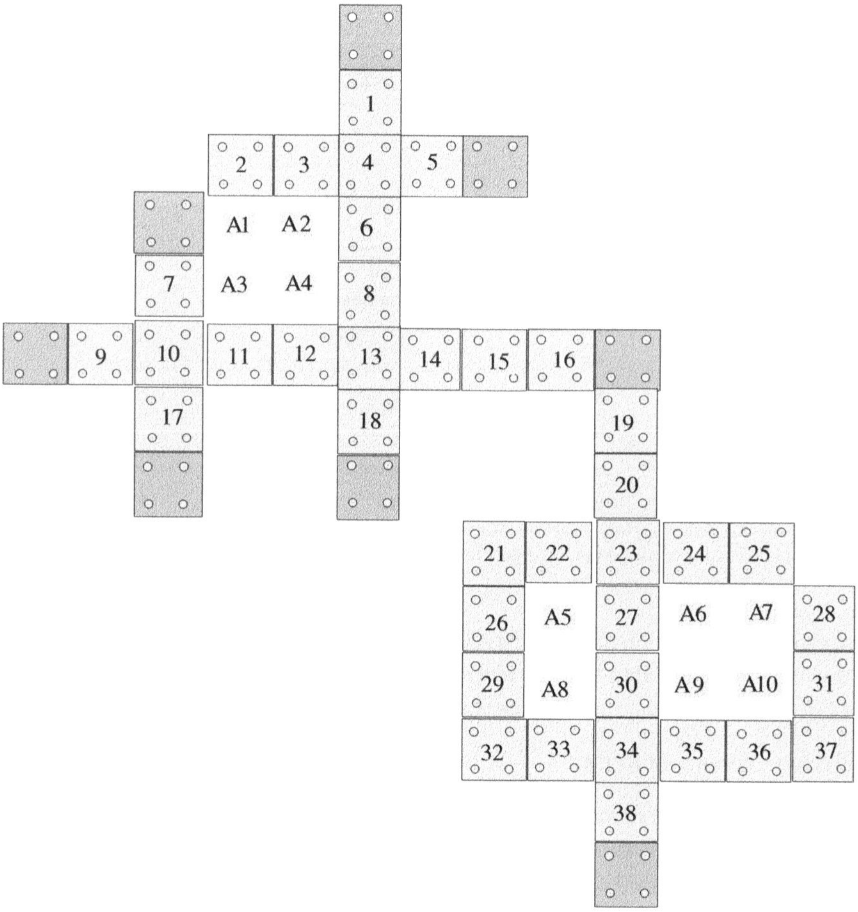

Figure 10.6 Cell numbering in BPIM2. Source: Dehbozorgi et al. [183]/with permission of Springer Nature.

Table 10.5 Cell omission fault evaluation in BPIM2.

Impact on output	Cell number	Impact on output	Cell number	Impact on output	Cell number	Impact on output	Cell number
No	34	No	23	No	12	Yes	1
No	35	No	24	Yes	13	Yes	2
No	36	No	25	No	14	Yes	3
Yes	37	Yes	26	No	15	Yes	4
Yes	38	No	27	No	16	Yes	5
		No	28	Yes	17	Yes	6
		No	29	Yes	18	Yes	7
		No	30	No	19	Yes	8
		No	31	No	20	No	9
		No	32	No	21	Yes	10
		Yes	33	No	22	No	11

Source: Dehbozorgi et al. [183]/with permission of Springer Nature.

Table 10.6 Extra-cell deposition fault evaluation in BPIM2.

Impact on output	Cell number	Impact on output	Cell number
No	6	Yes	1
Yes	7	No	2
No	8	No	3
No	9	No	4
No	10	No	5

Source: Dehbozorgi et al. [183]/with permission of Springer Nature.

To evaluate the extra-cell deposition defect, cells A1–A10 are added in the empty positions of the proposed structure. According to Table 10.6, correct output is obtained for 8 out of 10 added cells. Therefore, using Eq. (10.3), fault tolerance against extra-cell deposition defect is equal to 80%.

$$\text{(Extra cell) fault tolerance of BPIM2} = \frac{8}{10} \times 100 = 80\%$$

Although the new proposed PIM model is more extended due to including both processing and memory units, the model's fault tolerance has increased. Cell count, delay, and occupied area in this model are less than those in BPIM1, so it is more optimal.

In general, BPIM2 is very similar to BPIM1, especially considering their memory unit and multiplexer, with the only difference being that the memory unit in BPIM2 is placed at the end of the circuit. Despite the great similarities between these structures, the fault tolerance of BPIM2 is higher than that of BPIM1. The reason for this difference can be identified by comparing the circuit layout, cell positions, and cell count of these structures. The numbers of majority gates and inverters as well as the clock are the same in both structures (five majority gates and two NOT gates), while the numbers of cells used in wires are not the same.

In BPIM1, nine cells are between the primary input and memory input, and 13 cells are between the primary input and multiplexer input. In BPIM2, the primary input is directly connected to the multiplexer input, and the main output is connected to the memory unit only through six cells. Therefore, the main cause of fault is the long distance between the primary input and the inputs to the memory and multiplexer in BPIM1. This long distance alters the electrostatic effect between the cells in the presence of cell omission and extra-cell deposition defects. Under this condition, the signal is not properly transmitted to the neighboring cells and the output waveform is affected.

10.3.1.3 Fault Evaluation of BPIM3

The numbering of BPIM3 cells is shown in Figure 10.7. In these PIM structures, it has been tried to optimally design the memory and processing units as well as to use fewer majority gates and cells as possible. In the following, the effects of these considerations on performance improvement, fault tolerance, and design optimization are investigated.

For this structure, individual cells are omitted one by one, and the output result is obtained. A cell omission that causes an error in the output waveform is specified in Table 10.7. According to Table 10.7, the total cell count is equal to 31, and there are 8 cells whose omission does not cause a fault. By inserting these values into Eq. (10.3), the fault tolerance against cell omission defect is 25.8%.

$$\text{Fault tolerance of BPIM3 (Cell omission)} = \frac{8}{31} \times 100 = 25.8\%$$

In the next step, 8 cells (A1–A8) are added in the empty positions according to Figure 10.7. The addition of five of these cells does not cause a fault. The fault tolerance against extra-cell deposition defect according to Eq. (10.3) is equal to 62.5% (Table 10.8).

$$\text{Fault tolerance of BPIM3 (extra cell)} = \frac{5}{8} \times 100 = 62.5\%$$

The values of fault tolerance against both defects are acceptable because, in this structure, a multiplexer is used instead of directly using a majority gate, which is more vulnerable to faults. It is also worth mentioning that although both memory and processing units are integrated into PIM, the cell count, occupied area, and cost functions are still relatively optimal in this architecture.

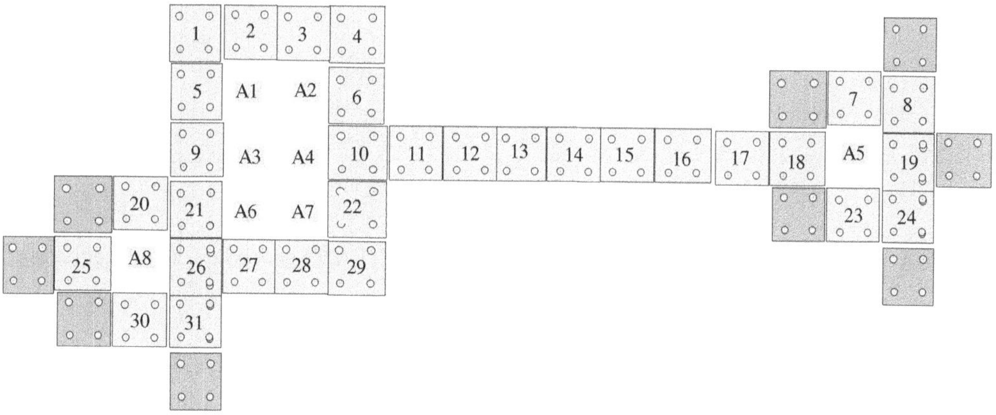

Figure 10.7 Cell numbering in BPIM3. Source: Dehbozorgi et al. [171]/with permission of Springer Nature.

Table 10.7 Cell omission fault evaluation in BPIM3.

Impact on output	Cell number	Impact on output	Cell number	Impact on output	Cell number
Yes	23	No	12	Yes	1
Yes	24	No	13	No	2
Yes	25	No	14	No	3
Yes	26	Yes	15	Yes	4
No	27	Yes	16	Yes	5
Yes	28	Yes	17	Yes	6
Yes	29	No	18	Yes	7
Yes	30	Yes	19	Yes	8
Yes	31	Yes	20	Yes	9
		Yes	21	Yes	10
		Yes	22	No	11

Source: Dehbozorgi et al. [171]/with permission of Springer Nature.

Table 10.8 Extra-cell deposition fault evaluation in BPIM3.

Impact on output	Cell number	Impact on output	Cell number	Impact on output	Cell number
No	7	No	4	No	1
Yes	8	No	5	No	2
		Yes	6	Yes	3

Source: Dehbozorgi et al. [171]/with permission of Springer Nature.

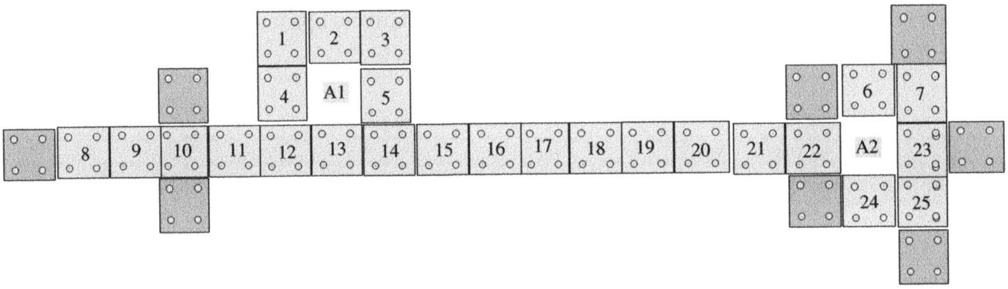

Figure 10.8 Cell numbering in BPIM4. Source: Dehbozorgi et al. [171]/with permission of Springer Nature.

10.3.1.4 Fault Evaluation of BPIM4

The cells in the BPIM4 structure are numbered, as shown in Figure 10.8.

To evaluate fault tolerance, individual cells are omitted one by one, and the output waveform is examined for different applied input states. The results of this evaluation are shown in Table 10.9.

Table 10.9 Cell omission fault evaluation in BPIM4.

Impact on output	Cell number	Impact on output	Cell number	Impact on output	Cell number
Yes	19	Yes	10	No	1
Yes	20	Yes	11	No	2
No	21	Yes	12	Yes	3
Yes	22	No	13	Yes	4
Yes	23	Yes	14	Yes	5
Yes	24	No	15	Yes	6
Yes	25	No	16	Yes	7
		No	17	Yes	8
		No	18	Yes	9

Source: Dehbozorgi et al. [171]/with permission of Springer Nature.

Table 10.10 Extra-cell deposition fault evaluation in BPIM4.

Impact on output	Cell number
No	1
No	2

Source: Dehbozorgi et al. [171]/with permission of Springer Nature.

By omitting 8 out of 25 cells, there is no change in the output. Finally, fault tolerance is calculated to be 32%, which is obtained by inserting the aforementioned values into Eq. (10.3).

$$\text{Fault tolerance of BPIM4 (Cell omission)} = \frac{8}{25} \times 100 = 32\%$$

To evaluate extra-cell deposition in BPIM4, there are only two empty positions for adding cells. Therefore, cells A1 and A2 are added to this structure, and the circuit output is examined. According to Table 10.10, adding each of the two cells does not affect the output, and fault tolerance against this defect is 100%, as obtained by Eq. (10.3). In other words, by adding cells in these empty positions, no fault is observed in the output.

$$\text{Fault tolerance of BPIM4 (extra cell)} = \frac{2}{2} \times 100 = 100\%$$

The performance of BPIM4 is optimal in terms of fault tolerance against both single-cell omission and extra-cell deposition defects compared to other examined PIM structures. Moreover, BPIM4 is 100% fault tolerant to extra-cell deposition defects. This can be attributed to the more compact design of BPIM4. There are very few empty positions in the structure, and the available area inside the feedback loop can accommodate only one cell. This structure uses three-input majority gates, which have relatively low fault tolerance, but compared to previous binary PIM structures, the memory loop is much smaller and more optimal, and a multiplexer is used instead of a majority gate in the output. Additionally, the distance between the input and output is reduced to the extent that the output waveform is not affected. In this way, the number of cells in wires is also reduced. Note that there is no special method to optimize QCA circuits, so circuit optimization or desired output waveform is achieved by changing the circuit layout, number of gates, and the location of the cells in the structure. Accordingly, besides the increase in fault tolerance, the cell count and occupied area are reduced compared to BPIM3, so it can be concluded that BPIM4 has better performance among all PIM structures.

11

Design Details of Binary Boolean Operators Using Basic PIM Cells: Proposing and Analyzing Basic-Extended Hypothesis

11.1 Introduction

In general, to evaluate smaller or basic structures, it is better to use them in designing extended circuits and gates. This is because if the proposed basic models can be used to design extended and more optimal circuits, it can be concluded that these models are appropriate and applicable. Therefore, to evaluate the basic processing in memory (PIM) cells proposed in this book, the binary and ternary Boolean operators AND, OR, and XOR are analyzed using the first proposed cell (BPIM1 and BPIM2), and the binary and ternary operators NAND and NOR are analyzed using the second proposed structure (BPIM3 and BPIM4). In all designs, the input values X and Y are assumed to be fixed; these values are respectively "0" and "1" in binary structures or A and B in ternary structures.

11.2 Basic-Extended Hypothesis

To better analyze the proposed structures, basic-extended hypothesis is proposed with the following statement:

> If it is possible to use primitive and basic PIM cells that are more optimal than similar cells to design extended circuits with more optimal performance compared to similar structures, then those primitive and basic PIM cells are suitable and more practical to design extended structures.

11.3 Designing Logic Gates with Binary QCA-Based PIM Capability

11.3.1 AND Gate Design Using PIM Basic Cells in Binary QCA

In this subsection, binary quantum dot cellular automata (QCA)-based AND gates BAND1 and BAND2 are introduced using the proposed binary basic cell models BPIM1 and BPIM2. Figure 11.1 shows the block diagram of the AND gate. This structure includes two basic PIM cells that are horizontally connected. To design this structure, the basic PIM cells are placed in an Akers array. Each PIM cell can store and process data. Therefore, by implementing these cells in the Akers array, the capability of storing and processing data is created and maintained for the entire structure. For the correct operation of the array, inputs X and Y carry fixed values of "0" and "1," respectively, and

Computing with Multi-Value Logic in Quantum Dot Cellular Automata, First Edition.
Reza Sabbaghi-Nadooshan, Reza Akbari-Hasanjani, Leila Dehbozorgi, Majid Haghparast, and Hamid Reza Akbari-Hasanjani.

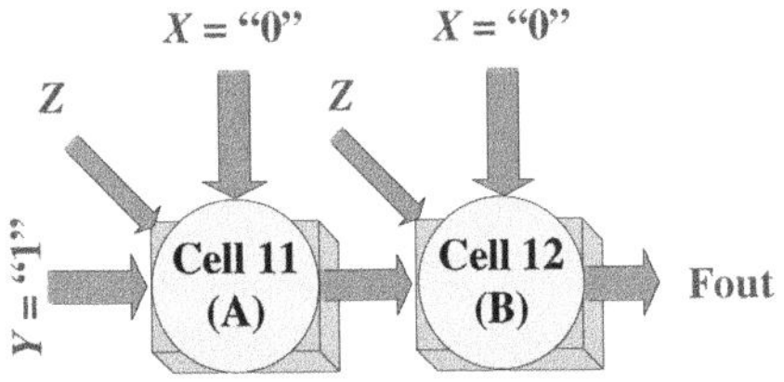

Figure 11.1 Block diagram of two-input AND gate using Akers array.

input Z controls cell operation [104]. Then, logic states are applied to each of these two Z inputs. Figure 11.2 shows the proposed structures for AND gate in binary QCA. Figure 11.3 shows the simulation results of BAND1 and BAND2 gates. If both inputs applied to a gate are equal to "1," the output signal is "1," otherwise it is "0."

The structure of BAND1 includes 145 cells, an occupied area of 0.27 μm^2, and delays of 1.5 clock cycles to the output of the last cell and 1.25 clock cycles to the output of the flip-flop. The second model, BAND2, consists of 108 cells with an occupied area of 0.19 μm^2 and delays of 1.5 clock cycles to the output of the last cell and 2.25 clock cycles to the flip-flop output.

By inserting the aforementioned values in Eqs. (10.1) and (10.2), the values of cost functions for each of the examined structures are calculated as follows:

BAND1:

$$\text{Cost}_{\text{A-T}} = A \times T = 0.27 \times 1.5 = 0.405$$

$$\text{Cost}_{\text{A/C}} = A/C = 0.27/145 = 0.00186 = 1.86 \times 10^{-3}$$

BAND2:

$$\text{Cost}_{\text{A-T}} = A \times T = 0.19 \times 1.5 = 0.285$$

$$\text{Cost}_{\text{A/C}} = A/C = 0.19/108 = 0.00175 = 1.75 \times 10^{-3}$$

The cost function values of both structures are relatively acceptable and optimal. The cost function values of the second structure are lower than those of the first structure. This is not surprising because BAND2 is based on BPIM2, which is more optimal than BPIM1 in terms of cost functions, cell count (13 cells), and occupied area (0.04 μm^2).

The basic-extended hypothesis, which was stated at the beginning of this section, is analyzed in detail in the following. By comparing the results of this section and those of the designs in Section (10.2), it can be seen that BPIM2 is more optimal than BPIM1 in terms of the occupied area, complexity, area–delay cost function, and area–complexity cost function. After designing the gates in this section, it is shown that the obtained parameter values of BAND2 are more optimal than those of BAND1. The basic-extended hypothesis is confirmed in this section; BAND2 has better performance because, according to this hypothesis, its basic cell, BPIM2, is more optimal and applicable.

11.3.2 OR Gate Design Using Binary QCA-Based Basic PIM Cells

In this subsection, the proposed basic cells BPIM1 and BPIM2 are used in the Akers array [78] to implement a binary QCA-based OR gate as shown in Figure 11.5. This structure includes two basic PIM cells that are vertically connected. Each cell can process and store data. Similar to the previous case, inputs X and Y are fixed and equal to logic values "0" and "1," while input Z controls the logic value of the PIM cell (Figure 11.4). According to simulation results shown in Figure 11.6, if one or both of the inputs are equal to the logic value "1," the output is equal to "1," otherwise it is "0."

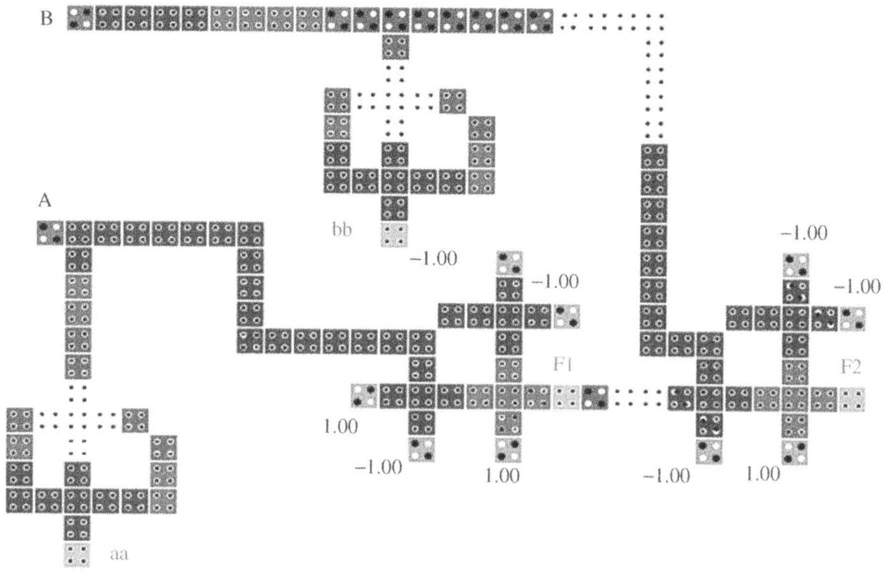

(a)

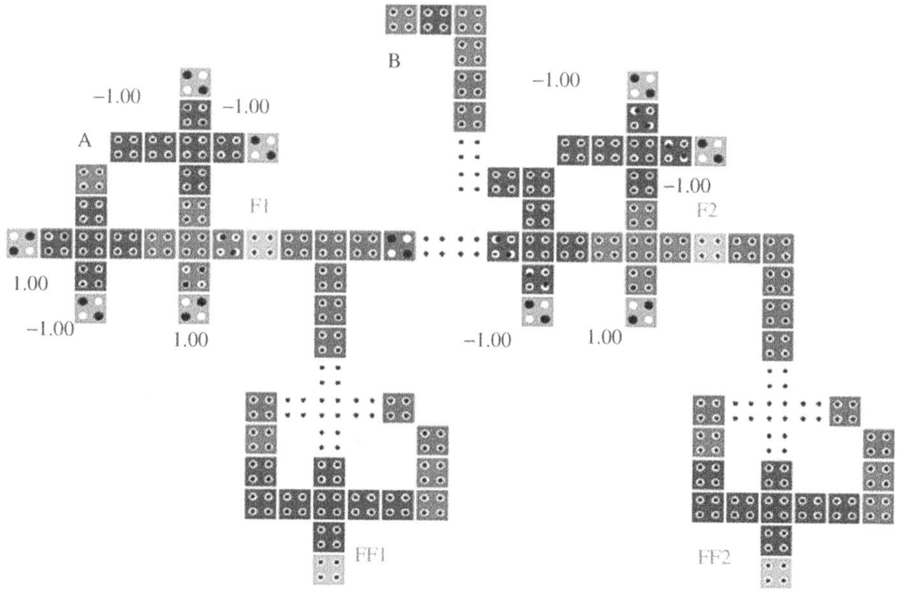

(b)

Figure 11.2 Two-input AND gate using PIM structure in binary QCA: (a) BAND1 and (b) BAND2. Source: Dehbozorgi et al. [183]/with permission of Springer Nature.

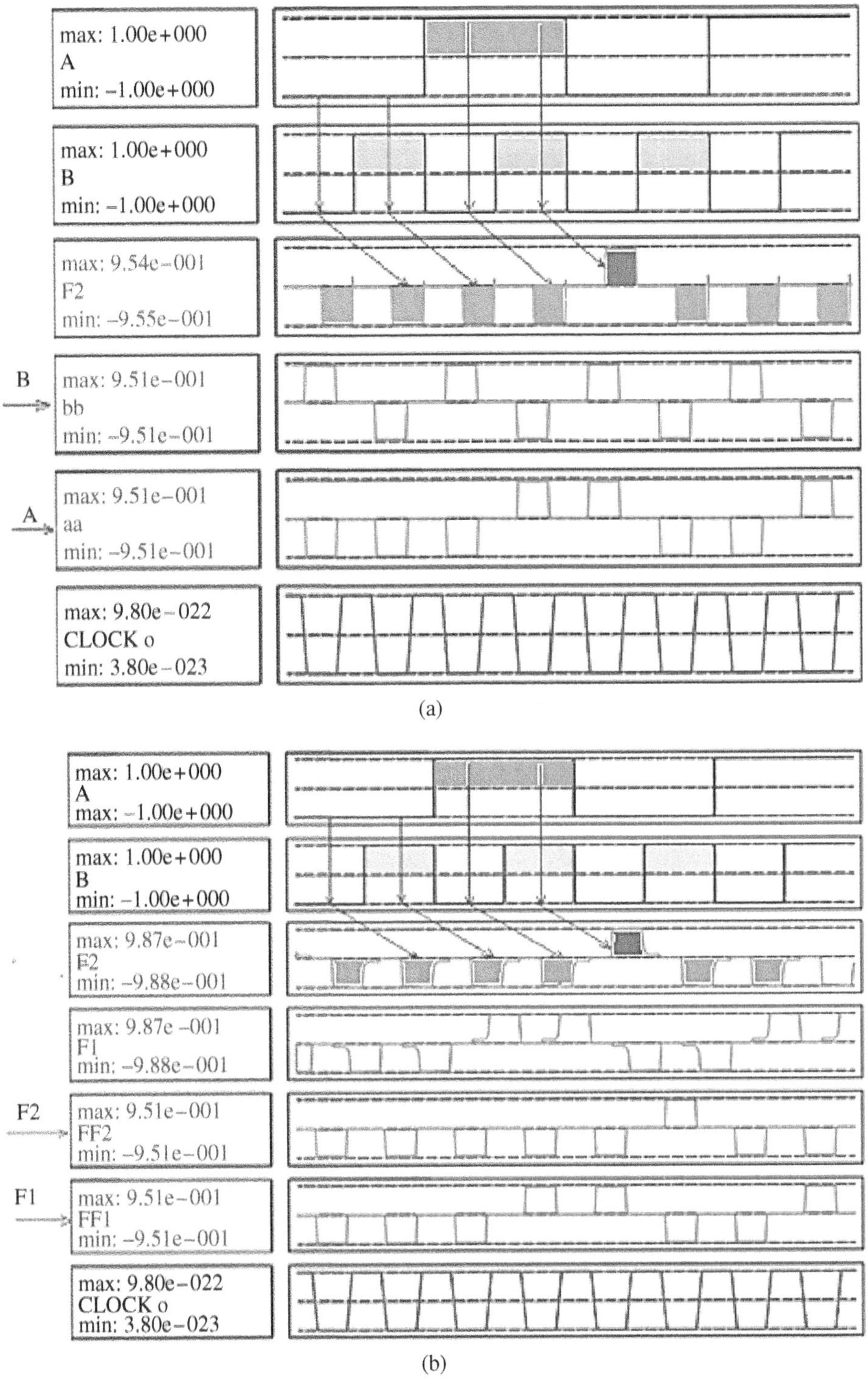

Figure 11.3 AND gate simulation results using binary QCA-based PIM structure for (a) BAND1 and (b) BAND2. Source: Dehbozorgi et al. [183]/with permission of Springer Nature.

Figure 11.4 Block diagram of two-input OR gate using Akers array. Source: Dehbozorgi et al. [183]/with permission of Springer Nature.

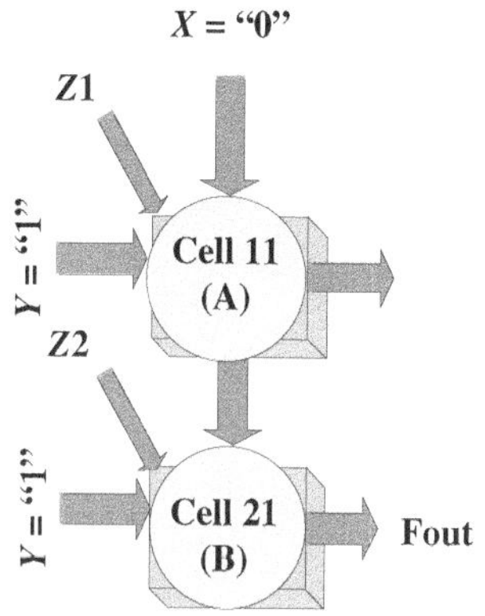

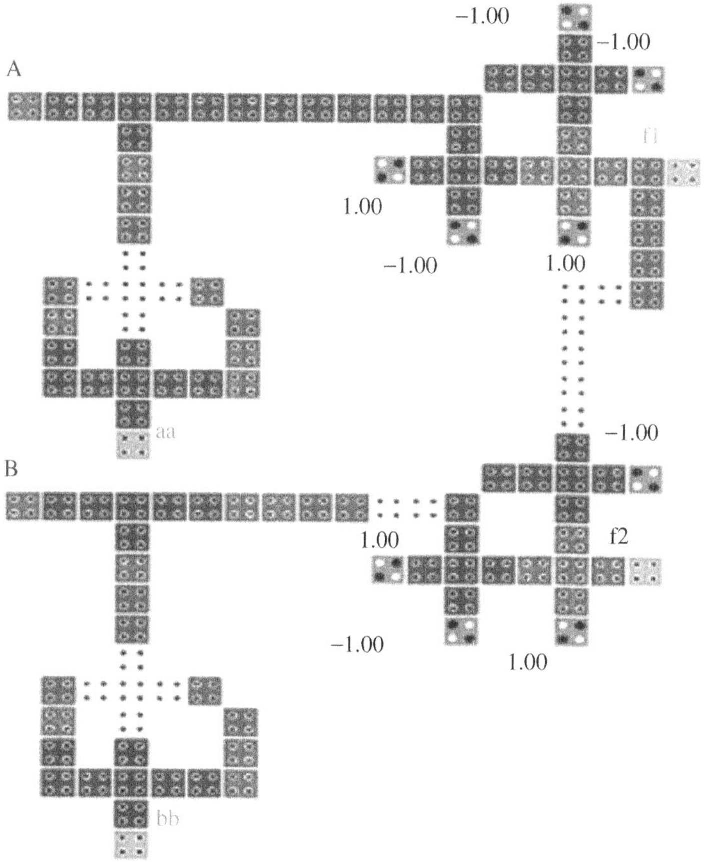

(a)

Figure 11.5 Two-input OR gate using binary QCA-based PIM structure: (a) BOR1 and (b) BOR2. Source: Dehbozorgi et al. [183]/with permission of Springer Nature.

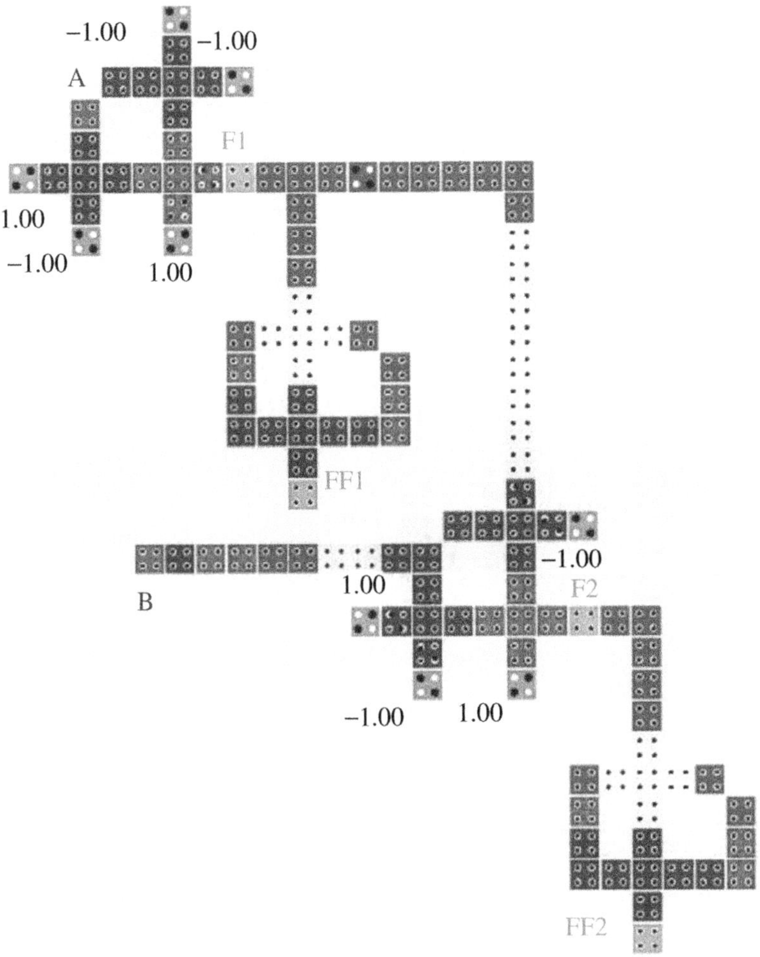

(b)

Figure 11.5 *(Continued)*

The first binary model, BOR1, includes 128 cells with an occupied area of 0.21 μm^2 and delays of 1.5 clock cycles to the output of the last cell and 1.25 clock cycles to the flip-flop output. The second binary model, BOR2, consists of 120 cells with an occupied area of 0.29 μm^2 and delays of 1.5 clock cycles to the output of the last cell and 2.25 clock cycles to the flip-flop output.

By inserting these values into Eqs. (10.1) and (10.2), the cost functions of each proposed structure are calculated as follows:

BOR1:

$$\text{Cost}_{\text{A-T}} = A \times T = 0.21 \times 1.5 = 0.315$$

$$\text{Cost}_{\text{A/C}} = A/C = 0.21/128 = 0.00164 = 1.64 \times 10^{-3}$$

BOR2:

$$\text{Cost}_{\text{A-T}} = A \times T = 0.29 \times 1.5 = 0.435$$

$$\text{Cost}_{\text{A/C}} = A/C = 0.29/120 = 0.00241 = 2.41 \times 10^{-3}$$

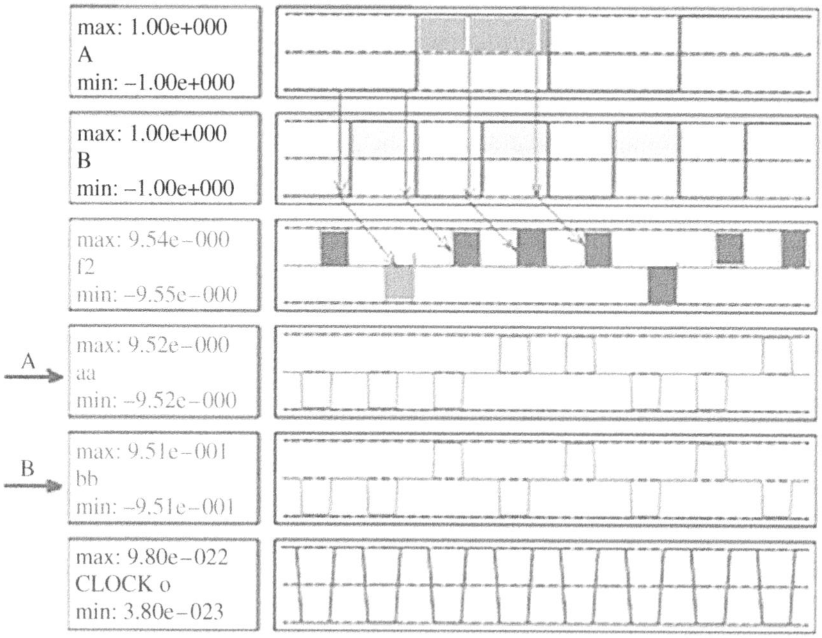

(a)

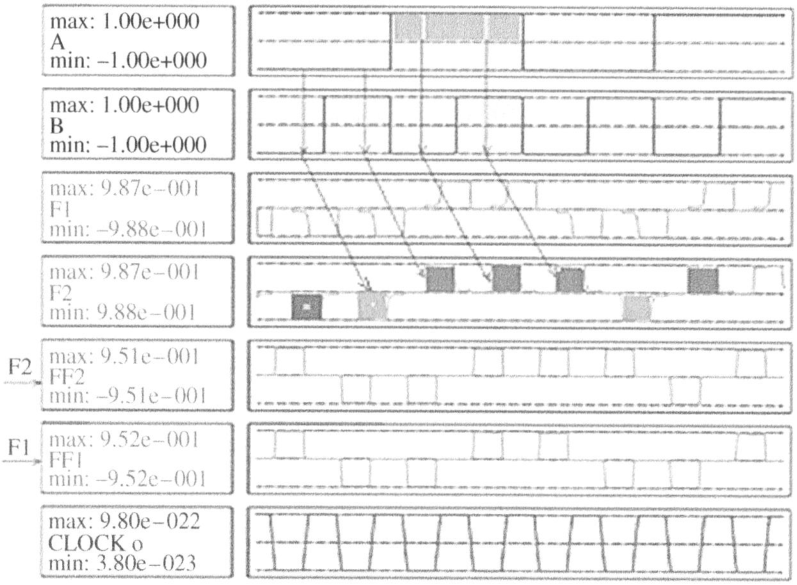

(b)

Figure 11.6 OR gate simulation results using binary QCA-based PIM: (a) BOR1 and (b) BOR2. Source: Dehbozorgi et al. [183]/with permission of Springer Nature.

Since BOR1 and BOR2 are designed using basic cells BPIM1 and BPIM2, they can be evaluated using the basic-extended hypothesis. In Section 10.2, the characteristics of BPIM1 and BPIM2 are determined. Because the cell count, occupied area, and cost functions of BPIM2 are smaller than those of BPIM1, BPIM2 is more optimal. Additionally, in this section, we prove BOR2 is more optimal than BOR1 in terms of complexity and delay, so the proposed hypothesis is confirmed. However, BOR1 shows better performance than BOR2 in terms of cost functions. This can be attributed to the larger occupied area of BOR2, which is approximately 0.08 μm^2 more extended than BOR1, as shown in Figure 11.5. Therefore, using Eqs. (10.1) and (10.2), larger values are obtained for the cost functions of BOR2. Consequently, in this analysis, BOR2 is superior to BOR1, with two optimal parameters out of the three main parameters. According to the basic-extended hypothesis, the basic cell of BOR2, BPIM2, is practical and appropriate.

11.3.3 NAND Gate Design Using Binary QCA-Based Basic PIM Cells

Operators BNAND1 and BNAND2 are designed based on BPIM3 and BPIM4 basic cells. These structures are designed in Akers array and consist of two cells that are connected horizontally, and a NOT gate is placed at the end. Figure 11.7 shows the block diagram of two-input NAND gate in Akers array. Each basic PIM cell can store and process data. According to Figure 11.8, in both BNAND1 and BNAND2, the input values X and Y are fixed and equal to "0" and "1", respectively. Unlike the proposed AND and OR structures, a new control input called "write/read" is added to BNAND1 and BNAND2 to improve the functionality of the circuit. The operation of the circuit can be set by this control input.

The simulation results of both proposed models are shown in Figure 11.9. It can be seen that when the write input is equal to "1," the result of the NAND operator is transferred to the output. Only when both inputs are "1," the output is "0," otherwise it returns "1."

The BNAND1 model has 124 cells with an occupied area of 0.19 μm^2 and a delay of 2.25 clock cycles. The BNAND2 model has 99 cells with an occupied area of 0.17 μm^2 and a delay of 2.75 clock cycles.

By inserting the related parameter values into Eqs. (10.1) and (10.2), the values of cost functions are calculated as follows:

BNAND1:

$$\text{Cost}_{\text{A-T}} = A \times T = 0.19 \times 2.25 = 0.4275$$

$$\text{Cost}_{\text{A/C}} = A/C = 0.19/124 = 0.00153 = 1.53 \times 10^{-3}$$

BNAND2:

$$\text{Cost}_{\text{A-T}} = A \times T = 0.17 \times 2.75 = 0.4675$$

$$\text{Cost}_{\text{A/C}} = A/C = 0.17/99 = 0.001717 = 1.717 \times 10^{-3}$$

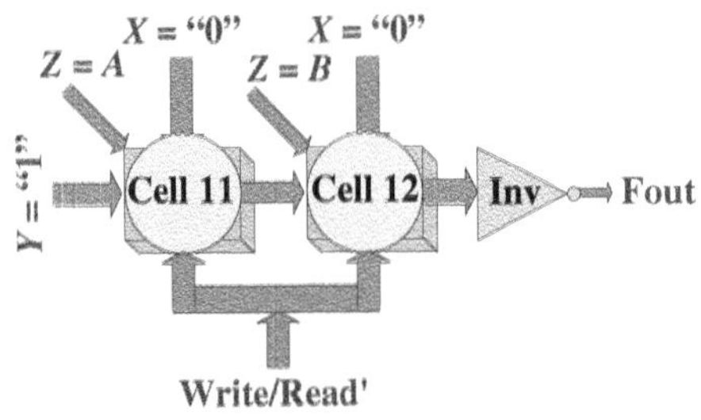

Figure 11.7 Block diagram of two-input NAND gate in Akers array. Source: Dehbozorgi et al. [171]/with permission of Springer Nature.

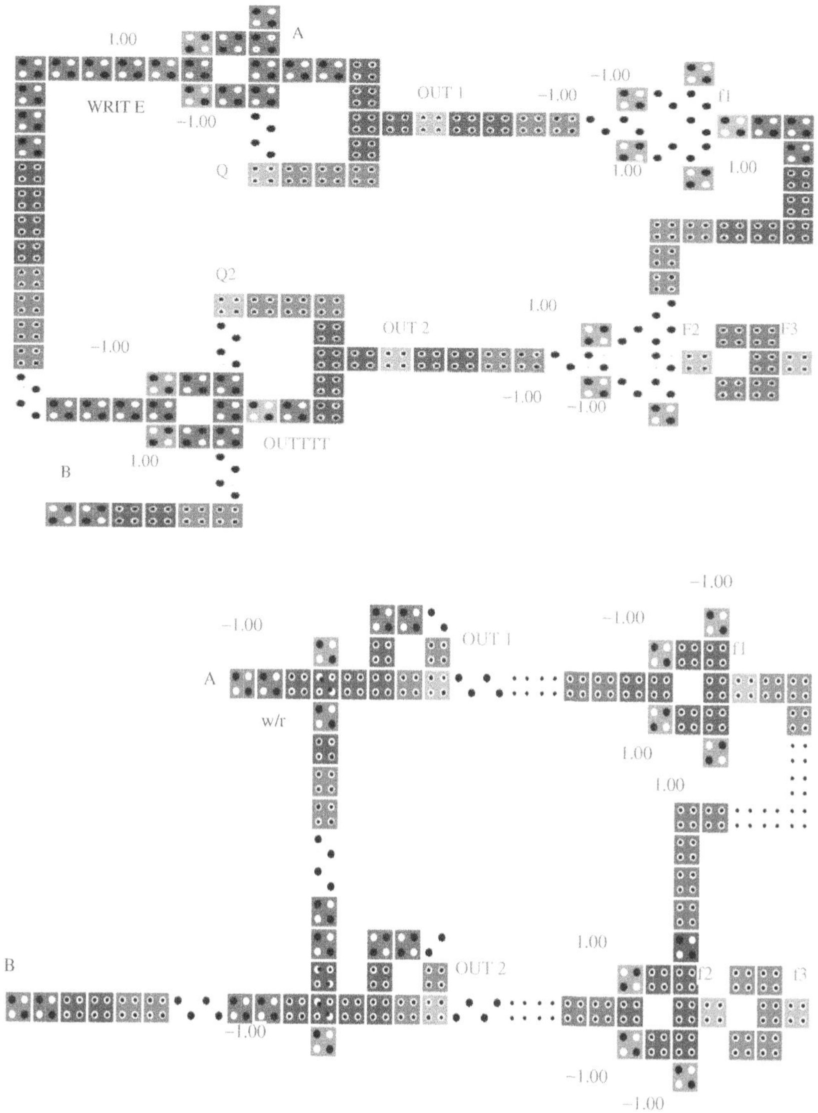

Figure 11.8 Two-input NAND gate based on PIM structure in binary QCA: (a) BNAND1 and (b) BNAND2. Source: Dehbozorgi et al. [171]/with permission of Springer Nature.

Furthermore, the values of cost functions are very similar. In other words, there is no noticeable difference in the occupied area dedicated to each circuit design, and the cost savings of both structures are almost identical.

By comparing the design parameters of BNAND1 and BNAND2, it is observed that the complexity and occupied area of BNAND2 are more optimal than those of BNAND1. Additionally, the cost functions of BNAND1 and BNAND2 are similar up to three decimal places. Therefore, the correctness of the basic-extended hypothesis is confirmed for these structures as well. This is because the basic structure of BPIM4 used in BNAND2 is more optimal than BPIM3. Therefore, it can be concluded that an extended and optimal structure, such as BNAND2, consists of an optimal basic structure, such as BPIM4.

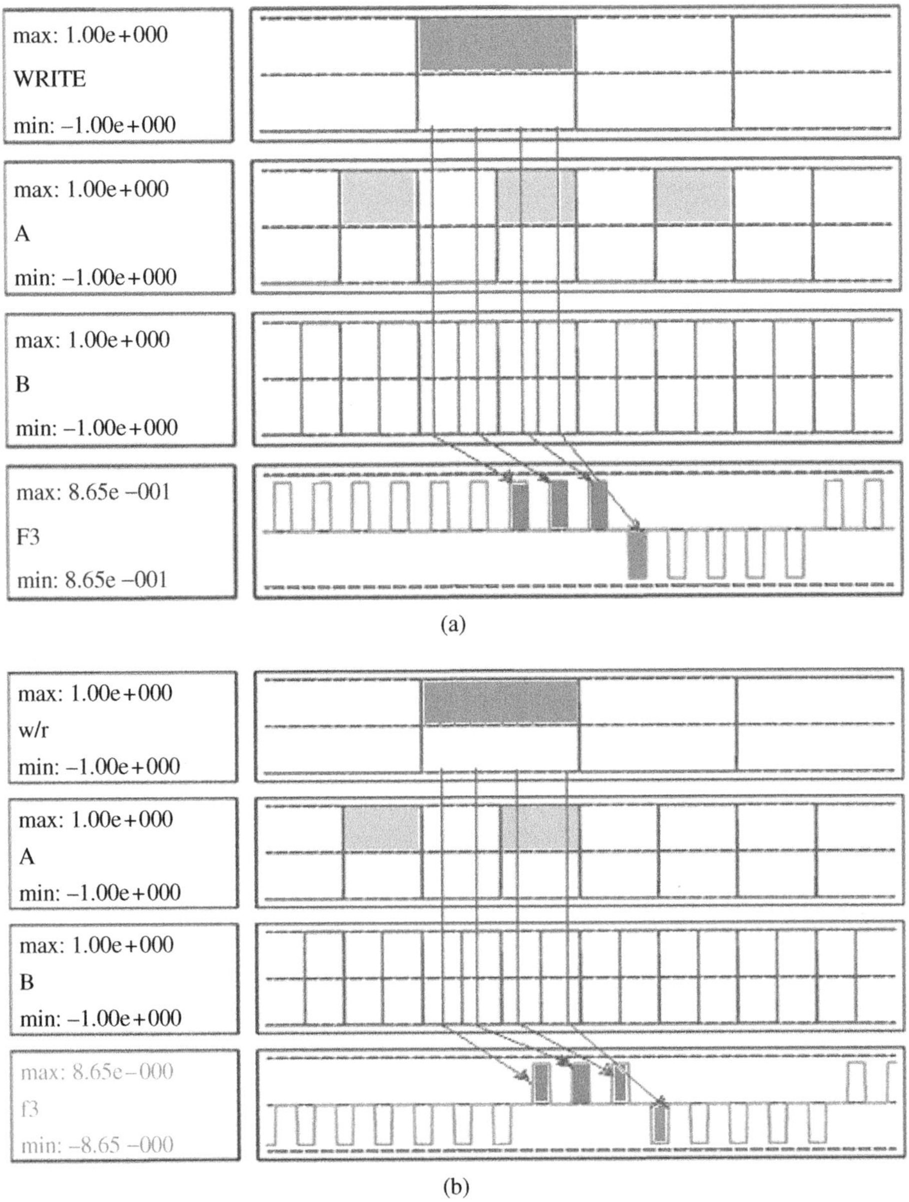

Figure 11.9 NAND simulation results based on PIM structure in binary QCA: (a) BNAND1 and (b) BNAND2. Source: Dehbozorgi et al. [171]/with permission of Springer Nature.

11.3.4 NOR Gate Using Basic PIM Cells in Binary QCA

The operators BNOR1 and BNOR2, which are implemented using Akers array, consist of two vertically connected cells and a NOT inverter at the end of the last cell (Figure 11.10). Each basic PIM cell can process and store data. As shown in Figure 11.11, BNOR1 and BNOR2 are composed of two basic cells BPIM3 and BPIM4. The values of X and Y are fixed and equal to "0" and "1," respectively. The control input Z changes with input states. In these structures, similar to NAND1

Figure 11.10 Block diagram of two-input NOR gate using Akers array. Source: Dehbozorgi et al. [171]/with permission of Springer Nature.

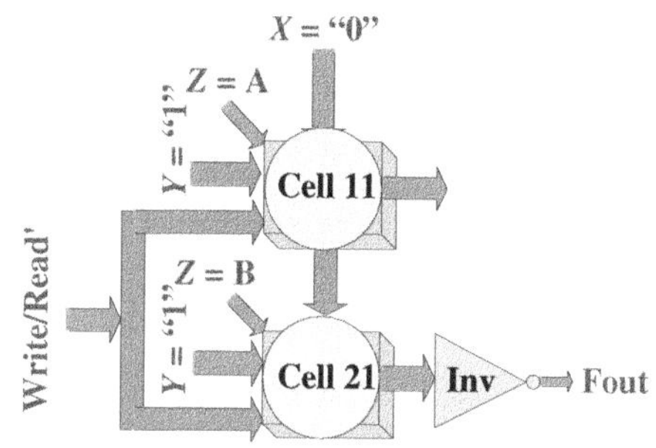

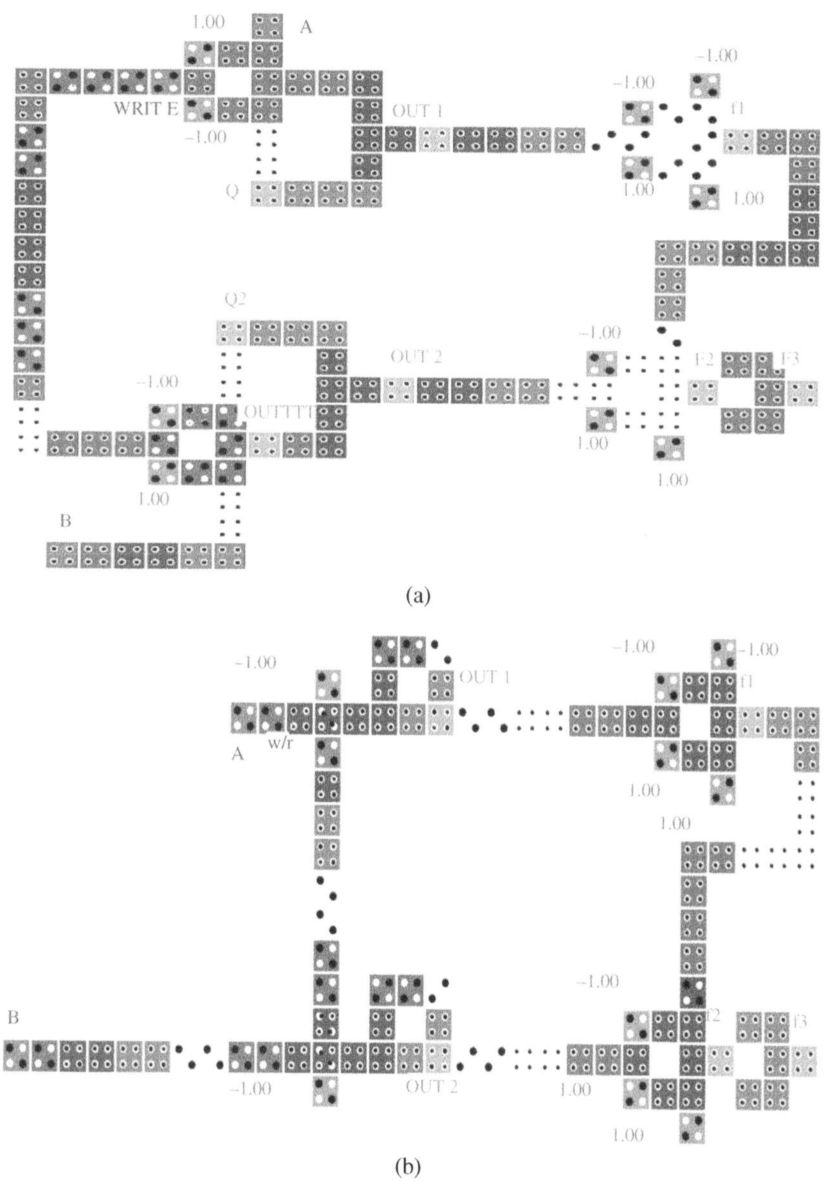

Figure 11.11 Two-input NOR gate using PIM structure in binary QCA: (a) BNOR1 and (b) BNOR2. Source: Dehbozorgi et al. [171]/with permission of Springer Nature.

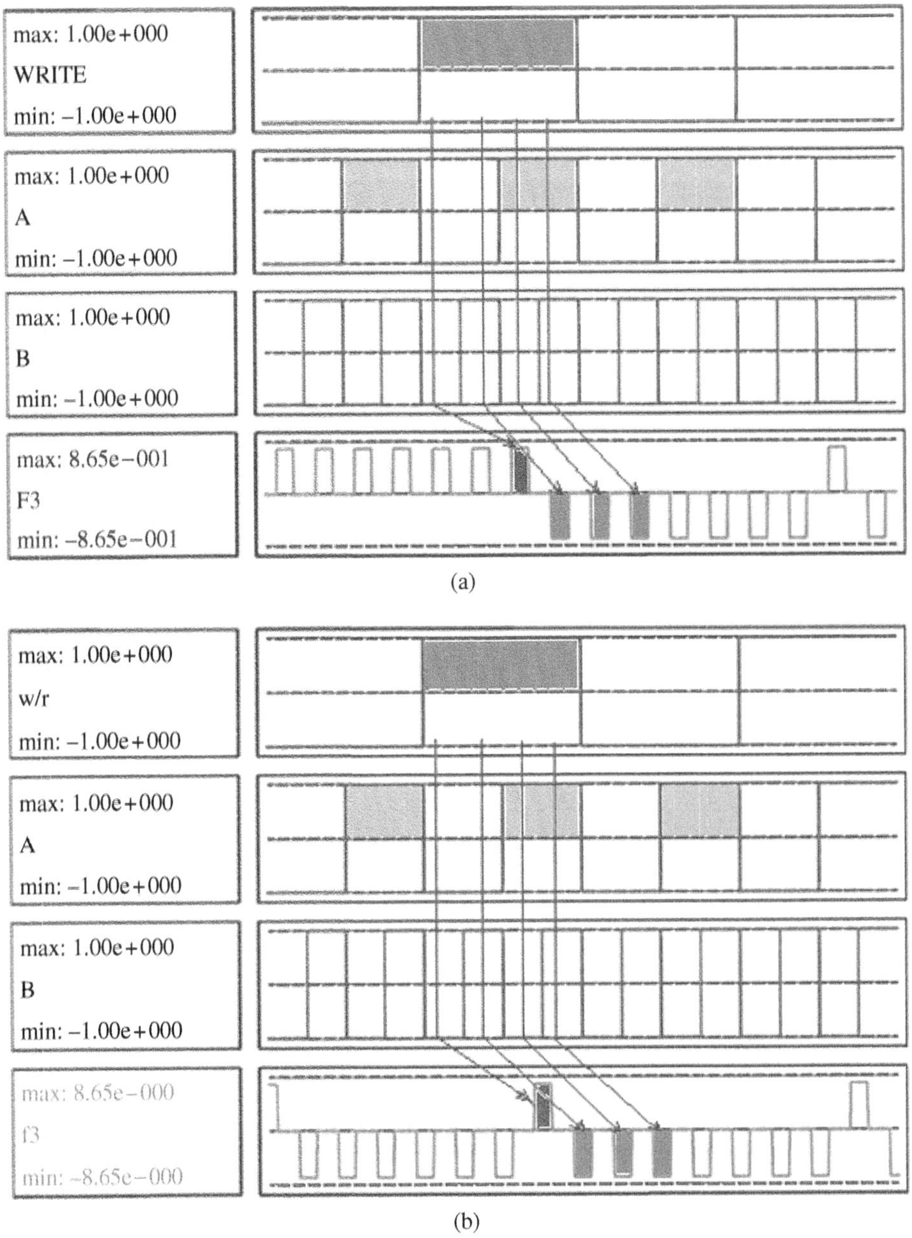

Figure 11.12 NOR simulation results based on PIM structure in binary QCA: (a) BNOR1 and (b) BNOR2. Source: Dehbozorgi et al. [171]/with permission of Springer Nature.

and NAND2, a new control input called WRITE/READ is added to control the operation of the circuit. Figure 11.12 shows the simulation results of BNOR1 and BNOR2. It is observed that when the write input is "1," the result of the NOR operator is transferred to the output. Only when both inputs are "0," the output is "1," otherwise it is "0." By comparing Figures 11.8 and 11.11, it is found that if only the values of fixed cells in the last multiplexer of the NAND structure are reversed and relocated from top to bottom, the new structure becomes a NOR gate.

The BNOR1 model consists of 124 cells with an occupied area of 0.19 μm^2 and a delay of 2.25 clock cycles. The BNOR2 model consists of 99 cells with an occupied area of 0.17 μm^2 and a delay of 2.75 clock cycles.

By inserting the aforementioned values into Eqs. (10.1) and (10.2), the cost functions are calculated as follows:

BNOR1:

$$\text{Cost}_{\text{A-T}} = A \times T = 0.19 \times 2.25 = 0.4275$$

$$\text{Cost}_{\text{A/C}} = A/C = 0.19/124 = 0.00153 = 1.53 \times 10^{-3}$$

BNOR2:

$$\text{Cost}_{\text{A-T}} = A \times T = 0.17 \times 2.75 = 0.4675$$

$$\text{Cost}_{\text{A/C}} = A/C = 0.17/99 = 0.001717 = 1.717 \times 10^{-3}$$

The values of the area–delay cost functions of BNOR2 and BNOR1 are the same up to one decimal place. The area–complexity cost function of BNOR2 slightly increases compared to BNOR1 because cell count decreases, but the occupied area does not decrease at the same rate, i.e., only by 0.02. However, by comparing the values of the area–complexity cost functions of BNOR1 and BNOR2 up to three decimal places, no significant difference is observed.

To investigate the basic-extended hypothesis for BNOR1 and BNOR2, the basic cells used in these structures are analyzed. According to parameter values obtained for BNOR2, this structure performs better than BNOR1 in terms of complexity and delay. Therefore, according to the basic-extended hypothesis, the basic structure of BNOR2, i.e., BPIM4, should perform better than its counterpart, BPIM3. This can be confirmed based on previous results, which indicate cell count, occupied area, and cost functions of BPIM4 are superior to those of BPIM3. Therefore, because BPIM4 is used in the BNOR2 structure, this structure has more desirable parameter values.

The simulations and analyses of the proposed AND, OR, NAND, and NOR structures show that any extended structure that performs better than its counterparts is implemented using optimal and more practical PIM cells, as also implied by the basic-extended hypothesis. Therefore, the correctness of this hypothesis is confirmed. Among the four PIM designs proposed until now, the best and most practical ones are BPIM4 and BPIM2. Next, the structure of BXOR is examined.

11.3.5 XOR Gate Design Using Basic PIM Cells in Binary QCA

The structure of basic PIM cells and their layout in the Akers array are shown in Figure 11.13a. The block diagram of the XOR gate and its structure in the Akers array is shown in Figure 11.13b. There are two rows and two columns in this structure. Each input and each row share a memory. For the multiplexers of the cells in the top row, the value of X is fixed and equal to "1," while the values of X in the cells of the second row are set and controlled by the outputs of the upper row. The input Y of the cells in the left column is equal to "1," while the values of Y of the cells in the second column are set and controlled by the outputs of the left column (see Figure 11.14).

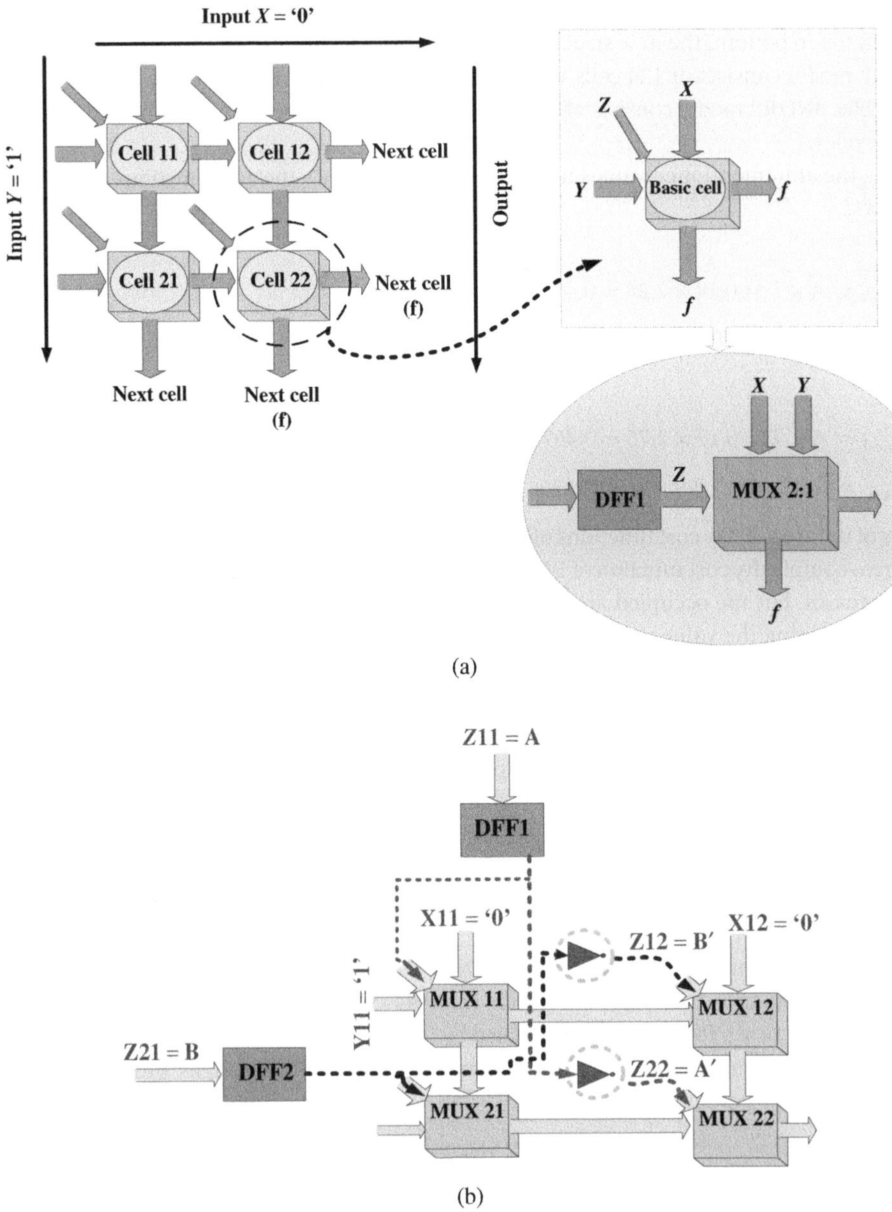

Figure 11.13 (a) structure of basic cells and their layout in Akers array and (b) block diagram of two-input XOR gate in Akers array. Source: Dehbozorgi et al. [183]/with permission of Springer Nature.

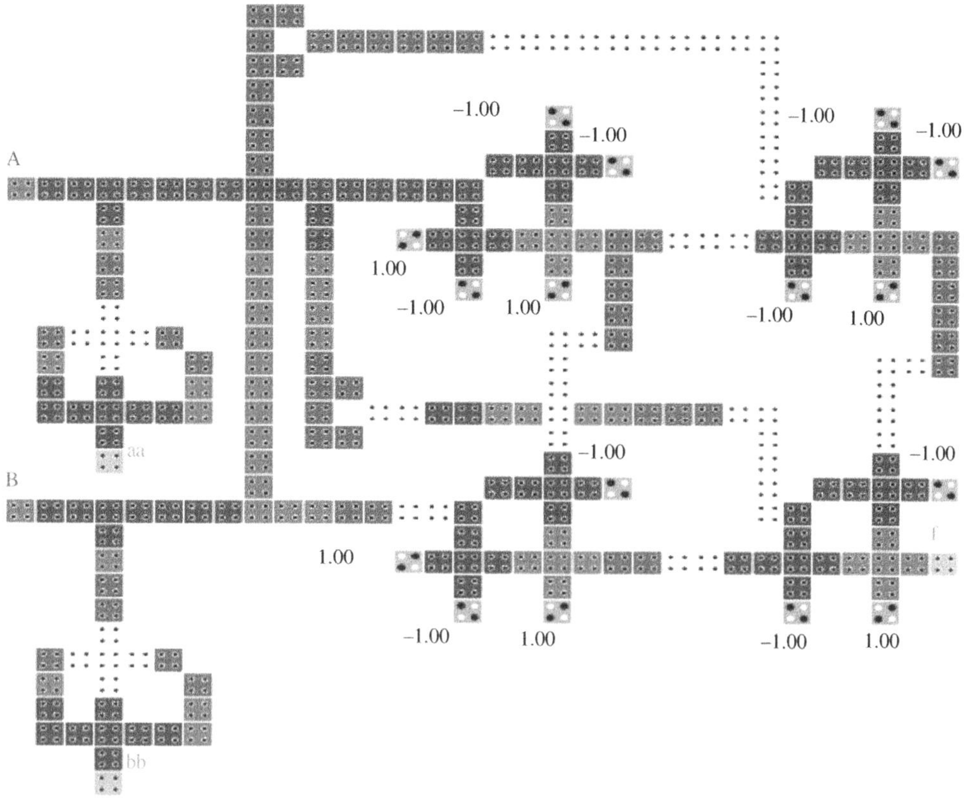

Figure 11.14 Two-input BXOR gate using PIM in binary QCA. Source: Dehbozorgi et al. [183]/with permission of Springer Nature.

When one of the two inputs is equal to logic "1," the output is "1," while for both inputs equal to logic "1" or "0," the output is logic "0" (Figure 11.15). In the block diagram, two outputs aa and bb are returned, which are equal to inputs A and B, respectively.

In fact, the DFFs at the beginning of the circuit are responsible for storing the input values while they perform data processing down the path within the array. The BXOR structure consists of 266 cells with an occupied area of 0.35 μm^2 and a delay of 2.5 clock cycles.

By inserting the aforementioned values into Eqs. (10.1) and (10.2), the values of cost functions can be calculated as follows:

$$\text{Cost}_{\text{A-T}} = A \times T = 0.35 \times 2.5 = 0.875$$

$$\text{Cost}_{\text{A/C}} = A/C = 0.35/266 = 0.001315 = 1.315 \times 10^{-3}$$

The values obtained for both cost functions are relatively favorable. Interestingly, the area–complexity cost function has the lowest value among the binary gates proposed in this chapter. This can be attributed to the use of only two memories in common for four multiplexers in the Akers array, and the previous criterion (i.e., one memory per BPIM1 cell) is not met. Therefore, the cost is lower, or in other words, the cell density increases as a result of placing more QCA cells on a given area. This circuit is proposed only to evaluate the use of common memory and its effect on cost functions.

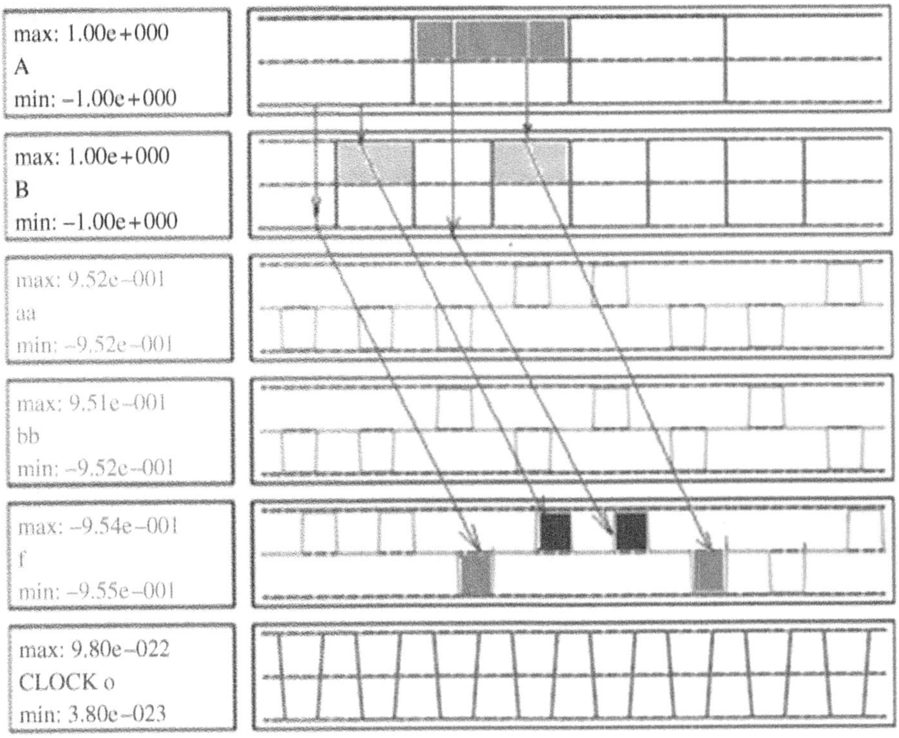

Figure 11.15 Simulation results of BXOR using PIM in binary QCA. Source: Dehbozorgi et al. [183]/with permission of Springer Nature.

12

Proposed PIM Cells in Ternary QCA and Their Fault Analysis

12.1 Introduction

Memory is a very important and fundamental structure whose optimization allows designing of larger high-performance structures. In this chapter and the next chapter, the ternary quantum dot cellular automata (QCA) technology is used to design loop-based memory structures. First, the design of basic processing in memory (PIM) cells in ternary QCA is analyzed. Then, in the next chapter, AND, OR, NAND, NOR, and XOR gates that use the presented PIM cell are designed, analyzed, and simulated. In the remainder of this chapter, the effects of cell omission and extra-cell deposition defects in the basic memory cells are investigated.

Initially, the first flip-flop model is proposed by simulating in MATLAB environment. Then, we discuss the proposed memory structures using the ternary quantum dot cellular automata (TQCA) Sim software version 1.0.11.1. In this chapter, to analyze and compare binary and ternary structures, we obtain the optimized values of their parameters using Eq. (12.1) [182].

$$\text{Improvement}\,(\%) = \left(1 - \frac{\text{our implemention result}}{\text{previous implemention result}}\right) \times 100 \tag{12.1}$$

12.2 Simulation of a Flip-Flop in Ternary QCA Using MATLAB

MATLAB is used to design a data flip-flop (DFF) in ternary QCA. Figure 12.1 shows the DFF design in ternary QCA. In this flip-flop, when the read/write input (R/W' or ENABLE) is equal to logic "1" or state B, the information of the flip-flop is transferred to the memory loop and output; otherwise, the previous information circulates in the memory loop.

In Figure 12.2, the proposed ternary QCA-based DFF is presented. This structure, which includes some junctions, is simulated by programming in the MATLAB 2016 environment. The first stage is initially designed, and then its output is fed into the second stage. Next, the output of the second stage is displayed and fed into the third stage. The result of the output is fed back to the previous stage, such that a continuous structure is realized by connecting the output of the previous stage to the input of the next stage. The diagrams in Figure 12.3 show the outputs of the stages, which are desirable according to the applied inputs.

In the aforementioned design, a basic DFF was simulated in MATLAB. However, this technique is very time-consuming and requires complex programming to design larger circuits. Therefore, in the remainder of this chapter, the proposed structures are presented for loop-based memory

Computing with Multi-Value Logic in Quantum Dot Cellular Automata, First Edition.
Reza Sabbaghi-Nadooshan, Reza Akbari-Hasanjani, Leila Dehbozorgi, Majid Haghparast, and Hamid Reza Akbari-Hasanjani.

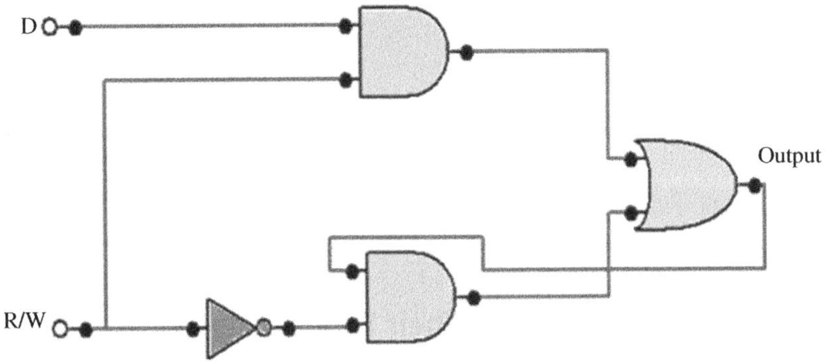

Figure 12.1 DFF structure.

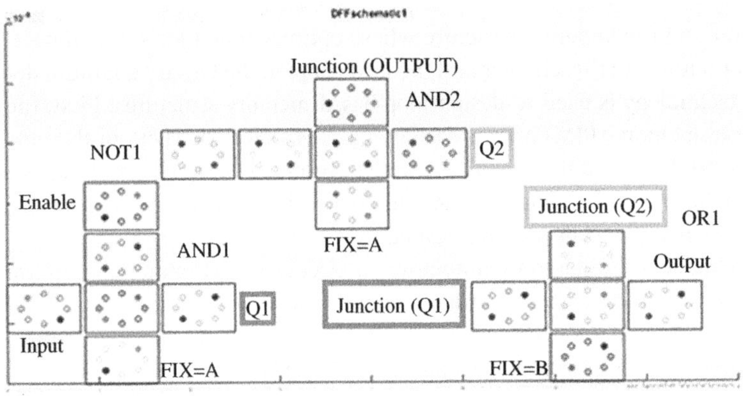

Figure 12.2 Proposed DFF implemented and simulated using MATLAB.

using TQCA Sim software version 1.0.11.1. The results from both MATLAB and TQCA Sim, which are presented in the following sections, are in good agreement. In this situation, the proper performance of the new simulator can also be examined.

12.3 Structures in Ternary QCA

12.3.1 First Model of PIM Cell (TPIM1)

The first ternary model for basic PIM cell is shown in Figure 12.4a. As mentioned in Chapter 9, to design the basic PIM cell, a relation such as Eq. (9.4) and Akers array can be used. The purpose of this structure is to determine the effect of the location of the flip-flop on improving the performance of the PIM cell. In TPIM1, similar to its binary counterpart, BPIM1, the DFF is placed at the beginning of the circuit to store the input information. Inputs X = "A" and Y = "B" are assumed to be fixed. If "B" is applied to the control input of the cell, Y is propagated to the output; otherwise, X is propagated. The output of the flip-flop is always equal to the primary input (Figure 12.4).

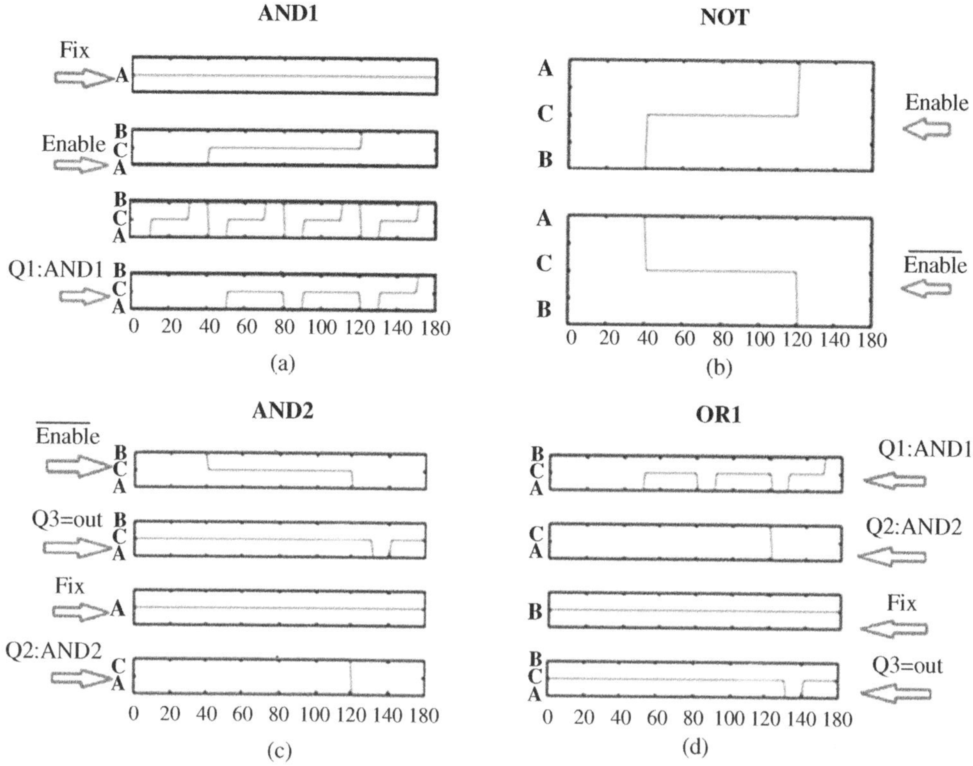

Figure 12.3 Output diagrams of the proposed ternary QCA-based DFF designed using MATLAB for (a) AND1, (b) NOT, (c) AND2, and (d) OR1 structures.

The TPIM1 cell contains 58 ternary QCA cells with an occupied area of 0.013399 μm^2 and delays of 0.5 clock cycles to the cell output and 1.25 clock cycles to the flip-flop output. The delay and occupied area of TPIM1 are much smaller than those of other QCA-based memory architectures.

By inserting these values of the design parameters into Eqs. (10.1) and (10.2), the cost functions are calculated as follows:

$$\text{Cost}_{\text{A-T}} = A \times T = 0.0133 \times 0.5 = 0.00665$$

$$\text{Cost}_{\text{A/C}} = A/C = 0.0133/58 = 0.001864 = 0.23086 \times 10^{-3}$$

According to these results, TPIM1 has an acceptable performance in terms of speed (a delay of 0.5 clock cycles), occupied area, and area–delay cost function.

For a more detailed analysis, the design parameter values of TPIM1 are inserted into Eq. (12.1), and a comparison is made with its binary counterpart, BPIM1. According to calculations, the occupied area, area–delay cost function, and area–complexity cost function of TPIM1 are optimized, respectively, by 88.18%, 88%, and 87.66% compared to BPIM1. This can be mainly attributed to the transformation from binary to ternary structure and reduction in the dimensions of QCA cell.

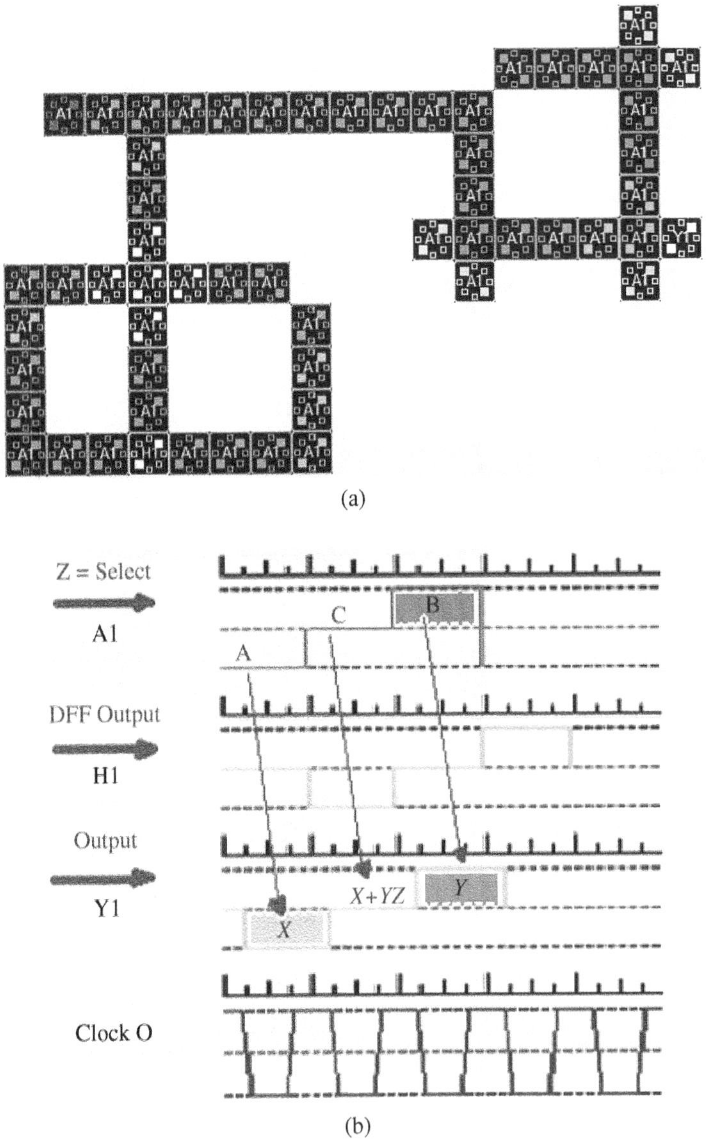

Figure 12.4 First model of PIM cell, TPIM1: (a) implementation in ternary QCA and (b) simulation results. Source: Dehbozorgi et al. [183]/with permission of Springer Nature.

Accordingly, TPIM1 has better performance and parameter values than BPIM1, so this model can be used to store input information in ternary circuits with PIM capability.

12.3.2 Second Model of PIM Cell (TPIM2)

Figure 12.5 shows the proposed ternary model of the basic cell TPIM2. In this structure, DFF is placed at the end of the circuit to evaluate its effect on improving the circuit design parameters. The flip-flop is responsible for storing the information at the output of the circuit. According to

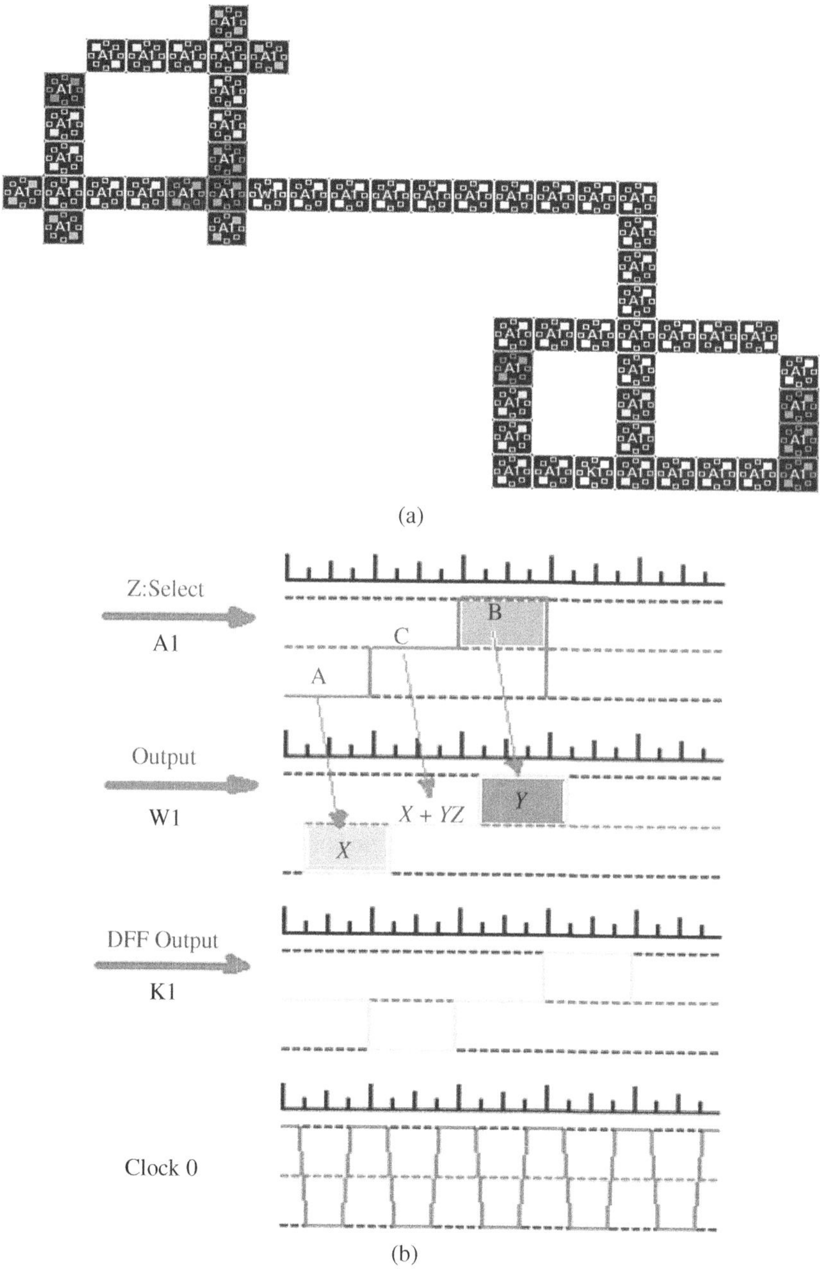

Figure 12.5 Second model of PIM cell, TPIM2: (a) implementation in ternary QCA and (b) simulation results. Source: Dehbozorgi et al. [183]/with permission of Springer Nature.

Figure 12.5, inputs $X =$ "A" and $Y =$ "B" are fixed. If "B" is applied to the control input of the cell, the value of Y is propagated to the output; otherwise, the value of X is propagated.

The TPIM2 cell includes 57 ternary QCA cells, which occupy 0.02008575 μm^2 with delays of 0.75 clock cycles to the main output and 1.25 clock cycles to the flip-flop output. Using Eqs. (10.1) and (10.2), cost functions are calculated as follows:

$$\text{Cost}_{\text{A-T}} = A \times T = 0.02 \times 0.75 = 0.015$$

$$\text{Cost}_{\text{A/C}} = A/C = 0.02/57 = 0.3508 \times 10^{-3}$$

Only the area–complexity cost function of TPIM2 is more optimal than that of TPIM1. To compare the performance of TPIM2 with BPIM2, the values of the design parameters should be inserted into Eq. (10.1). According to the calculations, TPIM2 is optimized compared to BPIM2 in terms of occupied area, area–delay cost function, and area–complexity cost function by 71.42%, 71.42%, and 76.97%, respectively. This can be mainly attributed to the transformation from binary to ternary structure and reduction in the dimensions of QCA cell.

Therefore, the ternary structure of TPIM2 has better performance and more optimal design parameters than the binary structure of BPIM2. This structure can be used to store the output information in ternary circuits with PIM capability.

12.3.3 Third Model of Ternary QCA-Based PIM Cell (TPIM3)

As mentioned earlier, no special rule is available to optimize the design of QCA circuits, so by only meeting some criteria, such as reducing cell count and occupied area as well as achieving a minimum number of clock zones by changing the structures, better results are achieved by trial and error. In the design of the ternary QCA-based basic PIM cell, it was tried to reduce the number of structural cells by varying the locations of QCA cells and connection between flip-flop and multiplexer (Figure 12.6). In the TPIM3 cell, a new control input is added to control the read/write operation. According to the Akers principle, the correct operation of the cell is ensured by applying the fixed values of "A" and "B" to the control inputs of the cell, i.e., X and Y, respectively.

Only when the read/write input is equal to "B" and the input information is propagated to the output and stored in the memory. The DFF output is propagated to the multiplexer as a control line. If the control line becomes equal to "A," the input value of X is propagated to the output, and if the control line becomes equal to "B," the input value of Y is propagated to the output. However, if the control line becomes equal to "C," the result of $X + YZ$, i.e., the value of "C" is propagated to the output. In this case, a write operation is performed.

If the Write/$\overline{\text{Read}}$ input is equal to "A," the previous value in the memory is propagated to the output of the cell, and a read operation is performed. If the value of the clock pulse is equal to "C," its value is ignored, and its output is not important. This is because two states are needed for the control input Write/$\overline{\text{Read}}$, but all three states are used for the primary input (INPUT). The proposed basic ternary PIM cell (TPIM3) shown in Figure 12.6 includes 54 cells with an occupied area of 0.014 μm^2 and a delay of 3.25 clock cycles.

This structure can store three states (A, B, and C) with less occupied area than its binary counterpart presented earlier. It is worth mentioning that, in TPIM3, cell count (complexity) and occupied area are reduced at the expense of increased delay.

By inserting the aforementioned values of the design parameters into Eqs. (10.1) and (10.2), the cost functions are calculated as follows:

$$\text{Cost}_{\text{A-T}} = A \times T = 0.014 \times 3.25 = 0.0455$$

$$\text{Cost}_{\text{A/C}} = A/C = 0.014/54 = 0.259 \times 10^{-3}$$

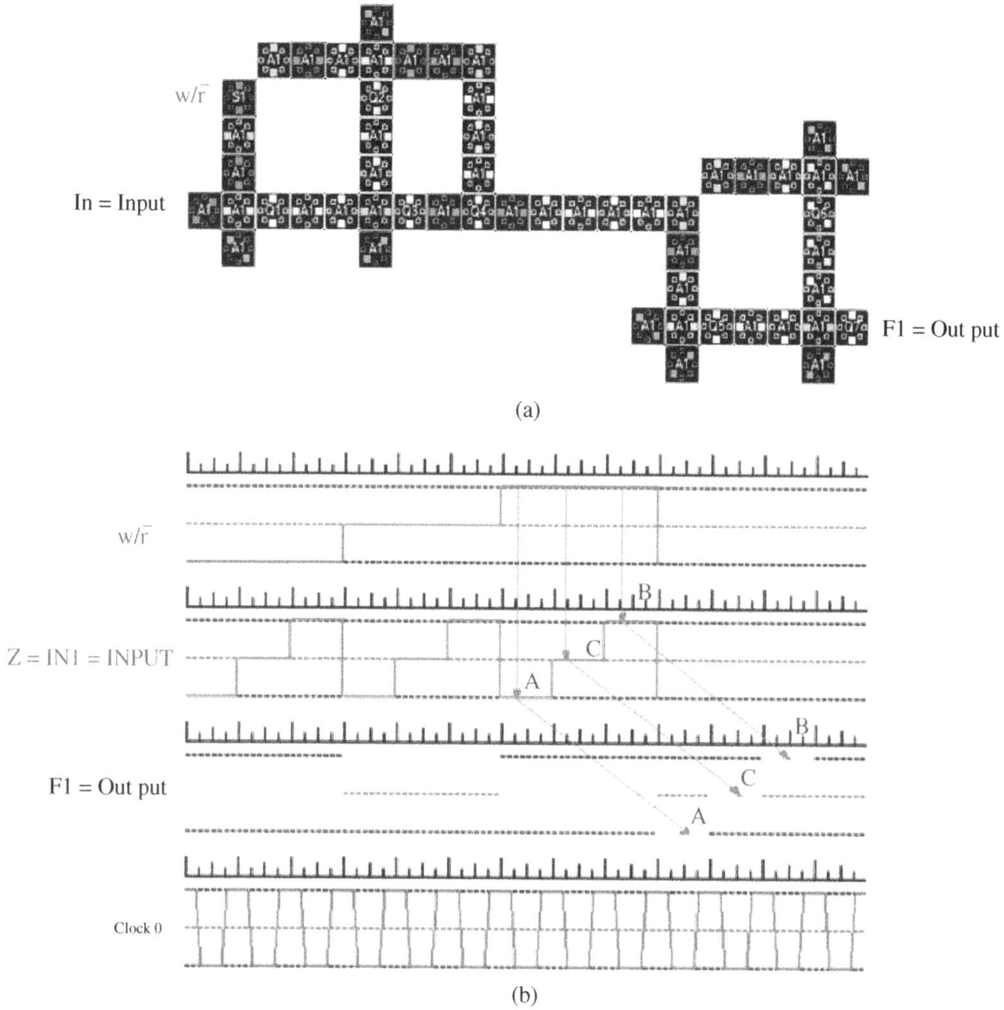

Figure 12.6 Third model of PIM cell (TPIM3): (a) implementation in ternary QCA and (b) simulation results. Source: Dehbozorgi et al. [171]/with permission of Springer Nature.

The area–complexity cost function of TPIM3 is not significantly different from that of TPIM1 or TPIM2. In general, the ternary QCA-based TPIM3 structure performs better than all binary PIM and memory structures in terms of cost functions and occupied area. However, here we specifically compare TPIM3 with BPIM3.

By inserting the design parameter values into Eq. (12.1), TPIM3 has optimized, compared to BPIM3, by 72%, 9%, and 79.264% in terms of occupied area, area–delay cost function, and area–complexity cost function, respectively. The main cause of this improvement is the migration from binary to ternary structure, which leads to smaller dimensions of the QCA cell. However, it can be noted that the binary structure of BPIM3 performs well to some extent because, in terms of the area–delay cost function, it can achieve relatively similar results to that of TPIM3.

Therefore, TPIM3 has better performance and optimal design parameters than BPIM3, and overall, compared to all binary structures. Another advantage of TPIM3 is the use of read/write control input, which is implemented as the first of its type among the proposed ternary PIM structures. Additionally, this structure can store input information.

In the remainder of this section, the performance of the switches in the Akers array is evaluated, and more careful attention is paid to the ternary QCA structure. In all presented models, the Akers array is used to design PIM in ternary QCA. In the Akers array, two switches are used. By applying the control input to the PIM cell, the operation of the switches can be controlled. For the correct design of gates and larger circuits that include memory, the Akers array is used to apply fixed values "0" and "1" to inputs "X" and "Y," respectively.

Now we discuss the implementation of a PIM cell using Akers structure in QCA. In Figure 12.7a, the QCA-based multiplexer structure, which performs switching in Akers array, is presented. Whether the flip-flop is located at the beginning or end of the PIM cell makes no difference. The operation of both switches must be regulated by two AND gates (Figure 12.7b). When the clock pulse "Z" becomes "1" (or ternary B), the corresponding AND gate operates as a closed switch, and the value of "Y" is propagated to the gate output. Moreover, the AND gate that is connected to the inverse of "Z" operates as an open switch, and if the clock pulse "Z" becomes "0" (or ternary A), the AND gate connected to it operates as an open switch, and the AND gate connected to the inverse of "Z" operates as a closed switch, then the value of "X" is propagated to the gate output (Figure 12.7c).

In the ternary design in Figure 12.7d, an additional state is considered for the switch, i.e., when the value of "Z" becomes C, both AND gates (switches) jointly determine the output state based on Eq. (7.4) and considering the values of X and Y. In this case, the off- or on-state of the switches cannot be clearly shown in the figure, but it can only be said that a third state occurs in which the output of the cell depends on both inputs X and Y, and based on the relationship between those inputs, the output is calculated. Therefore, the ternary cell has three modes of operation.

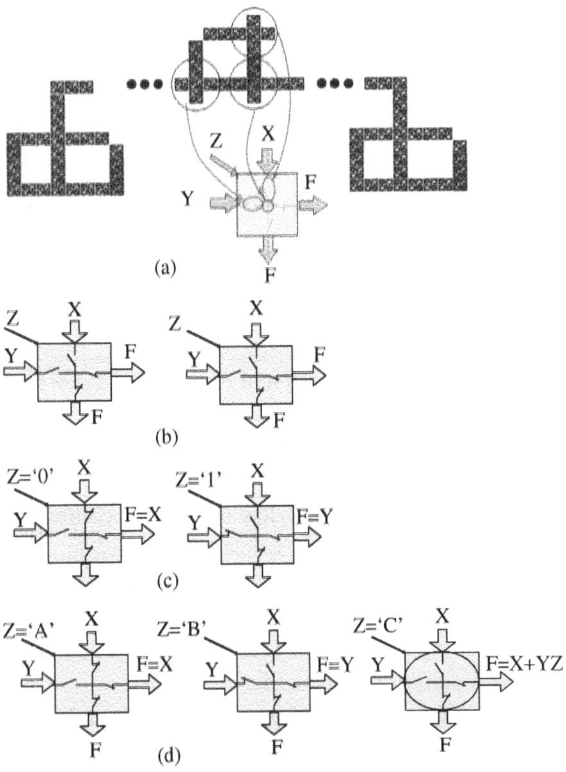

Figure 12.7 (a) Diagram and operating principle of switches in the QCA-based Akers basic cell, (b) operating principle of switches in the Akers cell, (c) operating principle of switches in the binary cell, and (d) operating principle of switches in the ternary cell. Source: Dehbozorgi et al. [183]/with permission of Springer Nature.

In this section, the proposed ternary circuits that are more similar in layout or design are compared in terms of different parameters such as area–delay and area–complexity cost functions, delay, complexity, and occupied area.

12.4 Fault Analysis

In the fabrication stage in QCA technology, faults may occur, as in other technologies. In QCA-based circuits, typical dimensions are in the nanometer range, so the possibility of fault occurrence increases. However, those faults that can be simulated with the available software packages and are in accordance with a realistic fabrication process can be analyzed.

As with fault analysis in binary structures, we focus on the faults that can be analyzed and simulated in both binary and ternary QCA structures. The fault tolerance of each circuit against the cell omission or extra-cell deposition defects is calculated according to Eq. (10.3).

12.4.1 Fault Tolerance for Cell Omission or Extra-Cell Deposition Defects in Proposed Models

All fault analysis criteria in ternary QCA are similar to those in binary QCA. First, the fault tolerance values of the proposed models against cell omission defect and then against extra-cell deposition defect in ternary structures are investigated. It is assumed that only one cell is added or omitted at a time during the manufacturing process. Subsequently, for all possible states applied to the input, it is checked whether the circuit output after fault occurrence is the same as the output of the fault-free circuit or not. For this purpose, the cells of each structure are numbered. Then, they are omitted one by one, and the effect of the omission or addition of each cell on the output waveform error can be evaluated.

In many studies on fault analysis in binary QCA, it has been mentioned that "by using a few more cells and some more area, the fault tolerance against manufacturing defects in conventional gates can be increased [177]." The main drawback of this scheme is the increase in complexity and the occupied area of the circuit. One of the main goals and innovations of this book is to present a different method, which has not been investigated so far, to increase fault tolerance against manufacturing defects. In this method, instead of adding more cells and increasing the occupied area, by transforming the circuit structure from binary QCA to ternary QCA, the fault tolerance against manufacturing defects increases, without increasing the occupied area.

In these structures, cell count may increase in some cases, but because of using ternary cells, which have smaller dimensions ($8 \times 8\ \text{nm}^2$ with a distance of 0.5 nm between two neighboring cells compared to the binary counterpart with an area of $18 \times 18\ \text{nm}^2$ and a distance of 2 nm between two neighboring cells), the issue of increasing the occupied area is resolved. As a result, with this method, the area–delay and (see Eq. (10.1)) area–complexity cost functions are also reduced (see Eq. (10.2)). This means that for a given occupied area, the cell count increases. In the following, the effect of fault in ternary structures is investigated.

In many studies on fault tolerance, some cells are added to parts of the structure during the design process to increase the fault tolerance of gates and circuits. However, in this book, a novel method is investigated so that instead of fundamental changes in the layout, number, and locations of the cells, the technology of QCA structure can be transformed from binary to ternary.

Accordingly, despite a possible slight increase in delay and complexity, the dimensions of the circuit are reduced, and fault tolerance against defects is increased. In ternary QCA, as mentioned, the cells are square-shaped with sides of 8 nm and a distance of 0.5 nm between the walls of a cell and an adjacent cell. Let us assume that three cells are placed next to each other on a ternary wire. If the middle cell is omitted due to a manufacturing defect, the distance between the walls of the first and third cells is 9 nm.

Now we perform the same calculations for three adjacent cells in a binary wire. Each binary cell is a square with sides of 18 nm and a distance of 2 nm between two neighboring cells. If the middle cell is omitted, the distance between the walls of the first and third cells is 22 nm.

Therefore, with a simple comparison, it can be concluded that the electrostatic interaction between two ternary cells, which are located at a distance of 9 nm from each other, is much stronger than that between two binary cells, which are located at a distance of more than twice, i.e., 22 nm, apart from each other. That is why the fault tolerance of the proposed ternary QCA structures is more than that of their binary counterparts.

Due to the smaller dimensions and closeness of the cells in the ternary structure, the electrostatic property of the preceding cell undergoes fewer changes even despite the omission or addition of the middle or surrounding cells, and its effect is propagated to the next cell with lower error probability.

12.4.1.1 Fault Analysis for First Model of PIM Cell (TPIM1)

Figure 12.8 shows the numbering of cells. This structure is designed based on the binary structure of BPIM1, and the memory is located on the input side.

By removing individual cells one by one, the fault occurrence is simulated. The simulation results are shown in Table 12.1. A total of 26 out of 50 cells result in correct output. By inserting these values

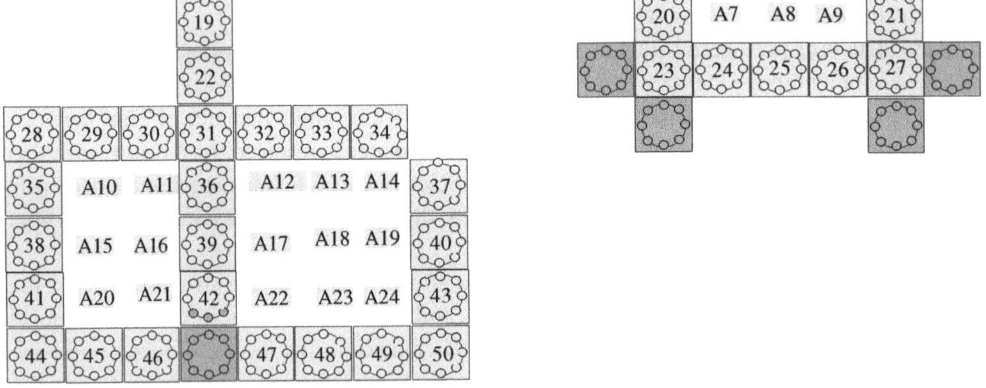

Figure 12.8 Numbering of cells in TPIM1. Source: Dehbozorgi et al. [183]/with permission of Springer Nature.

Table 12.1 Cell omission defect analysis for TPIM1.

Impact on output	Cell number	Impact on output	Cell number	Impact on output	Cell number	Impact on output	Cell number	Impact on output	Cell number
No	45	No	34	Yes	23	Yes	12	No	1
No	46	No	35	Yes	24	Yes	13	No	2
No	47	Yes	36	Yes	25	Yes	14	No	3
No	48	No	37	Yes	26	No	15	Yes	4
No	49	No	38	Yes	27	Yes	16	Yes	5
No	50	Yes	39	No	28	Yes	17	Yes	6
		No	40	No	29	No	18	Yes	7
		No	41	No	30	Yes	19	Yes	8
		Yes	42	No	31	Yes	20	Yes	9
		No	43	No	32	No	21	Yes	10
		No	44	No	33	Yes	22	Yes	11

Source: Dehbozorgi et al. [183]/with permission of Springer Nature.

Table 12.2 Extra-cell deposition defect analysis for TPIM1.

Impact on output	Cell number	Impact on output	Cell number	Impact on output	Cell number	Impact on output	Cell number	Impact on output	Cell number
No	21	No	16	No	11	No	6	No	1
No	22	No	17	No	12	No	7	No	2
No	23	No	18	No	13	Yes	8	Yes	3
No	24	No	19	No	14	No	9	No	4
		No	20	No	15	No	10	No	5

Source: Dehbozorgi et al. [183]/with permission of Springer Nature.

into Eq. (10.3), the fault tolerance against the cell omission defect is equal to 52%.

$$\text{(Cell omission) fault tolerance of TPIM1} = \frac{26}{50} \times 100 = 52\%$$

Moreover, cells A1–A24 are added in the empty positions, and the output results for different input states are shown in Table 12.2. Out of 24 added cells, 22 cells do not cause a change in the output waveform. Using Eq. (10.3), the fault tolerance against extra-cell deposition defect is equal to 91.6%.

$$\text{(Extra cell) fault tolerance of TPIM1} = \frac{22}{24} \times 100 = 91.6\%$$

The fault tolerance against extra-cell deposition and cell omission defects in the ternary structure of TPIM1 increases compared to BPIM1 because the manufacturing technology of binary QCA is different from that of ternary QCA. The dimensions of the cells and inter-cell distance are smaller in ternary QCA. This reduces the effects of electrostatic interaction between cells due to the addition or omission of the cells. As a result, the logic states of circuit cells propagate with less fault. Furthermore, the occupied area and complexity are also reduced in this structure. Therefore, ternary structures perform better than binary ones. Of course, in addition to all these advantages, it

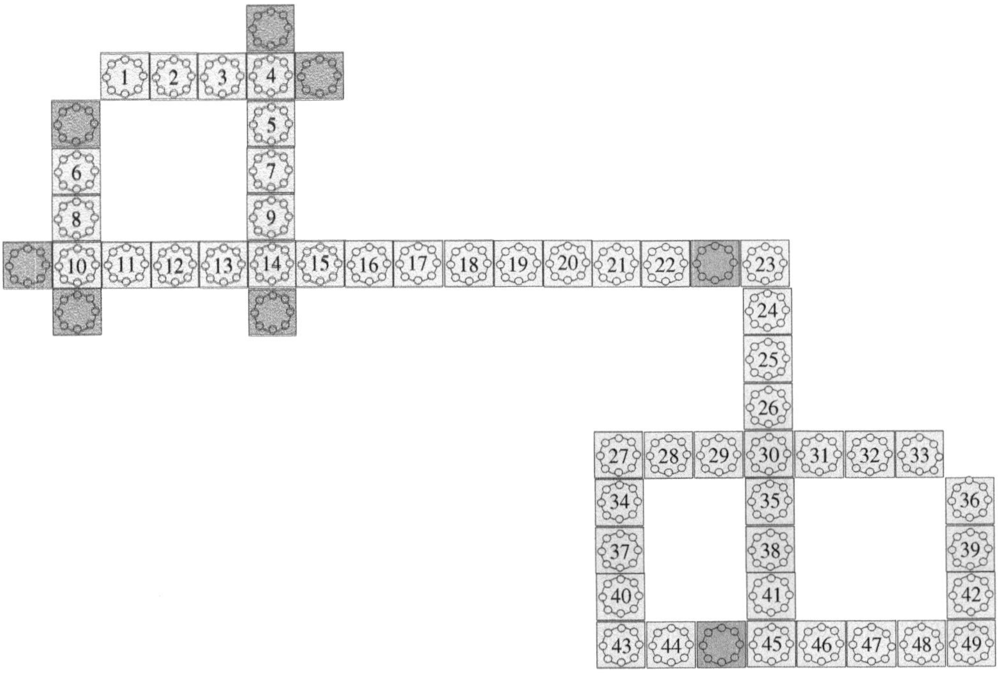

Figure 12.9 Numbering of cells in TPIM2. Source: Dehbozorgi et al. [183]/with permission of Springer Nature.

should be remembered that a ternary QCA structure can process and propagate higher information density than a binary QCA structure.

12.4.1.2 Fault Analysis for Second Model of PIM Cell (TPIM2)

Figure 12.9 shows the numbering of cells in TPIM2. This structure is designed in ternary QCA based on the binary structure of BPIM2.

To determine fault tolerance, first, cells are removed one by one, and the output waveform corresponding to the fault is compared with the correct output. The results of this comparison are shown in Table 12.3. The total number of evaluated cells is equal to 49. The omission of 25 cells does not affect the output waveform. The fault tolerance is 51.02%, obtained by inserting the aforementioned values into Eq. (10.3).

$$\text{(Cell omission) fault tolerance of TPIM2} = \frac{25}{49} \times 100 = 51.02\%$$

To evaluate the extra-cell deposition defect, cells A1–A24 are added to the empty positions of the circuit. Among 24 evaluated cells, 22 cells did not lead to faulty output waveform. By inserting these values into Eq. (10.3), the fault tolerance against extra-cell deposition defect becomes 91.66% (Table 12.4).

$$\text{(Extra cell) fault tolerance of TPIM2} = \frac{22}{24} \times 100 = 91.6\%$$

Compared to BPIM2, the ternary structure TPIM2 is more fault tolerant against extra-cell deposition defect and occupies more area. Moreover, the fault tolerance of TPIM2 is the same as that of TPIM1 due to the same number of cells in empty positions. The higher fault tolerance against

Table 12.3 Cell omission defect analysis for TPIM2.

Impact on output	Cell number	Impact on output	Cell number	Impact on output	Cell number	Impact on output	Cell number	Impact on output	Cell number
Yes	45	No	34	Yes	23	Yes	12	No	1
No	46	Yes	35	Yes	24	Yes	13	No	2
No	47	No	36	Yes	25	Yes	14	No	3
No	48	No	37	Yes	26	Yes	15	Yes	4
No	49	Yes	38	No	27	Yes	16	No	5
		No	39	No	28	Yes	17	Yes	6
		No	40	No	29	Yes	18	No	7
		Yes	41	No	30	Yes	19	Yes	8
		No	42	No	31	Yes	20	No	9
		No	43	No	32	Yes	21	Yes	10
		No	44	No	33	Yes	22	Yes	11

Source: Dehbozorgi et al. [183]/with permission of Springer Nature.

Table 12.4 Extra-cell deposition defect analysis for TPIM2.

Impact on output	Cell number	Impact on output	Cell number	Impact on output	Cell number	Impact on output	Cell number	Impact on output	Cell number
No	21	No	16	No	11	No	6	No	1
No	22	No	17	No	12	No	7	No	2
No	23	No	18	No	13	Yes	8	Yes	3
No	24	No	19	No	14	No	9	No	4
		No	20	No	15	No	10	No	5

Source: Dehbozorgi et al. [183]/with permission of Springer Nature.

extra-cell deposition defect of TPIM2 compared to BPIM2 can be attributed to the smaller dimensions of the ternary cells. Therefore, in case of fault occurrence, the electrostatic effect of a cell state is more likely to be transferred to the next cell because the cells are closer to each other (so there is less disturbance), and the fault probability is reduced.

Although the distance from the output to the memory input in TPIM2 is longer than in BPIM2 (13 cells in the ternary structure and 6 cells in the binary structure), the fault tolerance against the cell omission defect in TPIM2 is only slightly lower than that in BPIM2. However, the slightly lower fault tolerance can be ignored because the occupied area of the ternary structure is smaller.

12.4.1.3 Fault Analysis for Third Model of PIM Cell (TPIM3)

Figure 12.10 shows the numbering of TPIM3 cells. This structure is designed for comparison with BPIM3 and BPIM4 binary structures.

Table 12.5 shows the results of removing individual cells one by one and the obtained correct and incorrect outputs. A total of 19 cells out of 43 evaluated cells result in correct output. Using Eq. (8.3), the fault tolerance against cell omission defect is obtained as 44.186%. In the empty positions, the cells A1–A24 are placed one by one, and the effect of extra-cell deposition on output is evaluated.

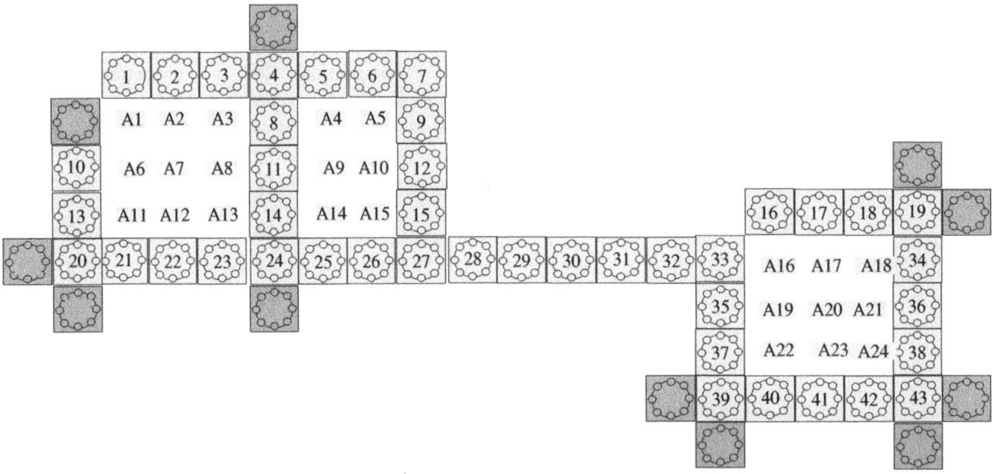

Figure 12.10 Numbering of cells in TPIM3. Source: Dehbozorgi et al. [171]/with permission of Springer Nature.

Table 12.5 Cell omission defect analysis for TPIM3.

Impact on output	Cell number	Impact on output	Cell number	Impact on output	Cell number	Impact on output	Cell number
Yes	34	Yes	23	No	12	Yes	1
Yes	35	Yes	24	Yes	13	Yes	2
Yes	36	No	25	No	14	No	3
Yes	37	No	26\	No	15	Yes	4
No	38	No	27	No	16	No	5
Yes	39	Yes	28	Yes	17	No	6
Yes	40	No	29	No	18	No	7
Yes	41	No	30	Yes	19	No	8
Yes	42	No	31	Yes	20	No	9
Yes	43	No	32	Yes	21	Yes	10
		Yes	33	Yes	22	Yes	11

Source: Dehbozorgi et al. [171]/with permission of Springer Nature.

The results of this evaluation are shown in Table 12.6. By adding 10 cells out of 24 evaluated cells, the output is unchanged and similar to the correct output. From Eq. (10.3), the fault tolerance against extra-cell deposition defect is calculated to be 41.66%.

As mentioned in the previous sections, a solution to improve fault tolerance against defects is to increase cell count. However, here we attempted to achieve more optimal structures by transforming binary QCA structures into ternary ones. However, due to the use of standard gates and wires that have an odd number of ternary QCA cells, the cell count in some circuits, including TPIM3, is greater than in similar structures. However, since ternary cells have smaller dimensions (8 nm) than binary ones (18 nm), the larger cell count can be ignored. This is because the fault tolerance against extra-cell deposition defect in TPIM3 is higher than that of BPIM3 and BPIM4, and the

Table 12.6 Extra-cell deposition defect analysis for TPIM3.

Impact on output	Cell number	Impact on output	Cell number	Impact on output	Cell number	Impact on output	Cell number	Impact on output	Cell number
No	21	Yes	16	No	11	No	6	Yes	1
No	22	Yes	17	Yes	12	Yes	7	No	2
Yes	23	No	18	Yes	13	Yes	8	No	3
No	24	Yes	19	Yes	14	Yes	9	No	4
		Yes	20	Yes	15	Yes	10	No	5

Source: Dehbozorgi et al. [171]/with permission of Springer Nature.

occupied area is reduced. The fault tolerance in this ternary structure is slightly reduced due to the reasons previously mentioned and the increase in the number of empty positions. This, of course, can be neglected considering the other advantages of ternary structures and the higher density of information carried in these structures compared to binary ones.

In ternary structures, TPIM1 and TPIM2, the fault tolerance against extra-cell deposition defect (91.66%) is very close to 100, and the fault tolerance against cell omission defect is higher than 50%. These results indicate the performance improvement of the ternary structures compared to the binary ones. However, in the remainder of this chapter, we compare the cost functions and other parameters of ternary structures and binary structures and prove the better performance of ternary structures.

In this section, the effect of defects on memory and PIM structures in ternary QCA was investigated, and some circuits and gates that could be designed using PIM capability in ternary QCA were then evaluated. Accordingly, the optimal design of the proposed basic cells and the feasibility of the realized structures were confirmed.

13

Design Details of Ternary Boolean Operators Using Basic PIM Cells and Analysis of Basic-Extended Hypothesis

13.1 Introduction

According to the basic-extended hypothesis introduced in Chapter 9, to evaluate smaller and basic structures, it is better to use them in the design of extended circuits and gates. This is because if extended and more optimal structures can be designed using the basic structures, then the proposed basic model is suitable and practical.

To verify this hypothesis in ternary quantum dot cellular automata (QCA), the ternary operators AND, OR, and XOR are examined using the proposed cells TPIM1 and TPIM2.

Since processing in memory (PIM) structures have attracted the attention of many researchers, in the following, the designs of gates and circuits with PIM capability are evaluated based on ternary QCA technology.

13.2 Designing Logic Gates With Ternary QCA-Based PIM Capability

13.2.1 AND Gate Design Using PIM Basic Cells in Ternary QCA

In this subsection, ternary QCA-based AND gates TAND1 and TAND2 are introduced using the proposed ternary basic cell models TPIM1 and TPIM2 (Figure 13.1). This structure includes two basic PIM cells horizontally connected in an Akers array. For the correct operation of the array, inputs X and Y carry fixed values of "A" and "B," respectively, and the main input Z controls cell operation [166] (Figure 13.1). Figure 13.2 shows the simulation results of the TAND1 and TAND2 gates. According to Eq. (13.1), if both inputs applied to a gate are equal to "B," the output signal is "B"; otherwise, it is "A."

$$\text{In ternary: } X \text{ AND } Y = X \cdot Y = \text{MAJ}(\text{"A"}, X, Y), \text{ "A" is a constant state} \tag{13.1}$$

The structure of TAND1 includes 135 cells with an occupied area of 0.04797 μm^2 and a delay of 2.25 clock cycles, while TAND2 consists of 109 cells with an occupied area of 0.026924 μm^2 and a delay of 1.75 clock cycles. By inserting the design parameters into Eqs. (10.1) and (10.2), the values of the cost functions of the proposed structures are calculated as follows:

TAND1:

$$\text{Cost}_{\text{A-T}} = A \times T = 0.04797 \times 2.25 = 0.1079325$$

$$\text{Cost}_{\text{A/C}} = A/C = 0.04797/135 = 0.00186 = 0.3553 \times 10^{-3}$$

Computing with Multi-Value Logic in Quantum Dot Cellular Automata, First Edition.
Reza Sabbaghi-Nadooshan, Reza Akbari-Hasanjani, Leila Dehbozorgi, Majid Haghparast, and Hamid Reza Akbari-Hasanjani.

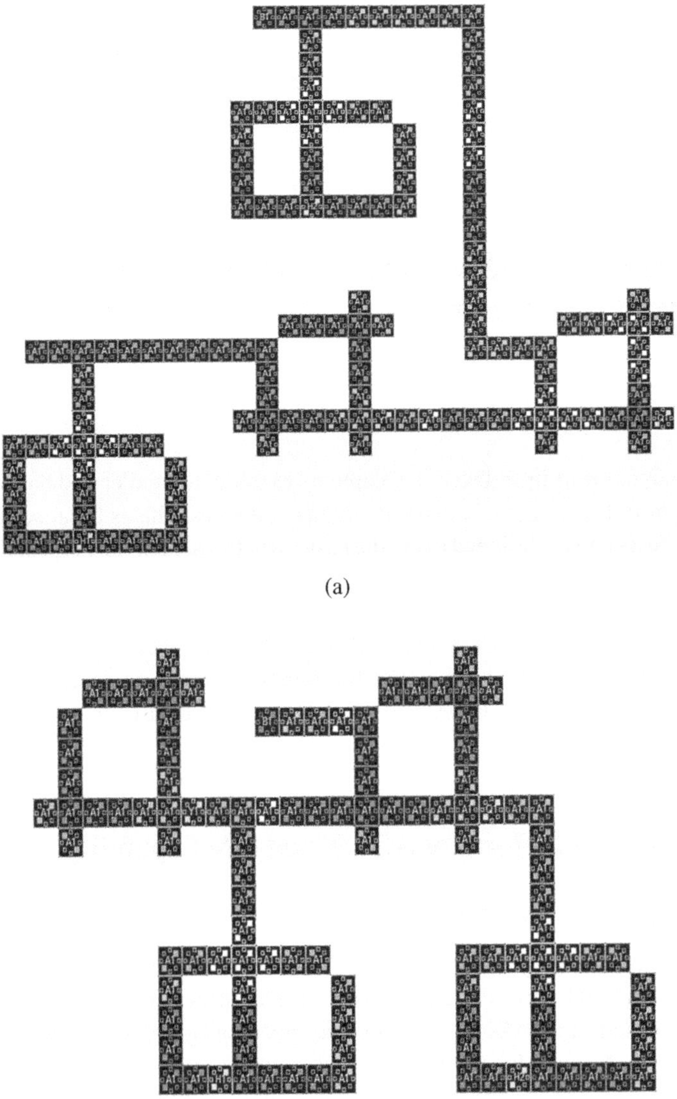

(a)

(b)

Figure 13.1 Two-input AND gate using PIM structure in ternary QCA: (a) TAND1 model and (b) TAND2 model. Source: Dehbozorgi et al. [183]/with permission of Springer Nature.

TAND2:

$$\text{Cost}_{\text{A-T}} = A \times T = 0.026924 \times 1.75 = 0.047117$$

$$\text{Cost}_{\text{A/C}} = A/C = 0.026924/109 = 0.247009 \times 10^{-3}$$

The obtained values of cost functions for both ternary structures are much more optimal than all extended binary structures, such as BAND1 and BAND2. To evaluate the basic-extended hypothesis, we start with BPIM1 and BPIM2, presented in Section 10.2. By comparing BPIM1 and BPIM2 with TPIM1 and TPIM2, it can be observed that TPIM1 and TPIM2 are more optimal than BPIM1 and BPIM2 in terms of complexity, occupied area, area–delay cost function, and area–complexity cost function. This time again, the more optimal occupied area of ternary structures is related to

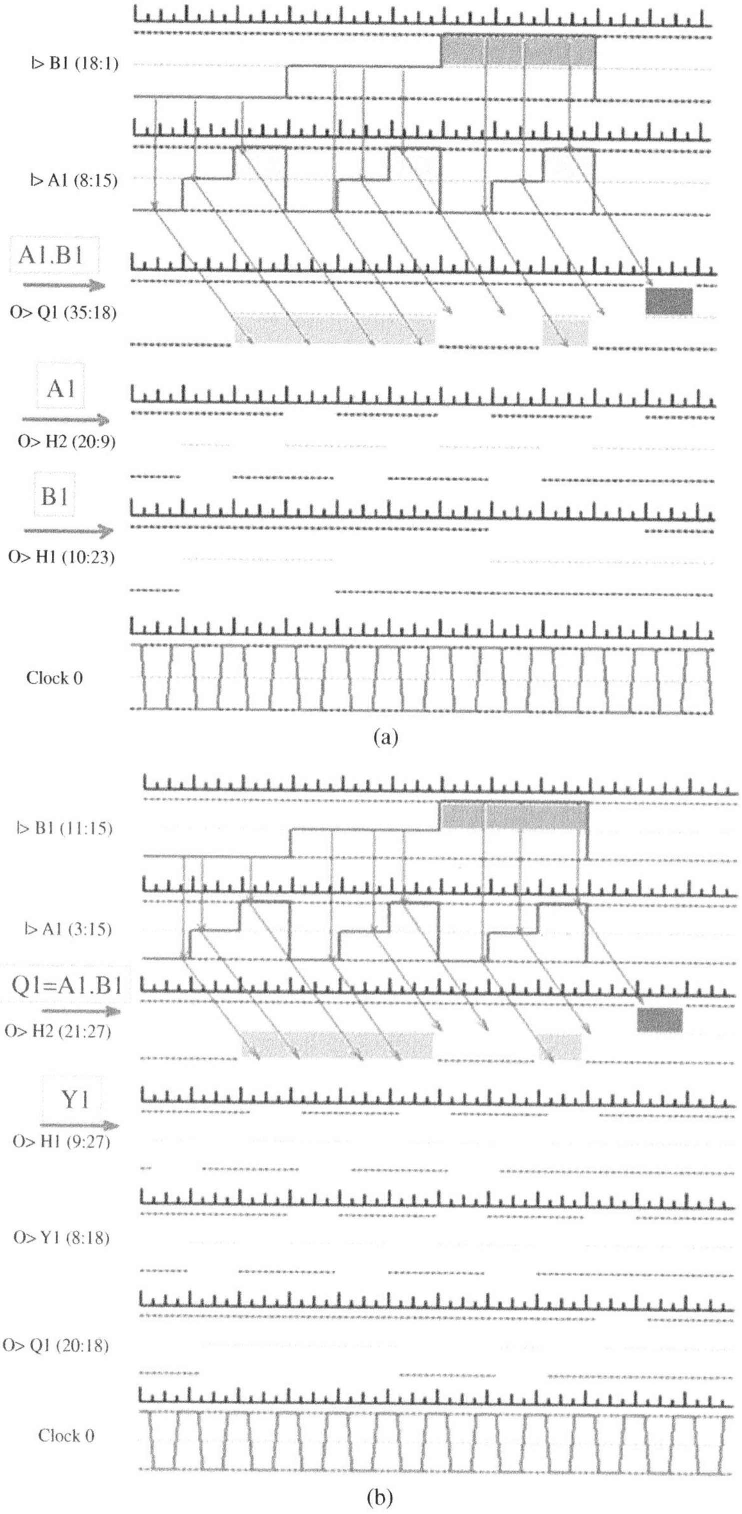

Figure 13.2 AND gate simulation results using ternary QCA-based PIM structure for (a) TAND1 and (b) TAND2. Source: Dehbozorgi et al. [183]/with permission of Springer Nature.

the smaller dimensions of ternary cells compared to binary ones. As mentioned, the basic-extended hypothesis is confirmed in this case too, because more optimal basic structures lead to more optimal extended circuits.

13.2.2 OR Gate Design Using Ternary QCA-Based Basic PIM Cells

The structures TOR1 and TOR2 are implemented in PIM-enabled ternary QCA by using the proposed basic cells TPIM1 and TPIM2 in the Akers array [104] (Figure 13.3). This structure includes two basic PIM cells that are vertically connected. Similar to the previous case, inputs X and Y are fixed and equal to logic values “A” and “B,” while the main input is applied to Z. Simulation results are shown in Figure 13.4. According to Eq. (13.2), if one or both inputs are equal to the logic value “B,” the output is equal to “B”; otherwise, it is “A.”

$$\text{In ternary: } X \text{ OR } Y = X + Y = \text{MAJ}(\text{“B”}, X, Y), \text{ “B” is a fixed state} \tag{13.2}$$

The first ternary model, TOR1, includes 122 cells with an occupied area of 20.02808 μm and a delay of 1.75 clock cycles. The second ternary model, TOR2, consists of 126 cells with an occupied area of 0.04299 μm^2 and a delay of 1.75 clock cycles. By inserting the aforementioned values into Eqs. (10.1) and (10.2), the values of the cost functions for each of the proposed structures are calculated as follows:

TOR1:

$$\text{Cost}_{\text{A-T}} = A \times T = 0.02808 \times 1.75 = 0.04914$$

$$\text{Cost}_{\text{A/C}} = A/C = 0.02808/122 = 0.23016 \times 10^{-3}$$

TOR2:

$$\text{Cost}_{\text{A-T}} = A \times T = 0.04299 \times 1.75 = 0.0752325$$

$$\text{Cost}_{\text{A/C}} = A/C = 0.04299/126 = 0.00241 = 0.34119 \times 10^{-3}$$

The design parameters and cost functions of TOR1 and TOR2 are more optimal than those of BOR1 and BOR2. If the basic structures of TOR1 and TOR2 are considered, the same conclusion can be drawn for TPIM1 and TPIM2 in comparison with BPIM1 and BPIM2. This, from the quantum point of view, can be attributed to the smaller dimensions of the ternary QCA cell compared to the binary one, as discussed in more detail earlier. Therefore, according to calculations, the basic-extended hypothesis can justify the optimal results of TOR1 and TOR2.

13.2.3 XOR Gate Design Using Ternary QCA-Based PIM Model

The ternary structure of the XOR gate is shown in Figure 13.5. In this structure, the number of input states is larger, and instead of two values, three values are applied to the inputs. Therefore, the density of processing and storage of information increases. Similar to its binary counterpart in the previous section, the Akers array is used to design this structure. The values of X inputs in the first row are fixed and equal to “A.” In the second row, the values of X inputs are determined from the output of the previous row. The Y input in the first column on the left is equal to “B,” but its value in the second column is determined from the output of the first (previous) column.

When one of the inputs is the state “C,” the output is “C.” If both inputs are the same, “A” or “B,” the output is “A.” Moreover, only if both inputs are different, i.e., equal to “A” and “B,” the output is “B” (Figure 13.5b). Two outputs H1 and H2 are also considered for storing the inputs, namely A1

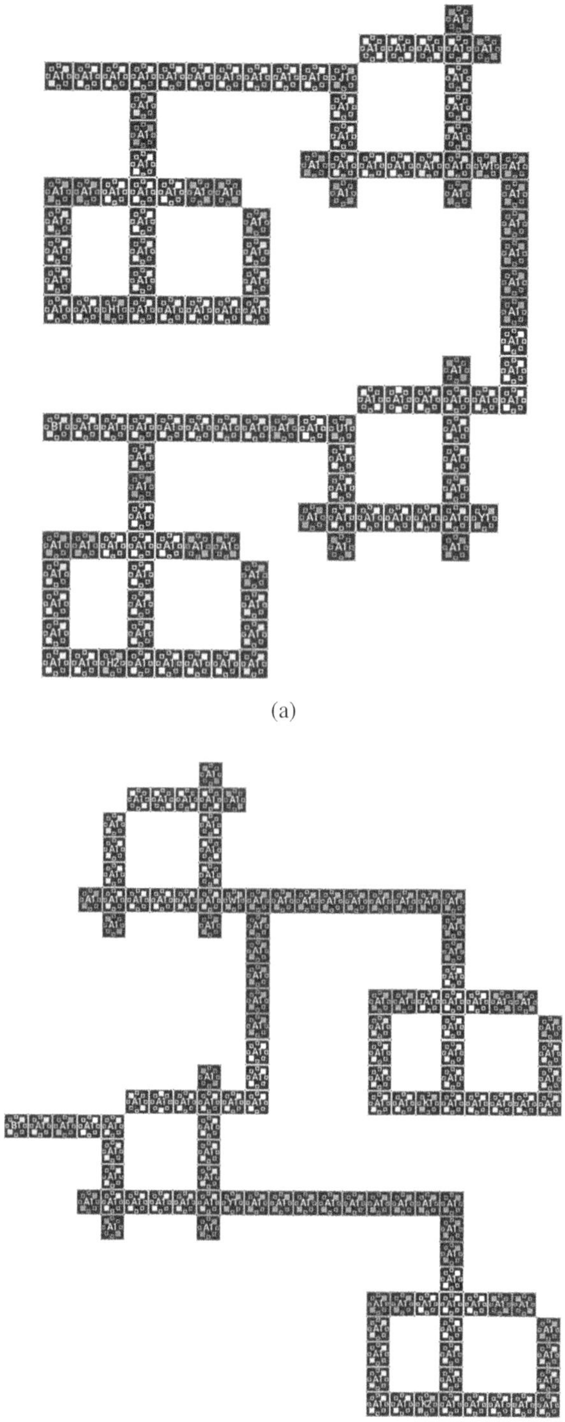

(a)

(b)

Figure 13.3 Two-input OR gate using PIM structure in ternary QCA: (a) TOR1 model and (b) TOR2 model. Source: Dehbozorgi et al. [183]/with permission of Springer Nature.

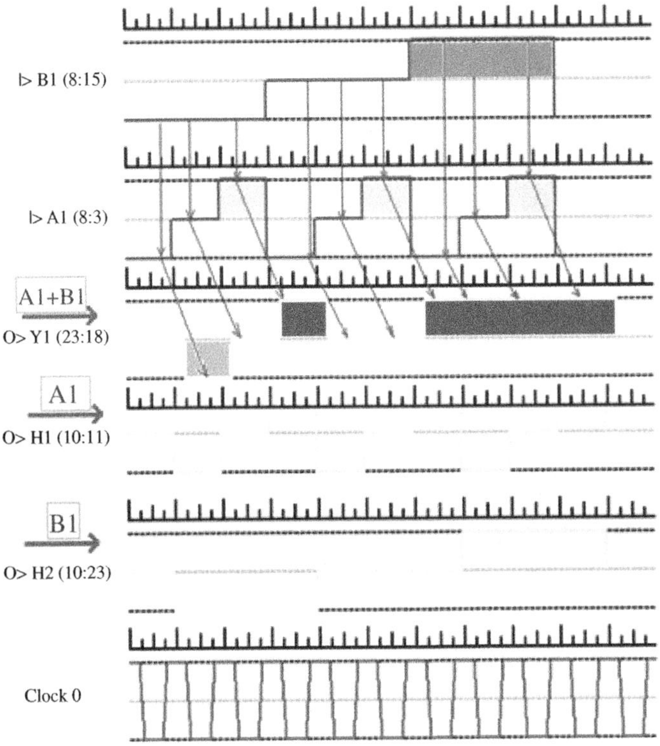

(a)

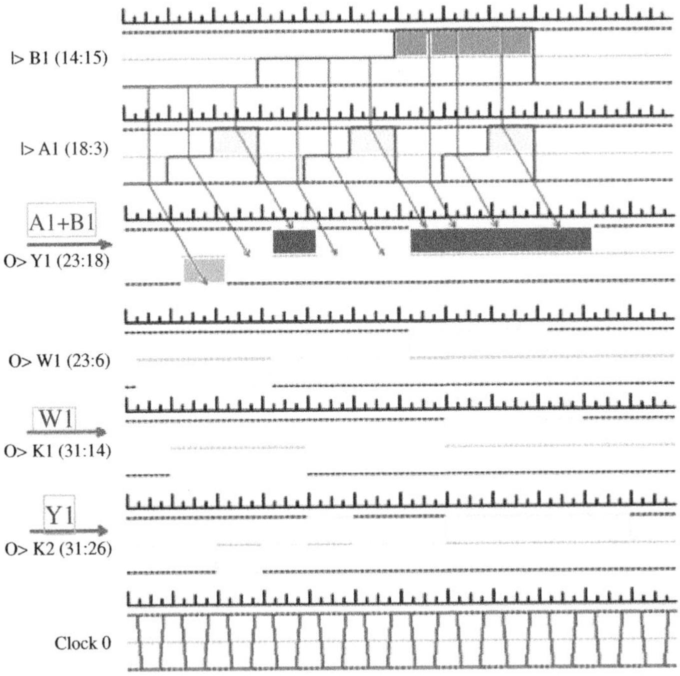

(b)

Figure 13.4 OR gate simulation results using ternary QCA-based PIM structure for (a) TOR1 and (b) TOR2. Source: Dehbozorgi et al. [183]/with permission of Springer Nature.

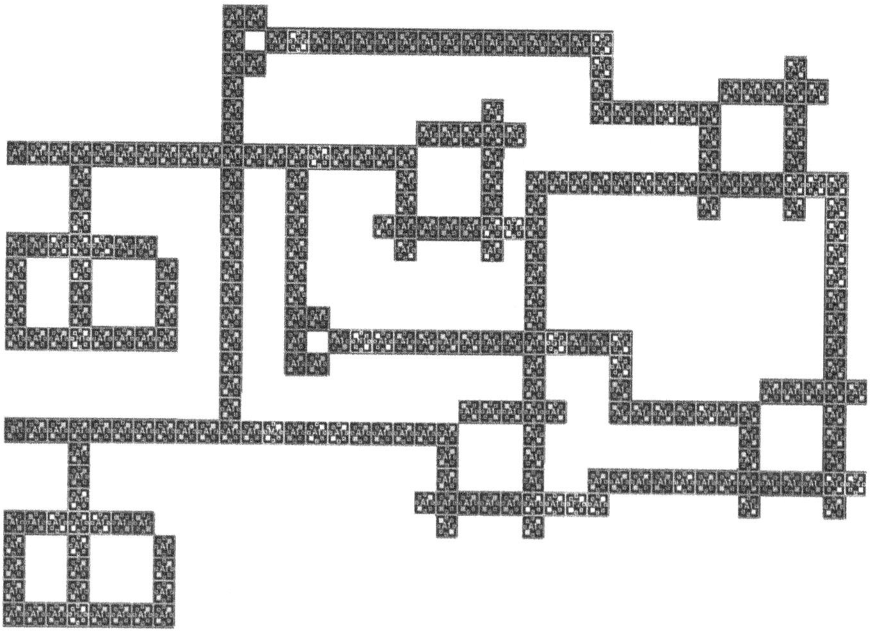

(a)

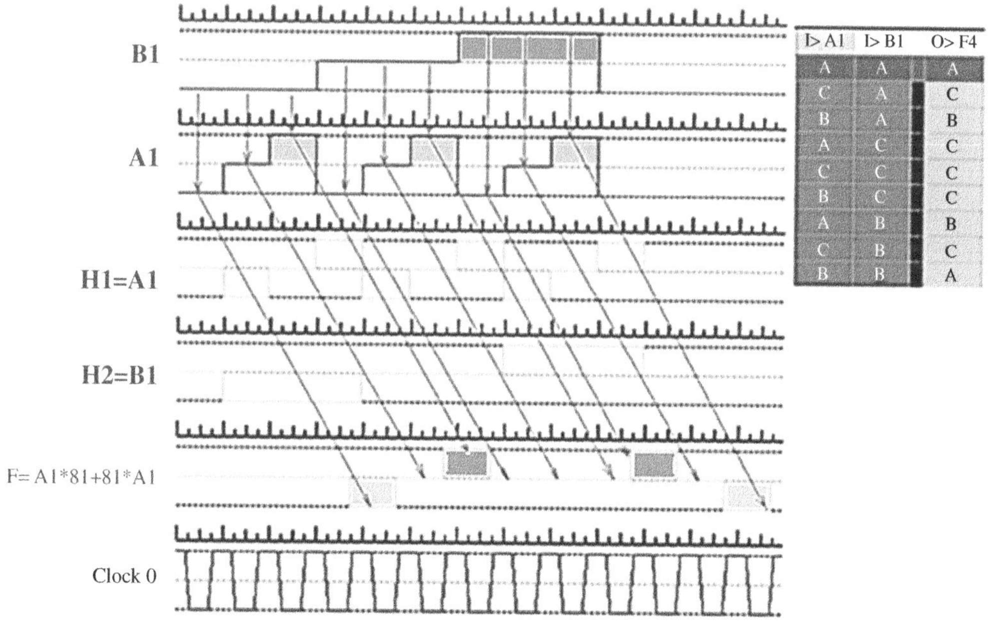

(b)

Figure 13.5 XOR gate using ternary QCA-based PIM: (a) TXOR model and (b) truth table and simulation results. Source: Dehbozorgi et al. [183]/with permission of Springer Nature.

and B1, as shown in Figure 13.5. In fact, the design of the XOR gate that uses ternary PIM structure leads to improved occupied area and more stored/processed pieces of information compared to its binary counterpart. The number of examined ternary cells in the XOR circuit is equal to 281 with an occupied area of 0.07774 μm^2 and a delay of 4 clock cycles.

By inserting the aforementioned values into Eqs. (10.1) and (10.2), the values of the cost functions of the proposed structures are calculated as follows:

TXOR:

$$\text{Cost}_{\text{A-T}} = A \times T = 0.0777 \times 4 = 0.3108$$

$$\text{Cost}_{\text{A/C}} = A/C = 0.0777/281 = 0.2765 \times 10^{-3}$$

The parameter values obtained for TXOR are more optimal than all proposed extended binary circuits, including BXOR. According to the basic-extended hypothesis, in this situation, the basic structure should be more optimal, which is exactly the case. Therefore, TPIM1, which is used in TXOR, has more balanced design parameters than BPIM1, which is used in BXOR. This can be attributed to the dimensions of the ternary QCA cell, which are smaller than the binary one.

Until this part of the book, we designed AND, OR, and XOR gates using the TPIM1 and TPIM2 models. Then by calculating cost functions and design parameters, we analyzed the proposed ternary structures and compared them with binary ones. The results indicated the better performance of the ternary structures. Moreover, in all ternary structures, the information-carrying density is higher than in binary structures, and the ternary circuits can collect more information from a system than binary circuits. Accordingly, the performance of ternary structures is closer to that of fuzzy and real systems. Ternary structures are more fault tolerant against manufacturing defects. Therefore, in general, the structures designed in ternary QCA have more optimal performance.

13.3 Evaluation, Analysis, and Comparison of Results

First, some models of basic random-access memory (RAM) and PIM cells in ternary QCA were evaluated. Then, the values of fault tolerance against extra-cell deposition and cell omission defects were calculated for the presented structures. Additionally, circuits and gates with ternary QCA-based PIM capability were proposed. For a more detailed comparison, the RAM and PIM design parameters are summarized in Table 13.1. In this table, the overall cost function is also presented as the product of "area/complexity" cost function and "area×delay" cost function.

As given in Table 13.1, TPIM1 performs better than other proposed structures. Moreover, the cost functions and design parameters of all ternary structures have lower values. The "occupied area" column in Table 13.1 shows that the proposed ternary structures occupy the smallest area. If the quantum aspects of the ternary QCA cells are considered, the reason for these results can be ascribed to the smaller dimensions of the ternary QCA cell (8×8 nm^2) compared to the binary one (18×18 nm^2).

The electrostatic interaction between two cells in the ternary structure is much stronger than in the binary structure. Accordingly, ternary structures show more fault tolerance against manufacturing defects, including extra-cell deposition and cell omission. Figure 13.6 shows that the minimum fault tolerance against cell omission defects in ternary and binary structures is equal to 42 and 23, respectively. Likewise, the minimum fault tolerance against extra-cell deposition defects in ternary and binary structures is equal to 41 and 40, respectively. By comparing the black and white 3D bar charts, the superiority of the ternary structure over the binary one can be concluded.

Table 13.1 Parameters of proposed binary and ternary RAM and PIM cells.

SRAM structure	Technology (nm)	Ternary/ Binary (T/B)	Coplanar wire crossing	Complexity (cell number)	Occupied area (μm^2)	Latency (clock cycles) (delay-s)	$Cost_{A-T}$ = (Area × Delay)	$Cost_{A/C}$ = Area/ Complexity ($\times 10^{-3}$)	Total cost = ($Cost_{A-T}$) × ($Cost_{A/C}$)	Fault tolerance (cell omission)%	Fault tolerance (extra cell)%
Proposed BPIM1	18	Binary	No	59	0.11	10–12 × 0.5	0.055	1.864	0.1025	23.52	40
Proposed BPIM2	18	Binary	No	46	0.07	10–12 × 0.75	0.0525	1.521	0.0798	57.8	80
Proposed BPIM3	18	Binary	No	40	0.05	10–12 × 1	0.05	1.25	0.0625	25.806	62.5
Proposed BPIM4	18	Binary	No	33	0.04	10–12 × 1.5	0.06	1.21	0.0726	32	100
Proposed TPIM1	8	Ternary	No	58	0.013	10–12 × 0.5	0.0066	0.23	0.0015	52	91.6
Proposed TPIM2	8	Ternary	No	57	0.02	10–12 × 0.75	0.015	0.35	0.0052	51.0204	91.66
Proposed TPIM3	8	Ternary	No	54	0.01	10–12 × 3.25	0.045	0.259	0.0116	44.186	41.66

Source: Dehbozorgi et al. [171, 183].

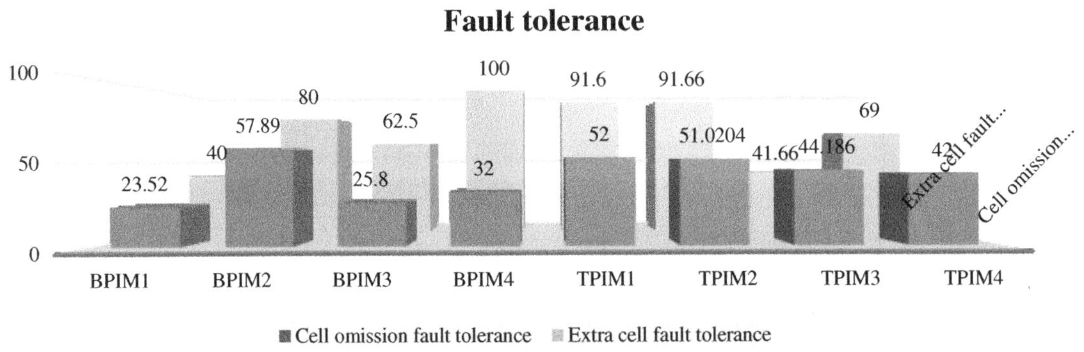

Figure 13.6 Comparison of fault tolerance against extra-cell deposition and cell omission defects for PIM structures in binary and ternary QCA. Source: Adapted from Dehbozorgi et al. [171, 183].

Although among the proposed binary structures, the fault tolerance against extra-cell deposition defect in BPIM4 is equal to 100, this structure does not show suitable fault tolerance against cell omission defects. Therefore, a structure that has a relatively good fault tolerance against both defects is more desirable, as proved for ternary structures. Furthermore, the overall cost function is more optimal in ternary structures than in binary ones.

Table 13.2 presents the design parameters of gates and extended circuits designed with basic binary PIM and memory cells.

Table 13.2 Design parameters of binary and ternary gates.

Model name	Technology (nm)	Ternary/ Binary (T/B)	Number of cells (complexity)	Occupied area (μm^2)	Latency to main output (clock cycle)	$Cost_{A-T}$ = (Area × Delay)	$Cost_{A/C}$ = Area/ Complexity ($\times 10^{-3}$)	Total cost = ($Cost_{A-T}$) × ($Cost_{A/C}$)
Proposed BAND1	18	Binary	145	0.27	1.5	0.0405	1.86	0.7533
Proposed BOR1	18	Binary	128	0.21	1.5	0.315	1.64	0.5166
Proposed BAND2	18	Binary	108	0.197	1.5	0.285	1.759	0.5013
Proposed BOR2	18	Binary	120	0.29	1.5	0.435	2.416	1.0509
Proposed BNAND1	18	Binary	124	0.19	2.25	0.4275	1.53	0.6540
Proposed BNOR1	18	Binary	124	0.19	2.25	0.4275	1.53	0.6540
Proposed BNAND2	18	Binary	99	0.17	2.75	0.4675	1.717	0.8026
Proposed BNOR2	18	Binary	99	0.17	2.75	0.4675	1.717	0.8026
Proposed TAND1	8	Ternary	135	0.04797	2.25	0.107	0.355	0.0379
Proposed TOR1	8	Ternary	122	0.02808	1.75	0.049	0.2301	0.0112
Proposed TAND2	8	Ternary	109	0.026924	1.75	0.0471	0.247	0.0116
Proposed TOR2	8	Ternary	126	0.042997	1.75	0.0752	0.341	0.0256

Binary QCA circuits are manufactured in 18 nm technology. Manufacturing technology has an important role in power consumption. In both binary and ternary QCA structures, gates with more optimal basic PIM cells in their structures have more optimized parameters.
Source: Adapted from Dehbozorgi et al. [171, 183].

Table 13.3 Comparison of binary and ternary XOR structures.

Circuit	Ternary/ Binary (T/B)	Technology (nm)	Latency (clock cycles)	Occupied area (μm^2)	Complexity (cell number)	$Cost_{A-T}$ = (Area × Delay)	$Cost_{A/C}$ = Area/ Complexity (×10^{-3})	Total cost = ($Cost_{A-T}$) × ($Cost_{A/C}$)
Proposed BXOR1	Binary	18	2.5	0.35	266	0.875	1.315	1.1506
Proposed TXOR1	Ternary	8	4	0.077	281	0.3108	0.2765	0.0859

Source: Adapted from Dehbozorgi et al. [171, 183].

Although cell count is not significantly reduced in ternary structures, the cost function values of these structures are smaller than all structures because they occupy less area.

Table 13.3 compares the extended circuits related to ternary and binary XOR structures. As given, in the ternary structure, the value of the cost function is less than 0.5. In the implementation of ternary XOR, due to the smaller dimensions of the ternary QCA cell, the occupied area of the designed XOR is reduced.

14

Conclusions and Suggestions

14.1 Introduction

In this chapter, the most significant achievements of this research are discussed, and some of its results are briefly mentioned. Finally, suggestions are made for researchers in this field to complete the activities carried out in this book and cases for which specific designs have yet to be presented.

14.2 Summary and Conclusion

In this book, for the first time, a new and comprehensive model is presented using quantum computations to design multi-value quantum dot cellular automata (QCA) for n-valued logic circuits based on this technology. Then, the energy calculations of these circuits were presented. Moreover, the quantum and electrostatic energy calculations for ternary, quaternary, and quinary cells were fully described. For cell structures with $n \geq 3$, the cell structure is designed based on the cells with smaller values of n (i.e., $n = 1$ or 2). The errors in the AND and OR gates with ternary cells are corrected in the proposed design.

To realize n-valued cells, the energies that affect ternary, quaternary, and quinary cells were calculated and studied in detail. This study showed that except for the external electrostatic energy that results from the effect of two adjacent cells, the amounts of other energies do not change as the polarization of the cells changes. Therefore, external electrostatic energy plays a vital role in the proposed models. Moreover, to confirm the effect of external electrostatic energy in the proposed cells, the interaction of two adjacent binary, ternary, quaternary, and quinary cells was individually evaluated using quantum computations in a nonlinear graph. The probability of polarization transfer from one cell to another was presented.

The basic logic gates were presented according to the QCA standard, using the proposed ternary, quaternary, and quinary cells. MATLAB was used to calculate the external electrostatic energy for the correct operation of the proposed gates. After calculating the external electrostatic energies, the total desired energy was obtained. We presented a model specific to the ternary, quaternary, and quinary logic. After presenting the multi-valued cells and using these cells for basic gates (ternary, quaternary, and quinary), the power consumption in each polarization state was calculated.

To avoid instability of the gates that are based on quaternary, quinary, or higher level cells, the cells are designed in two layers and according to the QCA standard. We provided drives for communication and the integration of input functions. In designing the drive for an input, a binary converter was provided for the ternary structure.

Computing with Multi-Value Logic in Quantum Dot Cellular Automata, First Edition.
Reza Sabbaghi-Nadooshan, Reza Akbari-Hasanjani, Leila Dehbozorgi, Majid Haghparast, and Hamid Reza Akbari-Hasanjani.

Additionally, for the model introduced in this book, TQCASim simulation software was modified based on the required calculations. A quaternary cell simulator, QQCAsim, was also presented. These simulators were used to simulate the gates, while gate simulations were also performed in MATLAB, and the simulation results in both cases were the same.

It should be noted that the models proposed in this book are comprehensive models that can be used to design *n*-valued circuits because the simulated gates have quite correct performance. Furthermore, a simulation tool has yet to be introduced to design ternary and quaternary circuits. These simulators can simulate various ternary and quaternary QCA circuits that have already been introduced in the literature.

Additionally, processing in memory (PIM) models in binary and ternary QCA were presented for the first time. These models were implemented by making changes in the basic structures of circuits, using more optimal binary basic gates, employing fewer gates and QCA cells, and relocating the memory part in PIM structures.

Based on calculations and simulations, and by omitting cells one by one in the first step or adding cells to the empty positions in the second step, the effects of cell omission and extra-cell deposition defects on memory and PIM structures in binary and ternary QCA were investigated for the first time. The fault tolerance against these defects was carefully calculated for each of the structures. For a more detailed comparison and analysis, both area–delay and area–complexity cost functions were calculated for all structures. In this book, the reason for higher fault tolerance of ternary structures compared to binary structures, in the case of manufacturing defects, is investigated. For this purpose, the exact external electrostatic interactions between two cells in binary and ternary QCA were calculated separately. This investigation indicates that the external electrostatic energy, which results from the effects of two adjacent cells on each other, is greater in ternary QCA than in binary QCA. Therefore, if a cell is added or omitted due to manufacturing faults, the stronger electrostatic energy in ternary QCA structures can transfer the effect of the previous cell to the next cell with less error probability.

Many studies on manufacturing defects have suggested that the deposition of more cells in the structure of basic gates can increase the fault tolerance against defects in these structures. This, however, has many disadvantages, such as the increase in complexity, delay, cost functions, and occupied area. Another innovation in this book is a new method proposed to reduce faults. In this method, the structure of the circuits is transformed from binary QCA into ternary QCA, and the fault tolerance against the manufacturing defects is improved without increasing the occupied area and cost functions.

It should be noted that the models proposed in this book are comprehensive and can be used to design extended circuits. This is because the simulated PIM cells have a completely correct and optimal response according to the truth tables and are compared with the results of similar designs in the literature.

To evaluate the performance of the proposed basic PIM cells and prove their applicability, different gates were designed using these cells. All structures had more desirable design parameters (delay, complexity, occupied area, and cost functions) than the designs in the literature. All the proposed ternary gates had a more optimal structure than the binary gates. Additionally, considering the more optimal design parameters of the PIM structure, it can even be suggested as a suitable alternative to random-access memory (RAM). Finally, it should be pointed out that the proposed ternary QCA-based structures are superior to binary ones.

14.3 Suggestions

- *Providing a Power Consumption Simulator*
  In this book, the power consumption of the proposed cells was calculated, but the circuit power loss needs to be calculated. One of the advantages of QCA technology is its very low power loss compared to complementary metal–oxide–semiconductor (CMOS) technology. In this book, the method of estimating power consumption in QCA technology is based on ternary, quaternary, and quinary values obtained by MATLAB using quantum mechanical analysis.
  QCAPro software was used to calculate the power loss in binary QCA circuits. It is suggested that to calculate the power loss of ternary, quaternary, and quinary QCA designs presented in this book, a simulation tool similar to QCAPro be designed.
- *Time-Dependent Performance Analysis*
  This book investigates steady-state polarization and power consumption based on time variations. It is suggested that the total energy be evaluated using a specific software for time-dependent quantum physics.
- *Designing Various Types of Circuits and Checking Errors in Multi-valued Circuits*
  This book presents a technique for designing n-valued cells and performing quantum calculations based on electron spin values and basic gates. Next, the designs of sequential and combinational circuits need to be examined, as well as performing error checks in these circuits.
- *Advance of Emulators*
  In this book, based on quantum calculations, the values of polarizations in each type of the proposed cells are presented. Based on the proposed models for the n-valued cells that were derived from binary and ternary cells with input and output integration drives, it is suggested to provide a comprehensive software package to simulate all different cells with different levels.

References

1 A. Henglein, "Small-particle research: physicochemical properties of extremely small colloidal metal and semiconductor particles," *Chemical Reviews*, vol. 89, no. 8, pp. 1861–1873, 1989.

2 K. Faaliyan, H. Abdoos, E. Borhani, and S. S. S. Afghahi, "Core-shell nanoparticles for medical applications: effects of surfactant concentration on the characteristics and magnetic properties of magnetite-silica nanoparticles," *Nanomedicine Journal*, vol. 6, no. 4, pp. 269–275, 2019.

3 A. Arafati, E. Borhani, S. M. S. Nourbakhsh, and H. Abdoos, "Synthesis and characterization of tetragonal/monoclinic mixed phases nanozirconia powders," *Ceramics International*, vol. 45, no. 10, pp. 12975–12982, 2019.

4 P. Tian, L. Tang, K. Teng, and S. Lau, "Graphene quantum dots from chemistry to applications," *Materials Today Chemistry*, vol. 10, pp. 221–258, 2018.

5 A. Ekimov and A. Onushchenko, "Quantum size effect in three-dimensional microscopic semiconductor crystals," *SPIE Milestone Series*, vol. 180, pp. 3–5, 2005.

6 L. Farzin, M. Shamsipur, L. Samandari, and S. Sheibani, "Recent advances in designing nanomaterial based biointerfaces for electrochemical biosensing cardiovascular biomarkers," *Journal of Pharmaceutical and Biomedical Analysis*, vol. 161, pp. 344–376, 2018.

7 S. K. Arumugasamy, G. Chellasamy, K. Yun, and J. Hyun, "Bio-quantum dots for electrochemical sensing of cardiac biomarkers of acute myocardial infarction," *Journal of Industrial and Engineering Chemistry*, vol. 129, pp. 488–498, 2024.

8 P. Walter *et al.*, "Early use of PbS nanotechnology for an ancient hair dyeing formula," *Nano Letters*, vol. 6, no. 10, pp. 2215–2219, 2006.

9 J. Cassidy and M. Zamkov, "Nanoshell quantum dots: quantum confinement beyond the exciton Bohr radius," *The Journal of Chemical Physics*, vol. 152, no. 11, p. 110902, 2020.

10 W. C. Chan, D. J. Maxwell, X. Gao, R. E. Bailey, M. Han, and S. Nie, "Luminescent quantum dots for multiplexed biological detection and imaging," *Current Opinion in Biotechnology*, vol. 13, no. 1, pp. 40–46, 2002.

11 C. Murray, D. J. Norris, and M. G. Bawendi, "Synthesis and characterization of nearly monodisperse CdE (E = sulfur, selenium, tellurium) semiconductor nanocrystallites," *Journal of the American Chemical Society*, vol. 115, no. 19, pp. 8706–8715, 1993.

12 J. Shen, Y. Zhu, X. Yang, J. Zong, J. Zhang, and C. Li, "One-pot hydrothermal synthesis of graphene quantum dots surface-passivated by polyethylene glycol and their photoelectric conversion under near-infrared light," *New Journal of Chemistry*, vol. 36, no. 1, pp. 97–101, 2012.

13 J. Robertson and E. O'reilly, "Electronic and atomic structure of amorphous carbon," *Physical Review B*, vol. 35, no. 6, p. 2946, 1987.

Computing with Multi-Value Logic in Quantum Dot Cellular Automata, First Edition
Reza Sabbaghi-Nadooshan, Reza Akbari-Hasanjani, Leila Dehbozorgi, Majid Haghparast, and Hamid Reza Akbari-Hasanjani.

14 D. Bera, L. Qian, T.-K. Tseng, and P. H. Holloway, "Quantum dots and their multimodal applications: a review," *Materials*, vol. 3, no. 4, pp. 2260–2345, 2010.

15 H. M. Kashani, T. Madrakian, A. Afkhami, F. Mahjoubi, and M. A. Moosavi, "Bottom-up and green-synthesis route of amino functionalized graphene quantum dot as a novel biocompatible and label-free fluorescence probe for *in vitro* cellular imaging of human ACHN cell lines," *Materials Science and Engineering B*, vol. 251, p. 114452, 2019.

16 Inamuddin, T. A. Rangreez, M. F. Ahmer, and R. Boddula; Editors: "*Quantum Dots: Properties and Applications* (Materials Research Foundations)," Materials Research Forum LLC, 360 pages, 2021. [ISBN-13: 978-1644901243].

17 D. Lee *et al.*, "Systematic Evolution of High-affinity ssDNA Sequence to Carbon Nanotube," 2023.

18 H.-Y. Xu, L. Huang, and Y.-C. Lai, "Relativistic quantum chaos in graphene," *Physics Today*, vol. 74, no. 2, pp. 44–49, 2021.

19 A. Ghaffarkhah *et al.*, "Synthesis, applications, and prospects of graphene quantum dots: a comprehensive review," *Small*, vol. 18, no. 2, p. 2102683, 2022.

20 L. T. Canham, "Silicon quantum wire array fabrication by electrochemical and chemical dissolution of wafers," *Applied Physics Letters*, vol. 57, no. 10, pp. 1046–1048, 1990.

21 V. Lehmann and U. Gösele, "Porous silicon formation: a quantum wire effect," *Applied Physics Letters*, vol. 58, no. 8, pp. 856–858, 1991.

22 M. Fujii, R. Fujii, M. Takada, and H. Sugimoto, "Silicon quantum dot supraparticles for fluorescence bioimaging," *ACS Applied Nano Materials*, vol. 3, no. 6, pp. 6099–6107, 2020.

23 Y. Xu *et al.*, "Group IV nanodots: synthesis, surface engineering and application in bioimaging and biotherapy," *Journal of Materials Chemistry B*, vol. 8, no. 45, pp. 10290–10308, 2020.

24 M. Marmiroli *et al.*, "Cadmium sulfide quantum dots, mitochondrial function and environmental stress: a mechanistic reconstruction through *in vivo* cellular approaches in *Saccharomyces cerevisiae*," *Nanomaterials*, vol. 13, no. 13, p. 1944, 2023.

25 S. Jain, S. Bharti, G. K. Bhullar, and S. Tripathi, "I-III-VI core/shell QDs: Synthesis, characterizations and applications," *Journal of Luminescence*, vol. 219, p. 116912, 2020.

26 A. Sahu and D. Kumar, "Core-shell quantum dots: a review on classification, materials, application, and theoretical modeling," *Journal of Alloys and Compounds*, vol. 924, p. 166508, 2022.

27 M. Safari, "Recent Advances in Quantum Dots-Based Biosensors," 2022.

28 P. Tandale *et al.*, "Fluorescent quantum dots: an insight on synthesis and potential biological application as drug carrier in cancer," *Biochemistry and Biophysics Reports*, vol. 26, p. 100962, 2021.

29 T. Torimoto, T. Kameyama, T. Uematsu, and S. Kuwabata, "Controlling optical properties and electronic energy structure of I–III–VI semiconductor quantum dots for improving their photofunctions," *Journal of Photochemistry and Photobiology C: Photochemistry Reviews*, vol. 54, p. 100569, 2023.

30 M. I. Ziane *et al.*, "Anisotropic optical properties of $Cu_2ZnSn(S_xSe_{1-x})_4$ solid solutions: first-principles calculations with TB-mBJ+U," *Optik*, vol. 243, p. 167490, 2021.

31 F. Ebrahim, O. Al-Hartomy, and S. Wageh, "Cadmium-based quantum dots alloyed structures: synthesis, properties, and applications," *Materials*, vol. 16, no. 17, p. 5877, 2023.

32 Y. Wang and N. Herron, "Nanometer-sized semiconductor clusters: materials synthesis, quantum size effects, and photophysical properties," *The Journal of Physical Chemistry*, vol. 95, no. 2, pp. 525–532, 1991.

33 V. Etacheri, R. Roshan, and V. Kumar, "Mg-doped ZnO nanoparticles for efficient sunlight-driven photocatalysis," *ACS Applied Materials & Interfaces*, vol. 4, no. 5, pp. 2717–2725, 2012.

34 P. P. Ingole, "A consolidated account of electrochemical determination of band structure parameters in II–VI semiconductor quantum dots: a tutorial review," *Physical Chemistry Chemical Physics*, vol. 21, no. 9, pp. 4695–4716, 2019.

35 J. D. Schiffman and R. G. Balakrishna, "Quantum dots as fluorescent probes: synthesis, surface chemistry, energy transfer mechanisms, and applications," *Sensors and Actuators B: Chemical*, vol. 258, pp. 1191–1214, 2018.

36 E. V. Shevchenko *et al.*, "Visualizing heterogeneity of monodisperse CdSe nanocrystals by their assembly into three-dimensional supercrystals," *ACS Nano*, vol. 14, no. 11, pp. 14989–14998, 2020.

37 S. S. Joglekar, H. M. Gholap, P. S. Alegaonkar, and A. A. Kale, "The interactions between CdTe quantum dots and proteins: understanding nano-bio interface," *AIMS Materials Science*, vol. 4, no. 1, pp. 209–222, 2017.

38 J. Lv, S. Liu, and Y. Miao, "Synthesis of biological quantum dots based on single-strand DNA and its application in melamine detection," *Spectrochimica Acta Part A: Molecular and Biomolecular Spectroscopy*, vol. 248, p. 119254, 2021.

39 M. Rafipoor *et al.*, "Strain engineering in InP/(Zn,Cd)Se core/shell quantum dots," *Chemistry of Materials*, vol. 30, no. 13, pp. 4393–4400, 2018.

40 J. Zhang, C. Li, J. Li, and X. Peng, "Synthesis of CdSe/ZnSe core/shell and CdSe/ZnSe/ZnS core/shell/shell nanocrystals: surface-ligand strain and CdSe–ZnSe lattice strain," *Chemistry of Materials*, vol. 35, no. 17, pp. 7049–7059, 2023.

41 J. Yu, M. Sharma, A. Sharma, S. Delikanli, H. Volkan Demir, and C. Dang, "All-optical control of exciton flow in a colloidal quantum well complex," *Light: Science & Applications*, vol. 9, no. 1, p. 27, 2020.

42 Y. Hu, Z. Wei, B. Wu, B. Shen, Q. Dai, and P. Feng, "Photoluminescence of ZnS: Mn quantum dot by hydrothermal method," *AIP Advances*, vol. 8, no. 1, p. 015014, 2018.

43 M. D. Villalva, V. Agarwal, M. Ulanova, P. S. Sachdev, and N. Braidy, "Quantum dots as a theranostic approach in Alzheimer's disease: a systematic review," *Nanomedicine*, vol. 16, no. 18, pp. 1595–1611, 2021.

44 Z. Yu, L. Guo, H. Du, T. Krauss, and J. Silcox, "Shell distribution on colloidal CdSe/ZnS quantum dots," *Nano Letters*, vol. 5, no. 4, pp. 565–570, 2005.

45 D. Van Thai, P. Van Ben, V. Q. Trung, B. H. Van, N. M. Nghia, and T. M. Thi, "Luminescence enhancement of ZnS: Mn and ZnS: Cu-PVP nanoparticles by laser radiation annealing," *Optical Materials*, vol. 135, p. 113020, 2023.

46 N. Órdenes-Aenishanslins, G. Anziani-Ostuni, C. P. Quezada, R. Espinoza-González, D. Bravo, and J. M. Pérez-Donoso, "Biological synthesis of CdS/CdSe core/shell nanoparticles and its application in quantum dot sensitized solar cells," *Frontiers in Microbiology*, vol. 10, p. 1587, 2019.

47 N. A. Hamizi *et al.*, "Investigation on surface properties of Mn-doped CdSe quantum dots studied by X-ray photoelectron spectroscopy," *Symmetry*, vol. 11, no. 10, p. 1250, 2019.

48 V. Renuga, C. N. Mohan, M. M. Jaabir, P. A. Prakash, and M. Navaneethan, "Synthesis and surface passivation of $CuInS_2$/MnS/ZnS core–multishell nanocrystals, their optical, structural, and morphological characterization, and their bioimaging applications," *Industrial and Engineering Chemistry Research*, vol. 57, no. 46, pp. 15703–15721, 2018.

49 A. I. Hassan and H. M. Saleh, "Quantum dots: properties and applications," *Quantum Dots – Properties Applications*, vol. 96, pp. 331–348, 2021.

50 E. Borovitskaya and M. S. Shur; Editors: "*Quantum Dots* (Selected Topics in Electronics and Systems)," World Scientific Publishing Company, 1st Edition, 216 pages, 2002. [ISBN-13: 978-9810249182].

51 D. De, S. Bhattacharaya, and K. P. Ghatak; Editors: "*Quantum Dots and Quantum Cellular Automata: Recent Trends and Applications* (Physics Research and Technology)," Nova Science Publishers, 1st Edition, 454 pages, 2023. [ISBN-13: 9781622579204].

52 B. A. Joyce, P. C. Kelires, A. G. Naumovets, and D. Vvedensky; Editors: "*Quantum Dots: Fundamentals, Applications, and Frontiers: Proceedings of the NATO ARW on Quantum Dots: Fundamentals, Applications, and Frontiers, Crete, Greece, 2003*," Springer, 2005 Edition, 412 pages, 2005. [ISBN-13: 978-1402033131].

53 K. Schmidt, G. Medeiros-Ribeiro, J. Garcia, and P. M. Petroff, "Size quantization effects in InAs self-assembled quantum dots," *Applied Physics Letters*, vol. 70, no. 13, pp. 1727–1729, 1997.

54 P. Das, S. R. Ahmed, S. Srinivasan, and A. R. Rajabzadeh, "4 Optical properties," in *Quantum Dots and Polymer Nanocomposites: Synthesis, Chemistry, and Applications*, CRC Press, p. 69, 2022.

55 X. Bai, F. Purcell-Milton, and Y. K. Gun'ko, "Optical properties, synthesis, and potential applications of Cu-based ternary or quaternary anisotropic quantum dots, polytypic nanocrystals, and core/shell heterostructures," *Nanomaterials*, vol. 9, no. 1, p. 85, 2019.

56 T. Kada *et al.*, "Two-step photon absorption in InAs/GaAs quantum-dot superlattice solar cells," *Physical Review B*, vol. 91, no. 20, p. 201303, 2015.

57 K. S. Thygesen, "Calculating excitons, plasmons, and quasiparticles in 2D materials and van der Waals heterostructures," *2D Materials*, vol. 4, no. 2, p. 022004, 2017.

58 S.-H. Choi, "Unique properties of graphene quantum dots and their applications in photonic/electronic devices," *Journal of Physics D: Applied Physics*, vol. 50, no. 10, p. 103002, 2017.

59 K. Singh and V. Kumar, *Carbonaceous Quantum Dots: Synthesis And Applications.* Bentham Science Publishers, 2023.

60 C. Carrillo-Carrión, S. Cárdenas, B. M. Simonet, and M. Valcárcel, "Quantum dots luminescence enhancement due to illumination with UV/Vis light," *Chemical Communications*, no. 35, pp. 5214–5226, 2009.

61 A. Kalluri, D. Debnath, B. Dharmadhikari, and P. Patra, "Graphene quantum dots: synthesis and applications," in *Methods in Enzymology*, vol. 609: Elsevier, 2018, pp. 335–354.

62 L. V. Goncharova and P. J. Simpson, "Ion beam modification for Si photonics," *Physics*, vol. 4, no. 2, pp. 383–93, 2022.

63 R. Palankar, N. Medvedev, A. Rong, and M. Delcea, "Fabrication of quantum dot microarrays using electron beam lithography for applications in analyte sensing and cellular dynamics," *ACS Nano*, vol. 7, no. 5, pp. 4617–4628, 2013.

64 Q. Yuan, T. Wang, P. Yu, H. Zhang, H. Zhang, and W. Ji, "A review on the electroluminescence properties of quantum-dot light-emitting diodes," *Organic Electronics*, vol. 90, p. 106086, 2021.

65 K. V. Chandekar *et al.*, "Improved linear and nonlinear optical properties of PbS thin films synthesized by spray pyrolysis technique for optoelectronics: an effect of Gd^{3+} doping concentrations," *Physica B: Condensed Matter*, vol. 641, p. 414099, 2022.

66 F. Weigert *et al.*, "Combining HR-TEM and XPS to elucidate the core–shell structure of ultra-bright CdSe/CdS semiconductor quantum dots," *Scientific Reports*, vol. 10, no. 1, p. 20712, 2020.

67 N. Byambasuren, J. Jo, H. Ko, B.-K. Ju, J. Y. Byun, and H. S. Jang, "Phosphine-free-synthesized ZnSe/ZnS Core/shell quantum dots for white light-emitting diodes," *Applied Sciences*, vol. 11, no. 21, p. 10060, 2021.

68 C. Hantschmann *et al.*, "Impact of dislocations in monolithic III-V lasers on silicon: a theoretical approach," in *Physics and Simulation of Optoelectronic Devices XXVIII*, 2020, vol. 11274: SPIE, pp. 64–72.

69 M. Volmer and A. Weber, "Keimbildung in übersättigten Gebilden," *Zeitschrift für Physikalische Chemie*, vol. 119, no. 1, pp. 277–301, 1926.

70 I. N. Stranski and L. V. Krastanow, "Sitzungsberichte der akademie der wissenschaften in wien," *Akad. Wiss. Lit. Mainz Math.-Natur. Kl. IIb*, vol. 146, p. 797, 1939.

71 T. Toujyou, T. Konishi, M. Hirayama, K. Yamaguchi, and S. Tsukamoto, "Intermittent growth for InAs quantum dot on GaAs(001)," *Journal of Crystal Growth*, vol. 551, p. 125891, 2020.

72 C. Lobo and R. Leon, "InGaAs island shapes and adatom migration behavior on (100), (110), (111), and (311) GaAs surfaces," *Journal of Applied Physics*, vol. 83, no. 8, pp. 4168–4172, 1998.

73 M. Hinojosa, I. Lombardero, C. Algora, and I. García, "Improved GaInP/GaAs/GaInAs inverted metamorphic triple-junction solar cells by reduction of Zn diffusion in the top subcell," *Solar Energy Materials and Solar Cells*, vol. 248, p. 112000, 2022.

74 S. Bereznev, I. Konovalov, A. Öpik, and J. Kois, "Hybrid $CuInS_2$/polypyrrole and $CuInS_2$/poly(3,4-ethylenedioxythiophene) photovoltaic structures," *Synthetic Metals*, vol. 152, no. 1–3, pp. 81–84, 2005.

75 C. Bernhardt, *Quantum Computing for Everyone*. MIT Press, 2019.

76 M. Gholami, R. Binaei, and M. Gholamnia Roshan, "Novel phase-frequency detector based on quantum-dot cellular automata nanotechnology," *International Journal of Engineering*, vol. 33, no. 2, pp. 269–276, 2020.

77 M. Parvane, E. Rahimi, and F. Jafarinejad, "Optimization of quantum cellular automata circuits by genetic algorithm," *International Journal of Engineering*, vol. 33, no. 2, pp. 229–236, 2020.

78 C. S. Lent, P. D. Tougaw, W. Porod, and G. H. Bernstein, "Quantum cellular automata," *Nanotechnology*, vol. 4, no. 1, p. 49, 1993.

79 C. S. Lent and P. D. Tougaw, "A device architecture for computing with quantum dots," *Proceedings of the IEEE*, vol. 85, no. 4, pp. 541–557, 1997.

80 Y. Zhang, H. Lv, H. Du, C. Huang, S. Liu, and G. Xie, "Modular design of QCA carry flow adders and multiplier with reduced wire crossing and number of logic gates," *International Journal of Circuit Theory and Applications*, vol. 44, no. 7, pp. 1351–1366, 2016.

81 T. Oya, T. Asai, T. Fukui, and Y. Amemiya, "A majority-logic nanodevice using a balanced pair of single-electron boxes," *Journal of Nanoscience and Nanotechnology*, vol. 2, no. 3–4, pp. 333–342, 2002.

82 T. Oya, T. Asai, T. Fukui, and Y. Amemiya, "A majority-logic device using an irreversible single-electron box," *IEEE Transactions on Nanotechnology*, vol. 2, no. 1, pp. 15–22, 2003.

83 A. Bachtold, P. Hadley, T. Nakanishi, and C. Dekker, "Logic circuits with carbon nanotube transistors," *Science*, vol. 294, no. 5545, pp. 1317–1320, 2001.

84 F. Sharifi, A. Panahi, H. Sharifi, K. Navi, N. Bagherzadeh, and H. Thapliyal, "Design of quaternary 4–2 and 5–2 compressors for nanotechnology," *Computers and Electrical Engineering*, vol. 56, pp. 64–74, 2016.

85 J. M. Tour, "Molecular electronics. Synthesis and testing of components," *Accounts of Chemical Research*, vol. 33, no. 11, pp. 791–804, 2000.

86 S. Erniyazov and J.-C. Jeon, "Carry save adder and carry look-ahead adder using inverter chain based coplanar QCA full adder for low energy dissipation," *Microelectronic Engineering*, vol. 211, pp. 37–43, 2019.

87 M. Zahmatkesh, S. Tabrizchi, S. Mohammadyan, K. Navi, and N. Bagherzadeh, "Robust coplanar full adder based on novel inverter in quantum cellular automata," *International Journal of Theoretical Physics*, vol. 58, pp. 639–655, 2019.

88 S. U. R. Qadri, Z. A. Bangi, M. Tariq Banday, and G. Mohiuddin Bhat, "Design and implementation of cryptographic element with low power dissipation in QCA," *Nanomaterials and Energy*, vol. 8, no. 1, pp. 96–106, 2019.

89 S.-S. Ahmadpour and M. Mosleh, "New designs of fault-tolerant adders in quantum-dot cellular automata," *Nano Communication Networks*, vol. 19, pp. 10–25, 2019.

90 F. Salimzadeh and S. R. Heikalabad, "Design of a novel reversible structure for full adder/subtractor in quantum-dot cellular automata," *Physica B: Condensed Matter*, vol. 556, pp. 163–169, 2019.

91 M. Mohammadi, A. Niknafs, and M. Eshghi, "Controlled gates for multi-level quantum computation," *Quantum Information Processing*, vol. 10, pp. 241–256, 2011.

92 I. L. Bajec, N. Zimic, and M. Mraz, "Towards the bottom-up concept: extended quantum-dot cellular automata," *Microelectronic Engineering*, vol. 83, no. 4–9, pp. 1826–1829, 2006.

93 I. L. Bajec, N. Zimic, and M. Mraz, "The ternary quantum-dot cell and ternary logic," *Nanotechnology*, vol. 17, no. 8, p. 1937, 2006.

94 H. Hosseinzadeh and S. R. Heikalabad, "A novel fault-tolerant majority gate in quantum-dot cellular automata to create a revolution in design of fault-tolerant nanostructures, with physical verification," *Microelectronic Engineering*, vol. 192, pp. 52–60, 2018.

95 M. Kianpour and R. Sabbaghi-Nadooshan, "A novel modular decoder implementation in quantum-dot cellular automata (QCA)," in *2011 International Conference on Nanoscience, Technology and Societal Implications*, 2011: IEEE, pp. 1–5.

96 A. Sadhu, K. Das, D. De, and M. R. Kanjilal, "SSTRNG: Self starved feedback SRAM based true random number generator using quantum cellular automata," *Microsystem Technologies*, vol. 26, pp. 2203–2215, 2020.

97 H. A. Mousavi, P. Keshavarzian, and A. S. Molahosseini, "A novel fast and small XOR-base full-adder in quantum-dot cellular automata," *Applied Nanoscience*, vol. 10, pp. 4037–4048, 2020.

98 J. C. Das and D. De, "Design of thermometer code-to-gray code converter circuit in quantum-dot cellular automata for nano-computing network," *Photonic Network Communications*, vol. 41, no. 3, pp. 259–273, 2021.

99 S. Ahmed, S. F. Naz, S. Sharma, and S. B. Ko, "Design of quantum-dot cellular automata-based communication system using modular N-bit binary to gray and gray to binary converters," *International Journal of Communication Systems*, vol. 34, no. 4, p. e4702, 2021.

100 J. C. Das and D. De, "Novel low power reversible binary incrementer design using quantum-dot cellular automata," *Microprocessors and Microsystems*, vol. 42, pp. 10–23, 2016.

101 S. M. Mohaghegh, R. Sabbaghi-Nadooshan, and M. Mohammadi, "Designing ternary quantum-dot cellular automata logic circuits based upon an alternative model," *Computers and Electrical Engineering*, vol. 71, pp. 43–59, 2018.

102 P. Pecar, M. Mraz, N. Zimic, M. Janez, and I. L. Bajec, "Solving the ternary quantum-dot cellular automata logic gate problem by means of adiabatic switching," *Japanese Journal of Applied Physics*, vol. 47, no. 6S, p. 5000, 2008.

103 P. Pecar, M. Janez, N. Zimic, M. Mraz, and I. L. Bajec, "The ternary quantum-dot cellular automata memorizing cell," in *2009 IEEE Computer Society Annual Symposium on VLSI*, 2009: IEEE, pp. 223–228.

104 M. Janez, I. L. Bajec, P. Pecar, A. Jazbec, N. Zimic, and M. Mraz, "Automatic design of optimal logic circuits based on ternary quantum-dot cellular automata," *WSEAS Transactions on Circuits and Systems*, vol. 7, pp. 919–928, 2008.

105 S. Mohammadi Mohaghegh, R. Sabbaghi-Nadooshan, and M. Mohammadi, "Design of a ternary QCA multiplier and multiplexer: a model-based approach," *Analog Integrated Circuits and Signal Processing*, vol. 101, pp. 23–29, 2019.

106 M. A. Tehrani and K. Navi, "A novel quantum dot cellular automata for implementation of multi-valued logic," in *Nano Today Conference, Elsevier*, 2009.

107 P. Pain, A. Sadhu, K. Das, and M. R. Kanjilal, "Physical proof and simulation of ternary logic gate in ternary quantum dot cellular automata," in *Computational Advancement in Communication Circuits and Systems: Proceedings of ICCACCS 2018*, 2020: Springer, pp. 375–385.

108 M. A. Tehrani, S. Bahrami, and K. Navi, "A novel ternary quantum-dot cell for solving majority voter gate problem," *Applied Nanoscience*, vol. 4, pp. 255–262, 2014.

109 S. Mohammadi Mohaghegh, R. Sabbaghi-Nadooshan, and M. Mohammadi, "Innovative model for ternary QCA gates," *IET Circuits, Devices and Systems*, vol. 12, no. 2, pp. 189–195, 2018.

110 A. Roohi, R. F. DeMara, and N. Khoshavi, "Design and evaluation of an ultra-area-efficient fault-tolerant QCA full adder," *Microelectronics Journal*, vol. 46, no. 6, pp. 531–542, 2015.

111 S. Hashemi and K. Navi, "Designing quantum-dot cellular automata circuits using a robust one layer crossover scheme," *The Journal of Engineering*, vol. 2014, no. 3, pp. 93–97, 2014.

112 J. S. Domingues, L. S. da Rosa, and F. de Souza Marques, "A straightforward methodology for QCA circuits design," in *2020 33rd Symposium on Integrated Circuits and Systems Design (SBCCI)*, 2020: IEEE, pp. 1–6.

113 R. Binaei and M. Gholami, "Using D flip-flop with reset terminal to design PFD in QCA nanotechnology," *International Journal of Electronics*, vol. 107, no. 12, pp. 1940–1962, 2020.

114 J. Pal, A. K. Pramanik, J. S. Sharma, A. K. Saha, and B. Sen, "An efficient, scalable, regular clocking scheme based on quantum dot cellular automata," *Analog Integrated Circuits and Signal Processing*, vol. 107, pp. 659–670, 2021.

115 I. Amlani, A. O. Orlov, R. K. Kummamuru, G. H. Bernstein, C. S. Lent, and G. L. Snider, "Experimental demonstration of a leadless quantum-dot cellular automata cell," *Applied Physics Letters*, vol. 77, no. 5, pp. 738–740, 2000.

116 K. Das, D. De, and M. De, "Realisation of semiconductor ternary quantum dot cellular automata," *Micro & Nano Letters*, vol. 8, no. 5, pp. 258–263, 2013.

117 K. Hennessy and C. S. Lent, "Clocking of molecular quantum-dot cellular automata," *Journal of Vacuum Science & Technology, B: Microelectronics and Nanometer Structures--Processing, Measurement, and Phenomena*, vol. 19, no. 5, pp. 1752–1755, 2001.

118 C. S. Lent, B. Isaksen, and M. Lieberman, "Molecular quantum-dot cellular automata," *Journal of the American Chemical Society*, vol. 125, no. 4, pp. 1056–1063, 2003.

119 J. F. Pulecio and S. Bhanja, "Reliability of bi-stable single domain nano magnets for cellular automata," in *2007 7th IEEE Conference on Nanotechnology (IEEE NANO)*, 2007: IEEE, pp. 782–786.

120 V. K. Sharma, "Optimal design for digital comparator using QCA nanotechnology with energy estimation," *International Journal of Numerical Modelling: Electronic Networks, Devices and Fields*, vol. 34, no. 2, p. e2822, 2021.

121 N. Safoev and J.-C. Jeon, "Design of high-performance QCA incrementer/decrementer circuit based on adder/subtractor methodology," *Microprocessors and Microsystems*, vol. 72, p. 102927, 2020.

122 J. Pal, M. Goswami, A. K. Saha, and B. Sen, "CFA: Toward the realization of conservative full adder in QCA with enhanced reliability," *Journal of Circuits, Systems and Computers*, vol. 30, no. 10, p. 2150172, 2021.

123 M. A. Mianroodi, R. Akbari-Hasanjani, and R. Sabbaghi-Nadooshan, "QQCAsim: A new simulation tool for quaternary logic in QCA technology," *Scientia Iranica*, in Press, 2023.

124 S. Yaqoob, S. Ahmed, S. F. Naz, S. Bashir, and S. Sharma, "Design of efficient N-bit shift register using optimized D flip flop in quantum dot cellular automata technology," *IET Quantum Communication*, vol. 2, no. 2, pp. 32–41, 2021.

125 J. Maharaj and S. Muthurathinam, "Effective RCA design using quantum dot cellular automata," *Microprocessors and Microsystems*, vol. 73, p. 102964, 2020.

126 E. AlKaldy, A. H. Majeed, M. S. Zainal, and D. M. Nor, "Optimum multiplexer design in quantum-dot cellular automata," *arXiv preprint arXiv:2002.00360*, 2020.

127 R. Laajimi and M. Niu, "Nanoarchitecture of quantum-dot cellular automata (QCA) using small area for digital circuits," in *Advanced Electronics Circuits–Principles, Architectures and Applications on Emerging Technologies*, IntechOpen, pp. 67–84, 2018.

128 D. Bhowmik, J. Pal, M. Goswami, P. Sen, A. K. Saha, and B. Sen, "Systematic cell placement in quantum-dot cellular automata embedding underlying regular clocking circuit," *IET Circuits, Devices and Systems*, vol. 15, no. 2), pp. 156–167, 2021.

129 R. Akbari-Hasanjani, R. Sabbaghi-Nadooshan, and M. R. Tanhayi, "New polarization and power calculations with error elimination in ternary QCA," *Computers and Electrical Engineering*, vol. 96, p. 107557, 2021.

130 R. Akbari-Hasanjani and R. Sabbaghi-Nadooshan, "Design and simulation of innovative QCA quaternary logic gates," *Advanced Theory and Simulations*, vol. 4, no. 9, p. 2100069, 2021.

131 R. Akbari-Hasanjani and R. Sabbaghi-Nadooshan, "Innovation quinary and n-value toward fuzzy logic QCA cell design," *Advanced Theory and Simulations*, vol. 5, no. 2, p. 2100304, 2022.

132 R. Akbari-Hasanjani and R. Sabbaghi-Nadooshan, "New design of binary to ternary converter," *IETE Journal of Research*, vol. 69, no. 4, pp. 2212–2223, 2021.

133 R. Akbari-Hasanjani, R. Sabbaghi-Nadooshan, and M. Haghparast, "Toward quaternary QCA: novel majority and XOR fuzzy gates," *IEEE Access*, vol. 10, pp. 38511–38522, 2022.

134 S. Srivastava and S. Bhanja, "Hierarchical probabilistic macromodeling for QCA circuits," *IEEE Transactions on Computers*, vol. 56, no. 2, pp. 174–190, 2007.

135 M. M. Arjmand, M. Soryani, K. Navi, and M. A. Tehrani, "A novel ternary-to-binary converter in quantum-dot cellular automata," in *2012 IEEE Computer Society Annual Symposium on VLSI*, 2012: IEEE, pp. 147–152.

136 J. Y. Wang *et al.*, "Direct and full-scale experimental verifications towards ground–satellite quantum key distribution," *Nature Photonics*, vol. 7, no. 5, pp. 387–393, 2013.

137 B. Robson and J. S. Clair, "Principles of quantum mechanics for artificial intelligence in medicine. Discussion with reference to the Quantum Universal Exchange Language (Q-UEL)," *Computers in Biology and Medicine*, vol. 143, p. 105323, 2022.

138 V. Vedral, "The role of relative entropy in quantum information theory," *Reviews of Modern Physics*, vol. 74, no. 1, p. 197, 2002.

139 D. Bouwmeester and A. Zeilinger, "The physics of quantum information: basic concepts," in *The Physics of Quantum Information: Quantum Cryptography, Quantum Teleportation, Quantum Computation*, 2000; Berlin, Heidelberg: Springer Berlin Heidelberg, pp. 1–14.

140 M. Biafore, "Can quantum computers have simple Hamiltonians?," in *Proceedings Workshop on Physics and Computation. PhysComp'94*, 1994: IEEE, pp. 63–68.

141 S. Çakmak, M. Kurt, and A. Gençten, "QFT based quantum arithmetic logic unit on IBM quantum computer," *arXiv preprint arXiv:2306.09560*, 2023.

142 D. Copsey *et al.*, "Toward a scalable, silicon-based quantum computing architecture," *IEEE Journal of Selected Topics in Quantum Electronics*, vol. 9, no. 6, pp. 1552–1569, 2003.

143 J. L. Brylinski and R. Brylinski, "Universal quantum gates," *Mathematics of Quantum Computation*, vol. 79, pp. 117–134, 2002.

144 Y. L. Ju, I. M. Tsai, and S. Y. Kuo, "Quantum circuit design and analysis for database search applications," *IEEE Transactions on Circuits and Systems I: Regular Papers*, vol. 54, no. 11, pp. 2552–2563, 2007.

145 N. Gisin and R. Thew, "Quantum communication," *Nature Photonics*, vol. 1, no. 3), pp. 165–171, 2007.

146 E. Andersson and P. Öhberg: "*Quantum Information and Coherence* (Scottish Graduate Series)," Springer, 2nd Edition, 295 pages, 2014. [ISBN-13: 978-3319040622].

147 H. Imai and M. Hayashi; Editors: "*Quantum Computation and Information: From Theory to Experiment*," Springer, 1st Edition, 295 pages, 2006. [ISBN-13: 978-3540331322].

148 P. D. Tougaw and C. S. Lent, "Logical devices implemented using quantum cellular automata," *Journal of Applied Physics*, vol. 75, no. 3, pp. 1818–1825, 1994.

149 I. Sturzu and M. Khatun, "Quantum calculation of thermal effect in quantum-dot cellular automata," *Complexity*, vol. 10, no. 4, pp. 73–78, 2005.

150 A. Palii *et al.*, "Mixed-valence magnetic molecular cell for quantum cellular automata: prospects of designing multifunctional devices through exploration of double exchange," *The Journal of Physical Chemistry C*, vol. 124, no. 46, pp. 25602–25614, 2020.

151 K. Walus, T. J. Dysart, G. A. Jullien, and R. A. Budiman, "QCADesigner: A rapid design and simulation tool for quantum-dot cellular automata," *IEEE Transactions on Nanotechnology*, vol. 3, no. 1, pp. 26–31, 2004.

152 J. Huang, M. Momenzadeh, M. B. Tahoori, and F. Lombardi, "Defect characterization for scaling of QCA devices [quantum dot cellular automata]," in *19th IEEE International Symposium on Defect and Fault Tolerance in VLSI Systems, 2004. DFT Proceedings*, 2004: IEEE, pp. 30–38.

153 S. Bhanja, M. Ottavi, F. Lombardi, and S. Pontarelli, "QCA circuits for robust coplanar crossing," *Journal of Electronic Testing*, vol. 23, pp. 193–210, 2007.

154 S. Srivastava, S. Sarkar, and S. Bhanja, "Power dissipation bounds and models for quantum-dot cellular automata circuits," in *2006 Sixth IEEE Conference on Nanotechnology*, 2006, vol. 1: IEEE, pp. 375–378.

155 J. Timler and C. S. Lent, "Power gain and dissipation in quantum-dot cellular automata," *Journal of Applied Physics*, vol. 91, no. 2, pp. 823–831, 2002.

156 S. Sheikhfaal, S. Angizi, S. Sarmadi, M. H. Moaiyeri, and S. Sayedsalehi, "Designing efficient QCA logical circuits with power dissipation analysis," *Microelectronics Journal*, vol. 46, no. 6, pp. 462–471, 2015.

157 I. Jahangir, D. M. N. Hasan, N. A. Siddique, S. Islam, and M. M. Hasan, "Design of quaternary sequential circuits using a newly proposed quaternary algebra," in *2009 12th International Conference on Computers and Information Technology*, 2009: IEEE, pp. 203–208.

158 S. N. Chowdhury, A. Faiyaz, and K. M. Ishtiak, "Formulation and design of useful logic gates using quaternary algebra," *International Journal of Scientific and Engineering Research*, vol. 6, no. 1, pp. 1–7, 2015.

159 I. Jahangir, A. Das, and M. Hasan, "Facile algebraic representation of a novel quaternary logic," *arXiv preprint arXiv:1108.5497*, 2011.

160 A. O. Orlov, I. Amlani, G. H. Bernstein, C. S. Lent, and G. L. Snider, "Realization of a functional cell for quantum-dot cellular automata," *Science*, vol. 277, no. 5328, pp. 928–930, 1997.

161 W. Chen and H. Ahmed, "Fabrication of 5–7 nm wide etched lines in silicon using 100 keV electron-beam lithography and polymethylmethacrylate resist," *Applied Physics Letters*, vol. 62, no. 13, pp. 1499–1501, 1993.

162 P. Gao *et al.*, "Plasmonic lithography for the fabrication of surface nanostructures with a feature size down to 9 nm," *Nanoscale*, vol. 12, no. 4, pp. 2415–2421, 2020.

163 B. Päivänranta, A. Langner, E. Kirk, C. David, and Y. Ekinci, "Sub-10 nm patterning using EUV interference lithography," *Nanotechnology*, vol. 22, no. 37, p. 375302, 2011.

164 M. Hofmann *et al.*, "Mix-and-match lithography and cryogenic etching for NIL template fabrication," *Microelectronic Engineering*, vol. 224, p. 111234, 2020.

165 R. Zhang, T. J. Emge, C. Zheng, and J. Li, "The first single-sized (~1 nm) and periodically ordered array of In_2Te_3 semiconductor quantum dots self-assembled in solution," *Journal of Materials Chemistry A*, vol. 1, no. 2, pp. 199–202, 2013.

166 J. P. Kulkarni, J. W. Tschanz, and V. K. De, "Energy efficient volatile memory circuits for the IoT Era," In: Alioto, M. (eds) *Enabling the Internet of Things*. Springer, Cham, 2017.

167 P. A. Pradeepa, G. B. Juliet, G. G. Priya, and S. K. Lakshmi, "Design of optimized QCA sequential circuits. SSRG," *International Journal of Electronics and Communication Engineering*, SSRG - IJECE, 2019.

168 P. Dutta and D. Mukhopadhyay, "New architecture for flip flops using quantum-dot cellular automata," in *Proceedings of the 48th Annual Convention of CSI - Volume II, Advances in Intelligent Systems and Computing*, volume 249, 2014.

169 S. Gupta, M. Imani, H. Kaur, and T. S. Rosing, "NNPIM: A processing in-memory architecture for neural network acceleration," *IEEE Transactions on Computers*, vol. 68, no. 9, pp. 1325–1337, 2019.

170 H. Kang, P. B. Gibbons, G. E. Blelloch, L. Dhulipala, Y. Gu, and C. McGuffey, "The processing-in-memory model," in *Proceedings of the 33rd ACM Symposium on Parallelism in Algorithms and Architectures*, 2021, pp. 295–306.

171 L. Dehbozorgi, R. Sabbaghi-Nadooshan, A. Kashaninia, "Realization of processing-in-memory using binary and ternary quantum-dot cellular automata," *The Journal of Supercomputing*, vol. 78, pp. 6846–6874, 2022. https://doi.org/10.1007/s11227-021-04152-1.

172 M. Kianpour and R. Sabbaghi-Nadooshan, "A novel design and simulation of 16 bits RAM implementation in quantum-dot cellular automata (QCA)," in *16th IEEE Mediterranean Electrotechnical Conference*, 2012: IEEE, pp. 637–640.

173 M. Ottavi, S. Pontarelli, V. Vankamamidi, A. Salsano, and F. Lombardi, "QCA memory with parallel read/serial write: design and analysis," *IEE Proceedings-Circuits, Devices and Systems*, vol. 153, no. 3, pp. 199–206, 2006.

174 A. S. Shamsabadi, B. S. Ghahfarokhi, K. Zamanifar, and N. Movahedinia, “Applying inherent capabilities of quantum-dot cellular automata to design: D flip-flop case study,” *Journal of Systems Architecture*, vol. 55, no. 3, pp. 180–187, 2009.

175 M. A. Dehkordi, A. S. Shamsabadi, B. S. Ghahfarokhi, and A. Vafaei, “Novel RAM cell designs based on inherent capabilities of quantum-dot cellular automata,” *Microelectronics Journal*, vol. 42, no. 5, pp. 701–708, 2011.

176 S. Hashemi and K. Navi, “New robust QCA D flip flop and memory structures,” *Microelectronics Journal*, vol. 43, no. 12, pp. 929–940, 2012.

177 S. Angizi, S. Sarmadi, S. Sayedsalehi, and K. Navi, “Design and evaluation of new majority gate-based RAM cell in quantum-dot cellular automata,” *Microelectronics Journal*, vol. 46, no. 1, pp. 43–51, 2015.

178 P. Chougule, B. Sen, R. Mukherjee, P. Patil, R. Kamat, and T. Dongale, “A processing in memory realization using quantum dot cellular automata (QCA): proposal and implementation,” *Journal of Nano- and Electronic Physics*, vol. 9, no. 1, p. 01021, 2017.

179 P. P. Chougule, B. Sen, and T. D. Dongale, “Realization of processing In-memory computing architecture using quantum dot cellular automata,” *Microprocessors and Microsystems*, vol. 52, pp. 49–58, 2017.

180 M. Goswami, J. Pal, M. R. Choudhury, P. P. Chougule, and B. Sen, “In memory computation using quantum-dot cellular automata,” *IET Computers & Digital Techniques*, vol. 14, no. 6, pp. 336–343, 2020.

181 S. Rani and T. N. Sasamal, “A new clocking scheme for quantum-dot cellular automata based designs with single or regular cells,” *Energy Procedia*, vol. 117, pp. 466–473, 2017.

182 Z. Taheri, A. Rezai, H. Rashidi, “Novel single layer fault tolerance RCA construction for QCA,” *Electronics and Energetics*, vol. 32, no 4, pp. 601–613, 2019.

183 L. Dehbozorgi, R. Sabbaghi-Nadooshan, and A. Kashaninia, “Novel fault-tolerant processing in memory cell in ternary quantum-dot cellular automata,” *Journal of Electronic Testing*, vol. 38, pp. 419–444, 2022. https://doi.org/10.1007/s10836-022-06018-7.

A

Matrices

A.1 Particle State Matrix in QQCA

$$S_1 = \begin{bmatrix}
\langle \frac{1}{2},\frac{1}{2}|S|\frac{1}{2},\frac{1}{2}\rangle & \langle \frac{1}{2},\frac{1}{2}|S|\frac{1}{2},\frac{1}{2}\rangle & \langle \frac{1}{2},\frac{1}{2}|S|\frac{1}{2},\frac{1}{2}\rangle & \langle \frac{1}{2},\frac{1}{2}|S|\frac{1}{2},\frac{1}{2}\rangle & \langle \frac{1}{2},\frac{1}{2}|S|\frac{1}{2},-\frac{1}{2}\rangle & \langle \frac{1}{2},\frac{1}{2}|S|\frac{1}{2},-\frac{1}{2}\rangle & \langle \frac{1}{2},\frac{1}{2}|S|\frac{1}{2},-\frac{1}{2}\rangle & \langle \frac{1}{2},\frac{1}{2}|S|\frac{1}{2},-\frac{1}{2}\rangle \\
\langle \frac{1}{2},\frac{1}{2}|S|\frac{1}{2},\frac{1}{2}\rangle & \langle \frac{1}{2},\frac{1}{2}|S|\frac{1}{2},\frac{1}{2}\rangle & \langle \frac{1}{2},\frac{1}{2}|S|\frac{1}{2},\frac{1}{2}\rangle & \langle \frac{1}{2},\frac{1}{2}|S|\frac{1}{2},\frac{1}{2}\rangle & \langle \frac{1}{2},\frac{1}{2}|S|\frac{1}{2},-\frac{1}{2}\rangle & \langle \frac{1}{2},\frac{1}{2}|S|\frac{1}{2},-\frac{1}{2}\rangle & \langle \frac{1}{2},\frac{1}{2}|S|\frac{1}{2},-\frac{1}{2}\rangle & \langle \frac{1}{2},\frac{1}{2}|S|\frac{1}{2},-\frac{1}{2}\rangle \\
\langle \frac{1}{2},\frac{1}{2}|S|\frac{1}{2},\frac{1}{2}\rangle & \langle \frac{1}{2},\frac{1}{2}|S|\frac{1}{2},\frac{1}{2}\rangle & \langle \frac{1}{2},\frac{1}{2}|S|\frac{1}{2},\frac{1}{2}\rangle & \langle \frac{1}{2},\frac{1}{2}|S|\frac{1}{2},\frac{1}{2}\rangle & \langle \frac{1}{2},\frac{1}{2}|S|\frac{1}{2},-\frac{1}{2}\rangle & \langle \frac{1}{2},\frac{1}{2}|S|\frac{1}{2},-\frac{1}{2}\rangle & \langle \frac{1}{2},\frac{1}{2}|S|\frac{1}{2},-\frac{1}{2}\rangle & \langle \frac{1}{2},\frac{1}{2}|S|\frac{1}{2},-\frac{1}{2}\rangle \\
\langle \frac{1}{2},\frac{1}{2}|S|\frac{1}{2},\frac{1}{2}\rangle & \langle \frac{1}{2},\frac{1}{2}|S|\frac{1}{2},\frac{1}{2}\rangle & \langle \frac{1}{2},\frac{1}{2}|S|\frac{1}{2},\frac{1}{2}\rangle & \langle \frac{1}{2},\frac{1}{2}|S|\frac{1}{2},\frac{1}{2}\rangle & \langle \frac{1}{2},\frac{1}{2}|S|\frac{1}{2},-\frac{1}{2}\rangle & \langle \frac{1}{2},\frac{1}{2}|S|\frac{1}{2},-\frac{1}{2}\rangle & \langle \frac{1}{2},\frac{1}{2}|S|\frac{1}{2},-\frac{1}{2}\rangle & \langle \frac{1}{2},\frac{1}{2}|S|\frac{1}{2},-\frac{1}{2}\rangle \\
\langle \frac{1}{2},-\frac{1}{2}|S|\frac{1}{2},\frac{1}{2}\rangle & \langle \frac{1}{2},-\frac{1}{2}|S|\frac{1}{2},\frac{1}{2}\rangle & \langle \frac{1}{2},-\frac{1}{2}|S|\frac{1}{2},\frac{1}{2}\rangle & \langle \frac{1}{2},-\frac{1}{2}|S|\frac{1}{2},\frac{1}{2}\rangle & \langle \frac{1}{2},-\frac{1}{2}|S|\frac{1}{2},-\frac{1}{2}\rangle & \langle \frac{1}{2},-\frac{1}{2}|S|\frac{1}{2},-\frac{1}{2}\rangle & \langle \frac{1}{2},-\frac{1}{2}|S|\frac{1}{2},-\frac{1}{2}\rangle & \langle \frac{1}{2},-\frac{1}{2}|S|\frac{1}{2},-\frac{1}{2}\rangle \\
\langle \frac{1}{2},-\frac{1}{2}|S|\frac{1}{2},\frac{1}{2}\rangle & \langle \frac{1}{2},-\frac{1}{2}|S|\frac{1}{2},\frac{1}{2}\rangle & \langle \frac{1}{2},-\frac{1}{2}|S|\frac{1}{2},\frac{1}{2}\rangle & \langle \frac{1}{2},-\frac{1}{2}|S|\frac{1}{2},\frac{1}{2}\rangle & \langle \frac{1}{2},-\frac{1}{2}|S|\frac{1}{2},-\frac{1}{2}\rangle & \langle \frac{1}{2},-\frac{1}{2}|S|\frac{1}{2},-\frac{1}{2}\rangle & \langle \frac{1}{2},-\frac{1}{2}|S|\frac{1}{2},-\frac{1}{2}\rangle & \langle \frac{1}{2},-\frac{1}{2}|S|\frac{1}{2},-\frac{1}{2}\rangle \\
\langle \frac{1}{2},-\frac{1}{2}|S|\frac{1}{2},\frac{1}{2}\rangle & \langle \frac{1}{2},-\frac{1}{2}|S|\frac{1}{2},\frac{1}{2}\rangle & \langle \frac{1}{2},-\frac{1}{2}|S|\frac{1}{2},\frac{1}{2}\rangle & \langle \frac{1}{2},-\frac{1}{2}|S|\frac{1}{2},\frac{1}{2}\rangle & \langle \frac{1}{2},-\frac{1}{2}|S|\frac{1}{2},-\frac{1}{2}\rangle & \langle \frac{1}{2},-\frac{1}{2}|S|\frac{1}{2},-\frac{1}{2}\rangle & \langle \frac{1}{2},-\frac{1}{2}|S|\frac{1}{2},-\frac{1}{2}\rangle & \langle \frac{1}{2},-\frac{1}{2}|S|\frac{1}{2},-\frac{1}{2}\rangle \\
\langle \frac{1}{2},-\frac{1}{2}|S|\frac{1}{2},\frac{1}{2}\rangle & \langle \frac{1}{2},-\frac{1}{2}|S|\frac{1}{2},\frac{1}{2}\rangle & \langle \frac{1}{2},-\frac{1}{2}|S|\frac{1}{2},\frac{1}{2}\rangle & \langle \frac{1}{2},-\frac{1}{2}|S|\frac{1}{2},\frac{1}{2}\rangle & \langle \frac{1}{2},-\frac{1}{2}|S|\frac{1}{2},-\frac{1}{2}\rangle & \langle \frac{1}{2},-\frac{1}{2}|S|\frac{1}{2},-\frac{1}{2}\rangle & \langle \frac{1}{2},-\frac{1}{2}|S|\frac{1}{2},-\frac{1}{2}\rangle & \langle \frac{1}{2},-\frac{1}{2}|S|\frac{1}{2},-\frac{1}{2}\rangle
\end{bmatrix}_{8\times 8} \tag{A.1}$$

Computing with Multi-Value Logic in Quantum Dot Cellular Automata, First Edition.
Reza Sabbaghi-Nadooshan, Reza Akbari-Hasanjani, Leila Dehbozorgi, Majid Haghparast, and Hamid Reza Akbari-Hasanjani.

$$S_2=\begin{bmatrix}
\langle \frac{1}{2},\frac{1}{2}|S|\frac{1}{2},\frac{1}{2}\rangle & \langle \frac{1}{2},\frac{1}{2}|S|\frac{1}{2},\frac{1}{2}\rangle & \langle \frac{1}{2},\frac{1}{2}|S|\frac{1}{2},-\frac{1}{2}\rangle & \langle \frac{1}{2},\frac{1}{2}|S|\frac{1}{2},-\frac{1}{2}\rangle & \langle \frac{1}{2},\frac{1}{2}|S|\frac{1}{2},\frac{1}{2}\rangle & \langle \frac{1}{2},\frac{1}{2}|S|\frac{1}{2},\frac{1}{2}\rangle & \langle \frac{1}{2},\frac{1}{2}|S|\frac{1}{2},-\frac{1}{2}\rangle & \langle \frac{1}{2},\frac{1}{2}|S|\frac{1}{2},-\frac{1}{2}\rangle \\
\langle \frac{1}{2},\frac{1}{2}|S|\frac{1}{2},\frac{1}{2}\rangle & \langle \frac{1}{2},\frac{1}{2}|S|\frac{1}{2},\frac{1}{2}\rangle & \langle \frac{1}{2},\frac{1}{2}|S|\frac{1}{2},-\frac{1}{2}\rangle & \langle \frac{1}{2},\frac{1}{2}|S|\frac{1}{2},-\frac{1}{2}\rangle & \langle \frac{1}{2},\frac{1}{2}|S|\frac{1}{2},\frac{1}{2}\rangle & \langle \frac{1}{2},\frac{1}{2}|S|\frac{1}{2},\frac{1}{2}\rangle & \langle \frac{1}{2},\frac{1}{2}|S|\frac{1}{2},-\frac{1}{2}\rangle & \langle \frac{1}{2},\frac{1}{2}|S|\frac{1}{2},-\frac{1}{2}\rangle \\
\langle \frac{1}{2},-\frac{1}{2}|S|\frac{1}{2},\frac{1}{2}\rangle & \langle \frac{1}{2},-\frac{1}{2}|S|\frac{1}{2},\frac{1}{2}\rangle & \langle \frac{1}{2},-\frac{1}{2}|S|\frac{1}{2},-\frac{1}{2}\rangle & \langle \frac{1}{2},-\frac{1}{2}|S|\frac{1}{2},-\frac{1}{2}\rangle & \langle \frac{1}{2},-\frac{1}{2}|S|\frac{1}{2},\frac{1}{2}\rangle & \langle \frac{1}{2},-\frac{1}{2}|S|\frac{1}{2},\frac{1}{2}\rangle & \langle \frac{1}{2},-\frac{1}{2}|S|\frac{1}{2},-\frac{1}{2}\rangle & \langle \frac{1}{2},-\frac{1}{2}|S|\frac{1}{2},-\frac{1}{2}\rangle \\
\langle \frac{1}{2},-\frac{1}{2}|S|\frac{1}{2},\frac{1}{2}\rangle & \langle \frac{1}{2},-\frac{1}{2}|S|\frac{1}{2},\frac{1}{2}\rangle & \langle \frac{1}{2},-\frac{1}{2}|S|\frac{1}{2},-\frac{1}{2}\rangle & \langle \frac{1}{2},-\frac{1}{2}|S|\frac{1}{2},-\frac{1}{2}\rangle & \langle \frac{1}{2},-\frac{1}{2}|S|\frac{1}{2},\frac{1}{2}\rangle & \langle \frac{1}{2},-\frac{1}{2}|S|\frac{1}{2},\frac{1}{2}\rangle & \langle \frac{1}{2},-\frac{1}{2}|S|\frac{1}{2},-\frac{1}{2}\rangle & \langle \frac{1}{2},-\frac{1}{2}|S|\frac{1}{2},-\frac{1}{2}\rangle \\
\langle \frac{1}{2},\frac{1}{2}|S|\frac{1}{2},\frac{1}{2}\rangle & \langle \frac{1}{2},\frac{1}{2}|S|\frac{1}{2},\frac{1}{2}\rangle & \langle \frac{1}{2},\frac{1}{2}|S|\frac{1}{2},-\frac{1}{2}\rangle & \langle \frac{1}{2},\frac{1}{2}|S|\frac{1}{2},-\frac{1}{2}\rangle & \langle \frac{1}{2},\frac{1}{2}|S|\frac{1}{2},\frac{1}{2}\rangle & \langle \frac{1}{2},\frac{1}{2}|S|\frac{1}{2},\frac{1}{2}\rangle & \langle \frac{1}{2},\frac{1}{2}|S|\frac{1}{2},-\frac{1}{2}\rangle & \langle \frac{1}{2},\frac{1}{2}|S|\frac{1}{2},-\frac{1}{2}\rangle \\
\langle \frac{1}{2},\frac{1}{2}|S|\frac{1}{2},\frac{1}{2}\rangle & \langle \frac{1}{2},\frac{1}{2}|S|\frac{1}{2},\frac{1}{2}\rangle & \langle \frac{1}{2},\frac{1}{2}|S|\frac{1}{2},-\frac{1}{2}\rangle & \langle \frac{1}{2},\frac{1}{2}|S|\frac{1}{2},-\frac{1}{2}\rangle & \langle \frac{1}{2},\frac{1}{2}|S|\frac{1}{2},\frac{1}{2}\rangle & \langle \frac{1}{2},\frac{1}{2}|S|\frac{1}{2},\frac{1}{2}\rangle & \langle \frac{1}{2},\frac{1}{2}|S|\frac{1}{2},-\frac{1}{2}\rangle & \langle \frac{1}{2},\frac{1}{2}|S|\frac{1}{2},-\frac{1}{2}\rangle \\
\langle \frac{1}{2},-\frac{1}{2}|S|\frac{1}{2},\frac{1}{2}\rangle & \langle \frac{1}{2},-\frac{1}{2}|S|\frac{1}{2},\frac{1}{2}\rangle & \langle \frac{1}{2},-\frac{1}{2}|S|\frac{1}{2},-\frac{1}{2}\rangle & \langle \frac{1}{2},-\frac{1}{2}|S|\frac{1}{2},-\frac{1}{2}\rangle & \langle \frac{1}{2},-\frac{1}{2}|S|\frac{1}{2},\frac{1}{2}\rangle & \langle \frac{1}{2},-\frac{1}{2}|S|\frac{1}{2},\frac{1}{2}\rangle & \langle \frac{1}{2},-\frac{1}{2}|S|\frac{1}{2},-\frac{1}{2}\rangle & \langle \frac{1}{2},-\frac{1}{2}|S|\frac{1}{2},-\frac{1}{2}\rangle \\
\langle \frac{1}{2},-\frac{1}{2}|S|\frac{1}{2},\frac{1}{2}\rangle & \langle \frac{1}{2},-\frac{1}{2}|S|\frac{1}{2},\frac{1}{2}\rangle & \langle \frac{1}{2},-\frac{1}{2}|S|\frac{1}{2},-\frac{1}{2}\rangle & \langle \frac{1}{2},-\frac{1}{2}|S|\frac{1}{2},-\frac{1}{2}\rangle & \langle \frac{1}{2},-\frac{1}{2}|S|\frac{1}{2},\frac{1}{2}\rangle & \langle \frac{1}{2},-\frac{1}{2}|S|\frac{1}{2},\frac{1}{2}\rangle & \langle \frac{1}{2},-\frac{1}{2}|S|\frac{1}{2},-\frac{1}{2}\rangle & \langle \frac{1}{2},-\frac{1}{2}|S|\frac{1}{2},-\frac{1}{2}\rangle
\end{bmatrix}_{8\times 8} \tag{A.2}$$

$$S_3=\begin{bmatrix}
\langle \frac{1}{2},\frac{1}{2}|S|\frac{1}{2},\frac{1}{2}\rangle & \langle \frac{1}{2},\frac{1}{2}|S|\frac{1}{2},-\frac{1}{2}\rangle & \langle \frac{1}{2},\frac{1}{2}|S|\frac{1}{2},\frac{1}{2}\rangle & \langle \frac{1}{2},\frac{1}{2}|S|\frac{1}{2},-\frac{1}{2}\rangle & \langle \frac{1}{2},\frac{1}{2}|S|\frac{1}{2},\frac{1}{2}\rangle & \langle \frac{1}{2},\frac{1}{2}|S|\frac{1}{2},-\frac{1}{2}\rangle & \langle \frac{1}{2},\frac{1}{2}|S|\frac{1}{2},\frac{1}{2}\rangle & \langle \frac{1}{2},\frac{1}{2}|S|\frac{1}{2},-\frac{1}{2}\rangle \\
\langle \frac{1}{2},-\frac{1}{2}|S|\frac{1}{2},\frac{1}{2}\rangle & \langle \frac{1}{2},-\frac{1}{2}|S|\frac{1}{2},-\frac{1}{2}\rangle & \langle \frac{1}{2},-\frac{1}{2}|S|\frac{1}{2},\frac{1}{2}\rangle & \langle \frac{1}{2},-\frac{1}{2}|S|\frac{1}{2},-\frac{1}{2}\rangle & \langle \frac{1}{2},-\frac{1}{2}|S|\frac{1}{2},\frac{1}{2}\rangle & \langle \frac{1}{2},-\frac{1}{2}|S|\frac{1}{2},-\frac{1}{2}\rangle & \langle \frac{1}{2},-\frac{1}{2}|S|\frac{1}{2},\frac{1}{2}\rangle & \langle \frac{1}{2},-\frac{1}{2}|S|\frac{1}{2},-\frac{1}{2}\rangle \\
\langle \frac{1}{2},\frac{1}{2}|S|\frac{1}{2},\frac{1}{2}\rangle & \langle \frac{1}{2},\frac{1}{2}|S|\frac{1}{2},-\frac{1}{2}\rangle & \langle \frac{1}{2},\frac{1}{2}|S|\frac{1}{2},\frac{1}{2}\rangle & \langle \frac{1}{2},\frac{1}{2}|S|\frac{1}{2},-\frac{1}{2}\rangle & \langle \frac{1}{2},\frac{1}{2}|S|\frac{1}{2},\frac{1}{2}\rangle & \langle \frac{1}{2},\frac{1}{2}|S|\frac{1}{2},-\frac{1}{2}\rangle & \langle \frac{1}{2},\frac{1}{2}|S|\frac{1}{2},\frac{1}{2}\rangle & \langle \frac{1}{2},\frac{1}{2}|S|\frac{1}{2},-\frac{1}{2}\rangle \\
\langle \frac{1}{2},-\frac{1}{2}|S|\frac{1}{2},\frac{1}{2}\rangle & \langle \frac{1}{2},-\frac{1}{2}|S|\frac{1}{2},-\frac{1}{2}\rangle & \langle \frac{1}{2},-\frac{1}{2}|S|\frac{1}{2},\frac{1}{2}\rangle & \langle \frac{1}{2},-\frac{1}{2}|S|\frac{1}{2},-\frac{1}{2}\rangle & \langle \frac{1}{2},-\frac{1}{2}|S|\frac{1}{2},\frac{1}{2}\rangle & \langle \frac{1}{2},-\frac{1}{2}|S|\frac{1}{2},-\frac{1}{2}\rangle & \langle \frac{1}{2},-\frac{1}{2}|S|\frac{1}{2},\frac{1}{2}\rangle & \langle \frac{1}{2},-\frac{1}{2}|S|\frac{1}{2},-\frac{1}{2}\rangle \\
\langle \frac{1}{2},\frac{1}{2}|S|\frac{1}{2},\frac{1}{2}\rangle & \langle \frac{1}{2},\frac{1}{2}|S|\frac{1}{2},-\frac{1}{2}\rangle & \langle \frac{1}{2},\frac{1}{2}|S|\frac{1}{2},\frac{1}{2}\rangle & \langle \frac{1}{2},\frac{1}{2}|S|\frac{1}{2},-\frac{1}{2}\rangle & \langle \frac{1}{2},\frac{1}{2}|S|\frac{1}{2},\frac{1}{2}\rangle & \langle \frac{1}{2},\frac{1}{2}|S|\frac{1}{2},-\frac{1}{2}\rangle & \langle \frac{1}{2},\frac{1}{2}|S|\frac{1}{2},\frac{1}{2}\rangle & \langle \frac{1}{2},\frac{1}{2}|S|\frac{1}{2},-\frac{1}{2}\rangle \\
\langle \frac{1}{2},-\frac{1}{2}|S|\frac{1}{2},\frac{1}{2}\rangle & \langle \frac{1}{2},-\frac{1}{2}|S|\frac{1}{2},-\frac{1}{2}\rangle & \langle \frac{1}{2},-\frac{1}{2}|S|\frac{1}{2},\frac{1}{2}\rangle & \langle \frac{1}{2},-\frac{1}{2}|S|\frac{1}{2},-\frac{1}{2}\rangle & \langle \frac{1}{2},-\frac{1}{2}|S|\frac{1}{2},\frac{1}{2}\rangle & \langle \frac{1}{2},-\frac{1}{2}|S|\frac{1}{2},-\frac{1}{2}\rangle & \langle \frac{1}{2},-\frac{1}{2}|S|\frac{1}{2},\frac{1}{2}\rangle & \langle \frac{1}{2},-\frac{1}{2}|S|\frac{1}{2},-\frac{1}{2}\rangle \\
\langle \frac{1}{2},\frac{1}{2}|S|\frac{1}{2},\frac{1}{2}\rangle & \langle \frac{1}{2},\frac{1}{2}|S|\frac{1}{2},-\frac{1}{2}\rangle & \langle \frac{1}{2},\frac{1}{2}|S|\frac{1}{2},\frac{1}{2}\rangle & \langle \frac{1}{2},\frac{1}{2}|S|\frac{1}{2},-\frac{1}{2}\rangle & \langle \frac{1}{2},\frac{1}{2}|S|\frac{1}{2},\frac{1}{2}\rangle & \langle \frac{1}{2},\frac{1}{2}|S|\frac{1}{2},-\frac{1}{2}\rangle & \langle \frac{1}{2},\frac{1}{2}|S|\frac{1}{2},\frac{1}{2}\rangle & \langle \frac{1}{2},\frac{1}{2}|S|\frac{1}{2},-\frac{1}{2}\rangle \\
\langle \frac{1}{2},-\frac{1}{2}|S|\frac{1}{2},\frac{1}{2}\rangle & \langle \frac{1}{2},-\frac{1}{2}|S|\frac{1}{2},-\frac{1}{2}\rangle & \langle \frac{1}{2},-\frac{1}{2}|S|\frac{1}{2},\frac{1}{2}\rangle & \langle \frac{1}{2},-\frac{1}{2}|S|\frac{1}{2},-\frac{1}{2}\rangle & \langle \frac{1}{2},-\frac{1}{2}|S|\frac{1}{2},\frac{1}{2}\rangle & \langle \frac{1}{2},-\frac{1}{2}|S|\frac{1}{2},-\frac{1}{2}\rangle & \langle \frac{1}{2},-\frac{1}{2}|S|\frac{1}{2},\frac{1}{2}\rangle & \langle \frac{1}{2},-\frac{1}{2}|S|\frac{1}{2},-\frac{1}{2}\rangle
\end{bmatrix}_{8\times 8} \tag{A.3}$$

A.2 Particle State Matrix in QQCA

$$
s_1 = \begin{bmatrix}
\langle \frac{1}{2},\frac{1}{2}|S|\frac{1}{2},\frac{1}{2}\rangle & \langle \frac{1}{2},\frac{1}{2}|S|\frac{1}{2},\frac{1}{2}\rangle & \langle \frac{1}{2},\frac{1}{2}|S|\frac{1}{2},\frac{1}{2}\rangle & \langle \frac{1}{2},\frac{1}{2}|S|\frac{1}{2},\frac{1}{2}\rangle & \langle \frac{1}{2},\frac{1}{2}|S|\frac{1}{2},\frac{1}{2}\rangle & \langle \frac{1}{2},\frac{1}{2}|S|\frac{1}{2},\frac{1}{2}\rangle & \langle \frac{1}{2},\frac{1}{2}|S|\frac{1}{2},\frac{1}{2}\rangle & \langle \frac{1}{2},\frac{1}{2}|S|\frac{1}{2},\frac{1}{2}\rangle & \langle \frac{1}{2},\frac{1}{2}|S|\frac{1}{2},-\frac{1}{2}\rangle & \langle \frac{1}{2},\frac{1}{2}|S|\frac{1}{2},-\frac{1}{2}\rangle & \langle \frac{1}{2},\frac{1}{2}|S|\frac{1}{2},-\frac{1}{2}\rangle & \langle \frac{1}{2},\frac{1}{2}|S|\frac{1}{2},-\frac{1}{2}\rangle & \langle \frac{1}{2},\frac{1}{2}|S|\frac{1}{2},-\frac{1}{2}\rangle & \langle \frac{1}{2},\frac{1}{2}|S|\frac{1}{2},-\frac{1}{2}\rangle & \langle \frac{1}{2},\frac{1}{2}|S|\frac{1}{2},-\frac{1}{2}\rangle & \langle \frac{1}{2},\frac{1}{2}|S|\frac{1}{2},-\frac{1}{2}\rangle \\
\langle \frac{1}{2},\frac{1}{2}|S|\frac{1}{2},\frac{1}{2}\rangle & \langle \frac{1}{2},\frac{1}{2}|S|\frac{1}{2},\frac{1}{2}\rangle & \langle \frac{1}{2},\frac{1}{2}|S|\frac{1}{2},\frac{1}{2}\rangle & \langle \frac{1}{2},\frac{1}{2}|S|\frac{1}{2},\frac{1}{2}\rangle & \langle \frac{1}{2},\frac{1}{2}|S|\frac{1}{2},\frac{1}{2}\rangle & \langle \frac{1}{2},\frac{1}{2}|S|\frac{1}{2},\frac{1}{2}\rangle & \langle \frac{1}{2},\frac{1}{2}|S|\frac{1}{2},\frac{1}{2}\rangle & \langle \frac{1}{2},\frac{1}{2}|S|\frac{1}{2},\frac{1}{2}\rangle & \langle \frac{1}{2},\frac{1}{2}|S|\frac{1}{2},-\frac{1}{2}\rangle & \langle \frac{1}{2},\frac{1}{2}|S|\frac{1}{2},-\frac{1}{2}\rangle & \langle \frac{1}{2},\frac{1}{2}|S|\frac{1}{2},-\frac{1}{2}\rangle & \langle \frac{1}{2},\frac{1}{2}|S|\frac{1}{2},-\frac{1}{2}\rangle & \langle \frac{1}{2},\frac{1}{2}|S|\frac{1}{2},-\frac{1}{2}\rangle & \langle \frac{1}{2},\frac{1}{2}|S|\frac{1}{2},-\frac{1}{2}\rangle & \langle \frac{1}{2},\frac{1}{2}|S|\frac{1}{2},-\frac{1}{2}\rangle & \langle \frac{1}{2},\frac{1}{2}|S|\frac{1}{2},-\frac{1}{2}\rangle \\
\langle \frac{1}{2},\frac{1}{2}|S|\frac{1}{2},\frac{1}{2}\rangle & \langle \frac{1}{2},\frac{1}{2}|S|\frac{1}{2},\frac{1}{2}\rangle & \langle \frac{1}{2},\frac{1}{2}|S|\frac{1}{2},\frac{1}{2}\rangle & \langle \frac{1}{2},\frac{1}{2}|S|\frac{1}{2},\frac{1}{2}\rangle & \langle \frac{1}{2},\frac{1}{2}|S|\frac{1}{2},\frac{1}{2}\rangle & \langle \frac{1}{2},\frac{1}{2}|S|\frac{1}{2},\frac{1}{2}\rangle & \langle \frac{1}{2},\frac{1}{2}|S|\frac{1}{2},\frac{1}{2}\rangle & \langle \frac{1}{2},\frac{1}{2}|S|\frac{1}{2},\frac{1}{2}\rangle & \langle \frac{1}{2},\frac{1}{2}|S|\frac{1}{2},-\frac{1}{2}\rangle & \langle \frac{1}{2},\frac{1}{2}|S|\frac{1}{2},-\frac{1}{2}\rangle & \langle \frac{1}{2},\frac{1}{2}|S|\frac{1}{2},-\frac{1}{2}\rangle & \langle \frac{1}{2},\frac{1}{2}|S|\frac{1}{2},-\frac{1}{2}\rangle & \langle \frac{1}{2},\frac{1}{2}|S|\frac{1}{2},-\frac{1}{2}\rangle & \langle \frac{1}{2},\frac{1}{2}|S|\frac{1}{2},-\frac{1}{2}\rangle & \langle \frac{1}{2},\frac{1}{2}|S|\frac{1}{2},-\frac{1}{2}\rangle & \langle \frac{1}{2},\frac{1}{2}|S|\frac{1}{2},-\frac{1}{2}\rangle \\
\langle \frac{1}{2},\frac{1}{2}|S|\frac{1}{2},\frac{1}{2}\rangle & \langle \frac{1}{2},\frac{1}{2}|S|\frac{1}{2},\frac{1}{2}\rangle & \langle \frac{1}{2},\frac{1}{2}|S|\frac{1}{2},\frac{1}{2}\rangle & \langle \frac{1}{2},\frac{1}{2}|S|\frac{1}{2},\frac{1}{2}\rangle & \langle \frac{1}{2},\frac{1}{2}|S|\frac{1}{2},\frac{1}{2}\rangle & \langle \frac{1}{2},\frac{1}{2}|S|\frac{1}{2},\frac{1}{2}\rangle & \langle \frac{1}{2},\frac{1}{2}|S|\frac{1}{2},\frac{1}{2}\rangle & \langle \frac{1}{2},\frac{1}{2}|S|\frac{1}{2},\frac{1}{2}\rangle & \langle \frac{1}{2},\frac{1}{2}|S|\frac{1}{2},-\frac{1}{2}\rangle & \langle \frac{1}{2},\frac{1}{2}|S|\frac{1}{2},-\frac{1}{2}\rangle & \langle \frac{1}{2},\frac{1}{2}|S|\frac{1}{2},-\frac{1}{2}\rangle & \langle \frac{1}{2},\frac{1}{2}|S|\frac{1}{2},-\frac{1}{2}\rangle & \langle \frac{1}{2},\frac{1}{2}|S|\frac{1}{2},-\frac{1}{2}\rangle & \langle \frac{1}{2},\frac{1}{2}|S|\frac{1}{2},-\frac{1}{2}\rangle & \langle \frac{1}{2},\frac{1}{2}|S|\frac{1}{2},-\frac{1}{2}\rangle & \langle \frac{1}{2},\frac{1}{2}|S|\frac{1}{2},-\frac{1}{2}\rangle \\
\langle \frac{1}{2},\frac{1}{2}|S|\frac{1}{2},\frac{1}{2}\rangle & \langle \frac{1}{2},\frac{1}{2}|S|\frac{1}{2},\frac{1}{2}\rangle & \langle \frac{1}{2},\frac{1}{2}|S|\frac{1}{2},\frac{1}{2}\rangle & \langle \frac{1}{2},\frac{1}{2}|S|\frac{1}{2},\frac{1}{2}\rangle & \langle \frac{1}{2},\frac{1}{2}|S|\frac{1}{2},\frac{1}{2}\rangle & \langle \frac{1}{2},\frac{1}{2}|S|\frac{1}{2},\frac{1}{2}\rangle & \langle \frac{1}{2},\frac{1}{2}|S|\frac{1}{2},\frac{1}{2}\rangle & \langle \frac{1}{2},\frac{1}{2}|S|\frac{1}{2},\frac{1}{2}\rangle & \langle \frac{1}{2},\frac{1}{2}|S|\frac{1}{2},-\frac{1}{2}\rangle & \langle \frac{1}{2},\frac{1}{2}|S|\frac{1}{2},-\frac{1}{2}\rangle & \langle \frac{1}{2},\frac{1}{2}|S|\frac{1}{2},-\frac{1}{2}\rangle & \langle \frac{1}{2},\frac{1}{2}|S|\frac{1}{2},-\frac{1}{2}\rangle & \langle \frac{1}{2},\frac{1}{2}|S|\frac{1}{2},-\frac{1}{2}\rangle & \langle \frac{1}{2},\frac{1}{2}|S|\frac{1}{2},-\frac{1}{2}\rangle & \langle \frac{1}{2},\frac{1}{2}|S|\frac{1}{2},-\frac{1}{2}\rangle & \langle \frac{1}{2},\frac{1}{2}|S|\frac{1}{2},-\frac{1}{2}\rangle \\
\langle \frac{1}{2},\frac{1}{2}|S|\frac{1}{2},\frac{1}{2}\rangle & \langle \frac{1}{2},\frac{1}{2}|S|\frac{1}{2},\frac{1}{2}\rangle & \langle \frac{1}{2},\frac{1}{2}|S|\frac{1}{2},\frac{1}{2}\rangle & \langle \frac{1}{2},\frac{1}{2}|S|\frac{1}{2},\frac{1}{2}\rangle & \langle \frac{1}{2},\frac{1}{2}|S|\frac{1}{2},\frac{1}{2}\rangle & \langle \frac{1}{2},\frac{1}{2}|S|\frac{1}{2},\frac{1}{2}\rangle & \langle \frac{1}{2},\frac{1}{2}|S|\frac{1}{2},\frac{1}{2}\rangle & \langle \frac{1}{2},\frac{1}{2}|S|\frac{1}{2},\frac{1}{2}\rangle & \langle \frac{1}{2},\frac{1}{2}|S|\frac{1}{2},-\frac{1}{2}\rangle & \langle \frac{1}{2},\frac{1}{2}|S|\frac{1}{2},-\frac{1}{2}\rangle & \langle \frac{1}{2},\frac{1}{2}|S|\frac{1}{2},-\frac{1}{2}\rangle & \langle \frac{1}{2},\frac{1}{2}|S|\frac{1}{2},-\frac{1}{2}\rangle & \langle \frac{1}{2},\frac{1}{2}|S|\frac{1}{2},-\frac{1}{2}\rangle & \langle \frac{1}{2},\frac{1}{2}|S|\frac{1}{2},-\frac{1}{2}\rangle & \langle \frac{1}{2},\frac{1}{2}|S|\frac{1}{2},-\frac{1}{2}\rangle & \langle \frac{1}{2},\frac{1}{2}|S|\frac{1}{2},-\frac{1}{2}\rangle \\
\langle \frac{1}{2},\frac{1}{2}|S|\frac{1}{2},\frac{1}{2}\rangle & \langle \frac{1}{2},\frac{1}{2}|S|\frac{1}{2},\frac{1}{2}\rangle & \langle \frac{1}{2},\frac{1}{2}|S|\frac{1}{2},\frac{1}{2}\rangle & \langle \frac{1}{2},\frac{1}{2}|S|\frac{1}{2},\frac{1}{2}\rangle & \langle \frac{1}{2},\frac{1}{2}|S|\frac{1}{2},\frac{1}{2}\rangle & \langle \frac{1}{2},\frac{1}{2}|S|\frac{1}{2},\frac{1}{2}\rangle & \langle \frac{1}{2},\frac{1}{2}|S|\frac{1}{2},\frac{1}{2}\rangle & \langle \frac{1}{2},\frac{1}{2}|S|\frac{1}{2},\frac{1}{2}\rangle & \langle \frac{1}{2},\frac{1}{2}|S|\frac{1}{2},-\frac{1}{2}\rangle & \langle \frac{1}{2},\frac{1}{2}|S|\frac{1}{2},-\frac{1}{2}\rangle & \langle \frac{1}{2},\frac{1}{2}|S|\frac{1}{2},-\frac{1}{2}\rangle & \langle \frac{1}{2},\frac{1}{2}|S|\frac{1}{2},-\frac{1}{2}\rangle & \langle \frac{1}{2},\frac{1}{2}|S|\frac{1}{2},-\frac{1}{2}\rangle & \langle \frac{1}{2},\frac{1}{2}|S|\frac{1}{2},-\frac{1}{2}\rangle & \langle \frac{1}{2},\frac{1}{2}|S|\frac{1}{2},-\frac{1}{2}\rangle & \langle \frac{1}{2},\frac{1}{2}|S|\frac{1}{2},-\frac{1}{2}\rangle \\
\langle \frac{1}{2},\frac{1}{2}|S|\frac{1}{2},\frac{1}{2}\rangle & \langle \frac{1}{2},\frac{1}{2}|S|\frac{1}{2},\frac{1}{2}\rangle & \langle \frac{1}{2},\frac{1}{2}|S|\frac{1}{2},\frac{1}{2}\rangle & \langle \frac{1}{2},\frac{1}{2}|S|\frac{1}{2},\frac{1}{2}\rangle & \langle \frac{1}{2},\frac{1}{2}|S|\frac{1}{2},\frac{1}{2}\rangle & \langle \frac{1}{2},\frac{1}{2}|S|\frac{1}{2},\frac{1}{2}\rangle & \langle \frac{1}{2},\frac{1}{2}|S|\frac{1}{2},\frac{1}{2}\rangle & \langle \frac{1}{2},\frac{1}{2}|S|\frac{1}{2},\frac{1}{2}\rangle & \langle \frac{1}{2},\frac{1}{2}|S|\frac{1}{2},-\frac{1}{2}\rangle & \langle \frac{1}{2},\frac{1}{2}|S|\frac{1}{2},-\frac{1}{2}\rangle & \langle \frac{1}{2},\frac{1}{2}|S|\frac{1}{2},-\frac{1}{2}\rangle & \langle \frac{1}{2},\frac{1}{2}|S|\frac{1}{2},-\frac{1}{2}\rangle & \langle \frac{1}{2},\frac{1}{2}|S|\frac{1}{2},-\frac{1}{2}\rangle & \langle \frac{1}{2},\frac{1}{2}|S|\frac{1}{2},-\frac{1}{2}\rangle & \langle \frac{1}{2},\frac{1}{2}|S|\frac{1}{2},-\frac{1}{2}\rangle & \langle \frac{1}{2},\frac{1}{2}|S|\frac{1}{2},-\frac{1}{2}\rangle \\
\langle \frac{1}{2},-\frac{1}{2}|S|\frac{1}{2},\frac{1}{2}\rangle & \langle \frac{1}{2},-\frac{1}{2}|S|\frac{1}{2},\frac{1}{2}\rangle & \langle \frac{1}{2},-\frac{1}{2}|S|\frac{1}{2},\frac{1}{2}\rangle & \langle \frac{1}{2},-\frac{1}{2}|S|\frac{1}{2},\frac{1}{2}\rangle & \langle \frac{1}{2},-\frac{1}{2}|S|\frac{1}{2},\frac{1}{2}\rangle & \langle \frac{1}{2},-\frac{1}{2}|S|\frac{1}{2},\frac{1}{2}\rangle & \langle \frac{1}{2},-\frac{1}{2}|S|\frac{1}{2},\frac{1}{2}\rangle & \langle \frac{1}{2},-\frac{1}{2}|S|\frac{1}{2},\frac{1}{2}\rangle & \langle \frac{1}{2},-\frac{1}{2}|S|\frac{1}{2},-\frac{1}{2}\rangle & \langle \frac{1}{2},-\frac{1}{2}|S|\frac{1}{2},-\frac{1}{2}\rangle & \langle \frac{1}{2},-\frac{1}{2}|S|\frac{1}{2},-\frac{1}{2}\rangle & \langle \frac{1}{2},-\frac{1}{2}|S|\frac{1}{2},-\frac{1}{2}\rangle & \langle \frac{1}{2},-\frac{1}{2}|S|\frac{1}{2},-\frac{1}{2}\rangle & \langle \frac{1}{2},-\frac{1}{2}|S|\frac{1}{2},-\frac{1}{2}\rangle & \langle \frac{1}{2},-\frac{1}{2}|S|\frac{1}{2},-\frac{1}{2}\rangle & \langle \frac{1}{2},-\frac{1}{2}|S|\frac{1}{2},-\frac{1}{2}\rangle \\
\langle \frac{1}{2},-\frac{1}{2}|S|\frac{1}{2},\frac{1}{2}\rangle & \langle \frac{1}{2},-\frac{1}{2}|S|\frac{1}{2},\frac{1}{2}\rangle & \langle \frac{1}{2},-\frac{1}{2}|S|\frac{1}{2},\frac{1}{2}\rangle & \langle \frac{1}{2},-\frac{1}{2}|S|\frac{1}{2},\frac{1}{2}\rangle & \langle \frac{1}{2},-\frac{1}{2}|S|\frac{1}{2},\frac{1}{2}\rangle & \langle \frac{1}{2},-\frac{1}{2}|S|\frac{1}{2},\frac{1}{2}\rangle & \langle \frac{1}{2},-\frac{1}{2}|S|\frac{1}{2},\frac{1}{2}\rangle & \langle \frac{1}{2},-\frac{1}{2}|S|\frac{1}{2},\frac{1}{2}\rangle & \langle \frac{1}{2},-\frac{1}{2}|S|\frac{1}{2},-\frac{1}{2}\rangle & \langle \frac{1}{2},-\frac{1}{2}|S|\frac{1}{2},-\frac{1}{2}\rangle & \langle \frac{1}{2},-\frac{1}{2}|S|\frac{1}{2},-\frac{1}{2}\rangle & \langle \frac{1}{2},-\frac{1}{2}|S|\frac{1}{2},-\frac{1}{2}\rangle & \langle \frac{1}{2},-\frac{1}{2}|S|\frac{1}{2},-\frac{1}{2}\rangle & \langle \frac{1}{2},-\frac{1}{2}|S|\frac{1}{2},-\frac{1}{2}\rangle & \langle \frac{1}{2},-\frac{1}{2}|S|\frac{1}{2},-\frac{1}{2}\rangle & \langle \frac{1}{2},-\frac{1}{2}|S|\frac{1}{2},-\frac{1}{2}\rangle \\
\langle \frac{1}{2},-\frac{1}{2}|S|\frac{1}{2},\frac{1}{2}\rangle & \langle \frac{1}{2},-\frac{1}{2}|S|\frac{1}{2},\frac{1}{2}\rangle & \langle \frac{1}{2},-\frac{1}{2}|S|\frac{1}{2},\frac{1}{2}\rangle & \langle \frac{1}{2},-\frac{1}{2}|S|\frac{1}{2},\frac{1}{2}\rangle & \langle \frac{1}{2},-\frac{1}{2}|S|\frac{1}{2},\frac{1}{2}\rangle & \langle \frac{1}{2},-\frac{1}{2}|S|\frac{1}{2},\frac{1}{2}\rangle & \langle \frac{1}{2},-\frac{1}{2}|S|\frac{1}{2},\frac{1}{2}\rangle & \langle \frac{1}{2},-\frac{1}{2}|S|\frac{1}{2},\frac{1}{2}\rangle & \langle \frac{1}{2},-\frac{1}{2}|S|\frac{1}{2},-\frac{1}{2}\rangle & \langle \frac{1}{2},-\frac{1}{2}|S|\frac{1}{2},-\frac{1}{2}\rangle & \langle \frac{1}{2},-\frac{1}{2}|S|\frac{1}{2},-\frac{1}{2}\rangle & \langle \frac{1}{2},-\frac{1}{2}|S|\frac{1}{2},-\frac{1}{2}\rangle & \langle \frac{1}{2},-\frac{1}{2}|S|\frac{1}{2},-\frac{1}{2}\rangle & \langle \frac{1}{2},-\frac{1}{2}|S|\frac{1}{2},-\frac{1}{2}\rangle & \langle \frac{1}{2},-\frac{1}{2}|S|\frac{1}{2},-\frac{1}{2}\rangle & \langle \frac{1}{2},-\frac{1}{2}|S|\frac{1}{2},-\frac{1}{2}\rangle \\
\langle \frac{1}{2},-\frac{1}{2}|S|\frac{1}{2},\frac{1}{2}\rangle & \langle \frac{1}{2},-\frac{1}{2}|S|\frac{1}{2},\frac{1}{2}\rangle & \langle \frac{1}{2},-\frac{1}{2}|S|\frac{1}{2},\frac{1}{2}\rangle & \langle \frac{1}{2},-\frac{1}{2}|S|\frac{1}{2},\frac{1}{2}\rangle & \langle \frac{1}{2},-\frac{1}{2}|S|\frac{1}{2},\frac{1}{2}\rangle & \langle \frac{1}{2},-\frac{1}{2}|S|\frac{1}{2},\frac{1}{2}\rangle & \langle \frac{1}{2},-\frac{1}{2}|S|\frac{1}{2},\frac{1}{2}\rangle & \langle \frac{1}{2},-\frac{1}{2}|S|\frac{1}{2},\frac{1}{2}\rangle & \langle \frac{1}{2},-\frac{1}{2}|S|\frac{1}{2},-\frac{1}{2}\rangle & \langle \frac{1}{2},-\frac{1}{2}|S|\frac{1}{2},-\frac{1}{2}\rangle & \langle \frac{1}{2},-\frac{1}{2}|S|\frac{1}{2},-\frac{1}{2}\rangle & \langle \frac{1}{2},-\frac{1}{2}|S|\frac{1}{2},-\frac{1}{2}\rangle & \langle \frac{1}{2},-\frac{1}{2}|S|\frac{1}{2},-\frac{1}{2}\rangle & \langle \frac{1}{2},-\frac{1}{2}|S|\frac{1}{2},-\frac{1}{2}\rangle & \langle \frac{1}{2},-\frac{1}{2}|S|\frac{1}{2},-\frac{1}{2}\rangle & \langle \frac{1}{2},-\frac{1}{2}|S|\frac{1}{2},-\frac{1}{2}\rangle \\
\langle \frac{1}{2},-\frac{1}{2}|S|\frac{1}{2},\frac{1}{2}\rangle & \langle \frac{1}{2},-\frac{1}{2}|S|\frac{1}{2},\frac{1}{2}\rangle & \langle \frac{1}{2},-\frac{1}{2}|S|\frac{1}{2},\frac{1}{2}\rangle & \langle \frac{1}{2},-\frac{1}{2}|S|\frac{1}{2},\frac{1}{2}\rangle & \langle \frac{1}{2},-\frac{1}{2}|S|\frac{1}{2},\frac{1}{2}\rangle & \langle \frac{1}{2},-\frac{1}{2}|S|\frac{1}{2},\frac{1}{2}\rangle & \langle \frac{1}{2},-\frac{1}{2}|S|\frac{1}{2},\frac{1}{2}\rangle & \langle \frac{1}{2},-\frac{1}{2}|S|\frac{1}{2},\frac{1}{2}\rangle & \langle \frac{1}{2},-\frac{1}{2}|S|\frac{1}{2},-\frac{1}{2}\rangle & \langle \frac{1}{2},-\frac{1}{2}|S|\frac{1}{2},-\frac{1}{2}\rangle & \langle \frac{1}{2},-\frac{1}{2}|S|\frac{1}{2},-\frac{1}{2}\rangle & \langle \frac{1}{2},-\frac{1}{2}|S|\frac{1}{2},-\frac{1}{2}\rangle & \langle \frac{1}{2},-\frac{1}{2}|S|\frac{1}{2},-\frac{1}{2}\rangle & \langle \frac{1}{2},-\frac{1}{2}|S|\frac{1}{2},-\frac{1}{2}\rangle & \langle \frac{1}{2},-\frac{1}{2}|S|\frac{1}{2},-\frac{1}{2}\rangle & \langle \frac{1}{2},-\frac{1}{2}|S|\frac{1}{2},-\frac{1}{2}\rangle \\
\langle \frac{1}{2},-\frac{1}{2}|S|\frac{1}{2},\frac{1}{2}\rangle & \langle \frac{1}{2},-\frac{1}{2}|S|\frac{1}{2},\frac{1}{2}\rangle & \langle \frac{1}{2},-\frac{1}{2}|S|\frac{1}{2},\frac{1}{2}\rangle & \langle \frac{1}{2},-\frac{1}{2}|S|\frac{1}{2},\frac{1}{2}\rangle & \langle \frac{1}{2},-\frac{1}{2}|S|\frac{1}{2},\frac{1}{2}\rangle & \langle \frac{1}{2},-\frac{1}{2}|S|\frac{1}{2},\frac{1}{2}\rangle & \langle \frac{1}{2},-\frac{1}{2}|S|\frac{1}{2},\frac{1}{2}\rangle & \langle \frac{1}{2},-\frac{1}{2}|S|\frac{1}{2},\frac{1}{2}\rangle & \langle \frac{1}{2},-\frac{1}{2}|S|\frac{1}{2},-\frac{1}{2}\rangle & \langle \frac{1}{2},-\frac{1}{2}|S|\frac{1}{2},-\frac{1}{2}\rangle & \langle \frac{1}{2},-\frac{1}{2}|S|\frac{1}{2},-\frac{1}{2}\rangle & \langle \frac{1}{2},-\frac{1}{2}|S|\frac{1}{2},-\frac{1}{2}\rangle & \langle \frac{1}{2},-\frac{1}{2}|S|\frac{1}{2},-\frac{1}{2}\rangle & \langle \frac{1}{2},-\frac{1}{2}|S|\frac{1}{2},-\frac{1}{2}\rangle & \langle \frac{1}{2},-\frac{1}{2}|S|\frac{1}{2},-\frac{1}{2}\rangle & \langle \frac{1}{2},-\frac{1}{2}|S|\frac{1}{2},-\frac{1}{2}\rangle \\
\langle \frac{1}{2},-\frac{1}{2}|S|\frac{1}{2},\frac{1}{2}\rangle & \langle \frac{1}{2},-\frac{1}{2}|S|\frac{1}{2},\frac{1}{2}\rangle & \langle \frac{1}{2},-\frac{1}{2}|S|\frac{1}{2},\frac{1}{2}\rangle & \langle \frac{1}{2},-\frac{1}{2}|S|\frac{1}{2},\frac{1}{2}\rangle & \langle \frac{1}{2},-\frac{1}{2}|S|\frac{1}{2},\frac{1}{2}\rangle & \langle \frac{1}{2},-\frac{1}{2}|S|\frac{1}{2},\frac{1}{2}\rangle & \langle \frac{1}{2},-\frac{1}{2}|S|\frac{1}{2},\frac{1}{2}\rangle & \langle \frac{1}{2},-\frac{1}{2}|S|\frac{1}{2},\frac{1}{2}\rangle & \langle \frac{1}{2},-\frac{1}{2}|S|\frac{1}{2},-\frac{1}{2}\rangle & \langle \frac{1}{2},-\frac{1}{2}|S|\frac{1}{2},-\frac{1}{2}\rangle & \langle \frac{1}{2},-\frac{1}{2}|S|\frac{1}{2},-\frac{1}{2}\rangle & \langle \frac{1}{2},-\frac{1}{2}|S|\frac{1}{2},-\frac{1}{2}\rangle & \langle \frac{1}{2},-\frac{1}{2}|S|\frac{1}{2},-\frac{1}{2}\rangle & \langle \frac{1}{2},-\frac{1}{2}|S|\frac{1}{2},-\frac{1}{2}\rangle & \langle \frac{1}{2},-\frac{1}{2}|S|\frac{1}{2},-\frac{1}{2}\rangle & \langle \frac{1}{2},-\frac{1}{2}|S|\frac{1}{2},-\frac{1}{2}\rangle \\
\langle \frac{1}{2},-\frac{1}{2}|S|\frac{1}{2},\frac{1}{2}\rangle & \langle \frac{1}{2},-\frac{1}{2}|S|\frac{1}{2},\frac{1}{2}\rangle & \langle \frac{1}{2},-\frac{1}{2}|S|\frac{1}{2},\frac{1}{2}\rangle & \langle \frac{1}{2},-\frac{1}{2}|S|\frac{1}{2},\frac{1}{2}\rangle & \langle \frac{1}{2},-\frac{1}{2}|S|\frac{1}{2},\frac{1}{2}\rangle & \langle \frac{1}{2},-\frac{1}{2}|S|\frac{1}{2},\frac{1}{2}\rangle & \langle \frac{1}{2},-\frac{1}{2}|S|\frac{1}{2},\frac{1}{2}\rangle & \langle \frac{1}{2},-\frac{1}{2}|S|\frac{1}{2},\frac{1}{2}\rangle & \langle \frac{1}{2},-\frac{1}{2}|S|\frac{1}{2},-\frac{1}{2}\rangle & \langle \frac{1}{2},-\frac{1}{2}|S|\frac{1}{2},-\frac{1}{2}\rangle & \langle \frac{1}{2},-\frac{1}{2}|S|\frac{1}{2},-\frac{1}{2}\rangle & \langle \frac{1}{2},-\frac{1}{2}|S|\frac{1}{2},-\frac{1}{2}\rangle & \langle \frac{1}{2},-\frac{1}{2}|S|\frac{1}{2},-\frac{1}{2}\rangle & \langle \frac{1}{2},-\frac{1}{2}|S|\frac{1}{2},-\frac{1}{2}\rangle & \langle \frac{1}{2},-\frac{1}{2}|S|\frac{1}{2},-\frac{1}{2}\rangle & \langle \frac{1}{2},-\frac{1}{2}|S|\frac{1}{2},-\frac{1}{2}\rangle
\end{bmatrix}_{16\times16} \quad \text{(A.4)}
$$

$$
S_2=\begin{bmatrix}
\langle\frac{1}{2},\frac{1}{2}|S|\frac{1}{2},\frac{1}{2}\rangle & \langle\frac{1}{2},\frac{1}{2}|S|\frac{1}{2},\frac{1}{2}\rangle & \langle\frac{1}{2},\frac{1}{2}|S|\frac{1}{2},\frac{1}{2}\rangle & \langle\frac{1}{2},\frac{1}{2}|S|\frac{1}{2},\frac{1}{2}\rangle & \langle\frac{1}{2},\frac{1}{2}|S|\frac{1}{2},-\frac{1}{2}\rangle & \langle\frac{1}{2},-\frac{1}{2}|S|\frac{1}{2},-\frac{1}{2}\rangle & \langle\frac{1}{2},\frac{1}{2}|S|\frac{1}{2},-\frac{1}{2}\rangle & \langle\frac{1}{2},\frac{1}{2}|S|\frac{1}{2},-\frac{1}{2}\rangle & \langle\frac{1}{2},\frac{1}{2}|S|\frac{1}{2},\frac{1}{2}\rangle & \langle\frac{1}{2},\frac{1}{2}|S|\frac{1}{2},\frac{1}{2}\rangle & \langle\frac{1}{2},\frac{1}{2}|S|\frac{1}{2},\frac{1}{2}\rangle & \langle\frac{1}{2},\frac{1}{2}|S|\frac{1}{2},\frac{1}{2}\rangle & \langle\frac{1}{2},\frac{1}{2}|S|\frac{1}{2},-\frac{1}{2}\rangle & \langle\frac{1}{2},\frac{1}{2}|S|\frac{1}{2},-\frac{1}{2}\rangle & \langle\frac{1}{2},\frac{1}{2}|S|\frac{1}{2},-\frac{1}{2}\rangle & \langle\frac{1}{2},\frac{1}{2}|S|\frac{1}{2},-\frac{1}{2}\rangle \\
\langle\frac{1}{2},\frac{1}{2}|S|\frac{1}{2},\frac{1}{2}\rangle & \langle\frac{1}{2},\frac{1}{2}|S|\frac{1}{2},\frac{1}{2}\rangle & \langle\frac{1}{2},\frac{1}{2}|S|\frac{1}{2},\frac{1}{2}\rangle & \langle\frac{1}{2},\frac{1}{2}|S|\frac{1}{2},\frac{1}{2}\rangle & \langle\frac{1}{2},\frac{1}{2}|S|\frac{1}{2},-\frac{1}{2}\rangle & \langle\frac{1}{2},\frac{1}{2}|S|\frac{1}{2},-\frac{1}{2}\rangle & \langle\frac{1}{2},\frac{1}{2}|S|\frac{1}{2},-\frac{1}{2}\rangle & \langle\frac{1}{2},\frac{1}{2}|S|\frac{1}{2},-\frac{1}{2}\rangle & \langle\frac{1}{2},\frac{1}{2}|S|\frac{1}{2},\frac{1}{2}\rangle & \langle\frac{1}{2},\frac{1}{2}|S|\frac{1}{2},\frac{1}{2}\rangle & \langle\frac{1}{2},\frac{1}{2}|S|\frac{1}{2},\frac{1}{2}\rangle & \langle\frac{1}{2},\frac{1}{2}|S|\frac{1}{2},\frac{1}{2}\rangle & \langle\frac{1}{2},\frac{1}{2}|S|\frac{1}{2},-\frac{1}{2}\rangle & \langle\frac{1}{2},\frac{1}{2}|S|\frac{1}{2},-\frac{1}{2}\rangle & \langle\frac{1}{2},\frac{1}{2}|S|\frac{1}{2},-\frac{1}{2}\rangle & \langle\frac{1}{2},\frac{1}{2}|S|\frac{1}{2},-\frac{1}{2}\rangle \\
\langle\frac{1}{2},\frac{1}{2}|S|\frac{1}{2},\frac{1}{2}\rangle & \langle\frac{1}{2},\frac{1}{2}|S|\frac{1}{2},\frac{1}{2}\rangle & \langle\frac{1}{2},\frac{1}{2}|S|\frac{1}{2},\frac{1}{2}\rangle & \langle\frac{1}{2},\frac{1}{2}|S|\frac{1}{2},\frac{1}{2}\rangle & \langle\frac{1}{2},\frac{1}{2}|S|\frac{1}{2},-\frac{1}{2}\rangle & \langle\frac{1}{2},\frac{1}{2}|S|\frac{1}{2},-\frac{1}{2}\rangle & \langle\frac{1}{2},\frac{1}{2}|S|\frac{1}{2},-\frac{1}{2}\rangle & \langle\frac{1}{2},\frac{1}{2}|S|\frac{1}{2},-\frac{1}{2}\rangle & \langle\frac{1}{2},\frac{1}{2}|S|\frac{1}{2},\frac{1}{2}\rangle & \langle\frac{1}{2},\frac{1}{2}|S|\frac{1}{2},\frac{1}{2}\rangle & \langle\frac{1}{2},\frac{1}{2}|S|\frac{1}{2},\frac{1}{2}\rangle & \langle\frac{1}{2},\frac{1}{2}|S|\frac{1}{2},\frac{1}{2}\rangle & \langle\frac{1}{2},\frac{1}{2}|S|\frac{1}{2},-\frac{1}{2}\rangle & \langle\frac{1}{2},\frac{1}{2}|S|\frac{1}{2},-\frac{1}{2}\rangle & \langle\frac{1}{2},\frac{1}{2}|S|\frac{1}{2},-\frac{1}{2}\rangle & \langle\frac{1}{2},\frac{1}{2}|S|\frac{1}{2},-\frac{1}{2}\rangle \\
\langle\frac{1}{2},\frac{1}{2}|S|\frac{1}{2},\frac{1}{2}\rangle & \langle\frac{1}{2},\frac{1}{2}|S|\frac{1}{2},\frac{1}{2}\rangle & \langle\frac{1}{2},\frac{1}{2}|S|\frac{1}{2},\frac{1}{2}\rangle & \langle\frac{1}{2},\frac{1}{2}|S|\frac{1}{2},\frac{1}{2}\rangle & \langle\frac{1}{2},\frac{1}{2}|S|\frac{1}{2},-\frac{1}{2}\rangle & \langle\frac{1}{2},\frac{1}{2}|S|\frac{1}{2},-\frac{1}{2}\rangle & \langle\frac{1}{2},\frac{1}{2}|S|\frac{1}{2},-\frac{1}{2}\rangle & \langle\frac{1}{2},\frac{1}{2}|S|\frac{1}{2},-\frac{1}{2}\rangle & \langle\frac{1}{2},\frac{1}{2}|S|\frac{1}{2},\frac{1}{2}\rangle & \langle\frac{1}{2},\frac{1}{2}|S|\frac{1}{2},\frac{1}{2}\rangle & \langle\frac{1}{2},\frac{1}{2}|S|\frac{1}{2},\frac{1}{2}\rangle & \langle\frac{1}{2},\frac{1}{2}|S|\frac{1}{2},\frac{1}{2}\rangle & \langle\frac{1}{2},\frac{1}{2}|S|\frac{1}{2},-\frac{1}{2}\rangle & \langle\frac{1}{2},\frac{1}{2}|S|\frac{1}{2},-\frac{1}{2}\rangle & \langle\frac{1}{2},\frac{1}{2}|S|\frac{1}{2},-\frac{1}{2}\rangle & \langle\frac{1}{2},\frac{1}{2}|S|\frac{1}{2},-\frac{1}{2}\rangle \\
\langle\frac{1}{2},-\frac{1}{2}|S|\frac{1}{2},\frac{1}{2}\rangle & \langle\frac{1}{2},-\frac{1}{2}|S|\frac{1}{2},\frac{1}{2}\rangle & \langle\frac{1}{2},-\frac{1}{2}|S|\frac{1}{2},\frac{1}{2}\rangle & \langle\frac{1}{2},-\frac{1}{2}|S|\frac{1}{2},\frac{1}{2}\rangle & \langle\frac{1}{2},-\frac{1}{2}|S|\frac{1}{2},-\frac{1}{2}\rangle & \langle\frac{1}{2},-\frac{1}{2}|S|\frac{1}{2},-\frac{1}{2}\rangle & \langle\frac{1}{2},-\frac{1}{2}|S|\frac{1}{2},-\frac{1}{2}\rangle & \langle\frac{1}{2},-\frac{1}{2}|S|\frac{1}{2},-\frac{1}{2}\rangle & \langle\frac{1}{2},-\frac{1}{2}|S|\frac{1}{2},\frac{1}{2}\rangle & \langle\frac{1}{2},-\frac{1}{2}|S|\frac{1}{2},\frac{1}{2}\rangle & \langle\frac{1}{2},-\frac{1}{2}|S|\frac{1}{2},\frac{1}{2}\rangle & \langle\frac{1}{2},-\frac{1}{2}|S|\frac{1}{2},\frac{1}{2}\rangle & \langle\frac{1}{2},-\frac{1}{2}|S|\frac{1}{2},-\frac{1}{2}\rangle & \langle\frac{1}{2},-\frac{1}{2}|S|\frac{1}{2},-\frac{1}{2}\rangle & \langle\frac{1}{2},-\frac{1}{2}|S|\frac{1}{2},-\frac{1}{2}\rangle & \langle\frac{1}{2},-\frac{1}{2}|S|\frac{1}{2},-\frac{1}{2}\rangle \\
\langle\frac{1}{2},-\frac{1}{2}|S|\frac{1}{2},\frac{1}{2}\rangle & \langle\frac{1}{2},-\frac{1}{2}|S|\frac{1}{2},\frac{1}{2}\rangle & \langle\frac{1}{2},-\frac{1}{2}|S|\frac{1}{2},\frac{1}{2}\rangle & \langle\frac{1}{2},-\frac{1}{2}|S|\frac{1}{2},\frac{1}{2}\rangle & \langle\frac{1}{2},-\frac{1}{2}|S|\frac{1}{2},-\frac{1}{2}\rangle & \langle\frac{1}{2},-\frac{1}{2}|S|\frac{1}{2},-\frac{1}{2}\rangle & \langle\frac{1}{2},-\frac{1}{2}|S|\frac{1}{2},-\frac{1}{2}\rangle & \langle\frac{1}{2},-\frac{1}{2}|S|\frac{1}{2},-\frac{1}{2}\rangle & \langle\frac{1}{2},-\frac{1}{2}|S|\frac{1}{2},\frac{1}{2}\rangle & \langle\frac{1}{2},-\frac{1}{2}|S|\frac{1}{2},\frac{1}{2}\rangle & \langle\frac{1}{2},-\frac{1}{2}|S|\frac{1}{2},\frac{1}{2}\rangle & \langle\frac{1}{2},-\frac{1}{2}|S|\frac{1}{2},\frac{1}{2}\rangle & \langle\frac{1}{2},-\frac{1}{2}|S|\frac{1}{2},-\frac{1}{2}\rangle & \langle\frac{1}{2},-\frac{1}{2}|S|\frac{1}{2},-\frac{1}{2}\rangle & \langle\frac{1}{2},-\frac{1}{2}|S|\frac{1}{2},-\frac{1}{2}\rangle & \langle\frac{1}{2},-\frac{1}{2}|S|\frac{1}{2},-\frac{1}{2}\rangle \\
\langle\frac{1}{2},-\frac{1}{2}|S|\frac{1}{2},\frac{1}{2}\rangle & \langle\frac{1}{2},-\frac{1}{2}|S|\frac{1}{2},\frac{1}{2}\rangle & \langle\frac{1}{2},-\frac{1}{2}|S|\frac{1}{2},\frac{1}{2}\rangle & \langle\frac{1}{2},-\frac{1}{2}|S|\frac{1}{2},\frac{1}{2}\rangle & \langle\frac{1}{2},-\frac{1}{2}|S|\frac{1}{2},-\frac{1}{2}\rangle & \langle\frac{1}{2},-\frac{1}{2}|S|\frac{1}{2},-\frac{1}{2}\rangle & \langle\frac{1}{2},-\frac{1}{2}|S|\frac{1}{2},-\frac{1}{2}\rangle & \langle\frac{1}{2},-\frac{1}{2}|S|\frac{1}{2},-\frac{1}{2}\rangle & \langle\frac{1}{2},-\frac{1}{2}|S|\frac{1}{2},\frac{1}{2}\rangle & \langle\frac{1}{2},-\frac{1}{2}|S|\frac{1}{2},\frac{1}{2}\rangle & \langle\frac{1}{2},-\frac{1}{2}|S|\frac{1}{2},\frac{1}{2}\rangle & \langle\frac{1}{2},-\frac{1}{2}|S|\frac{1}{2},\frac{1}{2}\rangle & \langle\frac{1}{2},-\frac{1}{2}|S|\frac{1}{2},-\frac{1}{2}\rangle & \langle\frac{1}{2},-\frac{1}{2}|S|\frac{1}{2},-\frac{1}{2}\rangle & \langle\frac{1}{2},-\frac{1}{2}|S|\frac{1}{2},-\frac{1}{2}\rangle & \langle\frac{1}{2},-\frac{1}{2}|S|\frac{1}{2},-\frac{1}{2}\rangle \\
\langle\frac{1}{2},-\frac{1}{2}|S|\frac{1}{2},\frac{1}{2}\rangle & \langle\frac{1}{2},-\frac{1}{2}|S|\frac{1}{2},\frac{1}{2}\rangle & \langle\frac{1}{2},-\frac{1}{2}|S|\frac{1}{2},\frac{1}{2}\rangle & \langle\frac{1}{2},-\frac{1}{2}|S|\frac{1}{2},\frac{1}{2}\rangle & \langle\frac{1}{2},-\frac{1}{2}|S|\frac{1}{2},-\frac{1}{2}\rangle & \langle\frac{1}{2},-\frac{1}{2}|S|\frac{1}{2},-\frac{1}{2}\rangle & \langle\frac{1}{2},-\frac{1}{2}|S|\frac{1}{2},-\frac{1}{2}\rangle & \langle\frac{1}{2},-\frac{1}{2}|S|\frac{1}{2},-\frac{1}{2}\rangle & \langle\frac{1}{2},-\frac{1}{2}|S|\frac{1}{2},\frac{1}{2}\rangle & \langle\frac{1}{2},-\frac{1}{2}|S|\frac{1}{2},\frac{1}{2}\rangle & \langle\frac{1}{2},-\frac{1}{2}|S|\frac{1}{2},\frac{1}{2}\rangle & \langle\frac{1}{2},-\frac{1}{2}|S|\frac{1}{2},\frac{1}{2}\rangle & \langle\frac{1}{2},-\frac{1}{2}|S|\frac{1}{2},-\frac{1}{2}\rangle & \langle\frac{1}{2},-\frac{1}{2}|S|\frac{1}{2},-\frac{1}{2}\rangle & \langle\frac{1}{2},-\frac{1}{2}|S|\frac{1}{2},-\frac{1}{2}\rangle & \langle\frac{1}{2},-\frac{1}{2}|S|\frac{1}{2},-\frac{1}{2}\rangle \\
\langle\frac{1}{2},\frac{1}{2}|S|\frac{1}{2},\frac{1}{2}\rangle & \langle\frac{1}{2},\frac{1}{2}|S|\frac{1}{2},\frac{1}{2}\rangle & \langle\frac{1}{2},\frac{1}{2}|S|\frac{1}{2},\frac{1}{2}\rangle & \langle\frac{1}{2},\frac{1}{2}|S|\frac{1}{2},\frac{1}{2}\rangle & \langle\frac{1}{2},\frac{1}{2}|S|\frac{1}{2},-\frac{1}{2}\rangle & \langle\frac{1}{2},\frac{1}{2}|S|\frac{1}{2},-\frac{1}{2}\rangle & \langle\frac{1}{2},\frac{1}{2}|S|\frac{1}{2},-\frac{1}{2}\rangle & \langle\frac{1}{2},\frac{1}{2}|S|\frac{1}{2},-\frac{1}{2}\rangle & \langle\frac{1}{2},\frac{1}{2}|S|\frac{1}{2},\frac{1}{2}\rangle & \langle\frac{1}{2},\frac{1}{2}|S|\frac{1}{2},\frac{1}{2}\rangle & \langle\frac{1}{2},\frac{1}{2}|S|\frac{1}{2},\frac{1}{2}\rangle & \langle\frac{1}{2},\frac{1}{2}|S|\frac{1}{2},\frac{1}{2}\rangle & \langle\frac{1}{2},\frac{1}{2}|S|\frac{1}{2},-\frac{1}{2}\rangle & \langle\frac{1}{2},\frac{1}{2}|S|\frac{1}{2},-\frac{1}{2}\rangle & \langle\frac{1}{2},\frac{1}{2}|S|\frac{1}{2},-\frac{1}{2}\rangle & \langle\frac{1}{2},\frac{1}{2}|S|\frac{1}{2},-\frac{1}{2}\rangle \\
\langle\frac{1}{2},\frac{1}{2}|S|\frac{1}{2},\frac{1}{2}\rangle & \langle\frac{1}{2},\frac{1}{2}|S|\frac{1}{2},\frac{1}{2}\rangle & \langle\frac{1}{2},\frac{1}{2}|S|\frac{1}{2},\frac{1}{2}\rangle & \langle\frac{1}{2},\frac{1}{2}|S|\frac{1}{2},\frac{1}{2}\rangle & \langle\frac{1}{2},\frac{1}{2}|S|\frac{1}{2},-\frac{1}{2}\rangle & \langle\frac{1}{2},\frac{1}{2}|S|\frac{1}{2},-\frac{1}{2}\rangle & \langle\frac{1}{2},\frac{1}{2}|S|\frac{1}{2},-\frac{1}{2}\rangle & \langle\frac{1}{2},\frac{1}{2}|S|\frac{1}{2},-\frac{1}{2}\rangle & \langle\frac{1}{2},\frac{1}{2}|S|\frac{1}{2},\frac{1}{2}\rangle & \langle\frac{1}{2},\frac{1}{2}|S|\frac{1}{2},\frac{1}{2}\rangle & \langle\frac{1}{2},\frac{1}{2}|S|\frac{1}{2},\frac{1}{2}\rangle & \langle\frac{1}{2},\frac{1}{2}|S|\frac{1}{2},\frac{1}{2}\rangle & \langle\frac{1}{2},\frac{1}{2}|S|\frac{1}{2},-\frac{1}{2}\rangle & \langle\frac{1}{2},\frac{1}{2}|S|\frac{1}{2},-\frac{1}{2}\rangle & \langle\frac{1}{2},\frac{1}{2}|S|\frac{1}{2},-\frac{1}{2}\rangle & \langle\frac{1}{2},\frac{1}{2}|S|\frac{1}{2},-\frac{1}{2}\rangle \\
\langle\frac{1}{2},\frac{1}{2}|S|\frac{1}{2},\frac{1}{2}\rangle & \langle\frac{1}{2},\frac{1}{2}|S|\frac{1}{2},\frac{1}{2}\rangle & \langle\frac{1}{2},\frac{1}{2}|S|\frac{1}{2},\frac{1}{2}\rangle & \langle\frac{1}{2},\frac{1}{2}|S|\frac{1}{2},\frac{1}{2}\rangle & \langle\frac{1}{2},\frac{1}{2}|S|\frac{1}{2},-\frac{1}{2}\rangle & \langle\frac{1}{2},\frac{1}{2}|S|\frac{1}{2},-\frac{1}{2}\rangle & \langle\frac{1}{2},\frac{1}{2}|S|\frac{1}{2},-\frac{1}{2}\rangle & \langle\frac{1}{2},\frac{1}{2}|S|\frac{1}{2},-\frac{1}{2}\rangle & \langle\frac{1}{2},\frac{1}{2}|S|\frac{1}{2},\frac{1}{2}\rangle & \langle\frac{1}{2},\frac{1}{2}|S|\frac{1}{2},\frac{1}{2}\rangle & \langle\frac{1}{2},\frac{1}{2}|S|\frac{1}{2},\frac{1}{2}\rangle & \langle\frac{1}{2},\frac{1}{2}|S|\frac{1}{2},\frac{1}{2}\rangle & \langle\frac{1}{2},\frac{1}{2}|S|\frac{1}{2},-\frac{1}{2}\rangle & \langle\frac{1}{2},\frac{1}{2}|S|\frac{1}{2},-\frac{1}{2}\rangle & \langle\frac{1}{2},\frac{1}{2}|S|\frac{1}{2},-\frac{1}{2}\rangle & \langle\frac{1}{2},\frac{1}{2}|S|\frac{1}{2},-\frac{1}{2}\rangle \\
\langle\frac{1}{2},\frac{1}{2}|S|\frac{1}{2},\frac{1}{2}\rangle & \langle\frac{1}{2},\frac{1}{2}|S|\frac{1}{2},\frac{1}{2}\rangle & \langle\frac{1}{2},\frac{1}{2}|S|\frac{1}{2},\frac{1}{2}\rangle & \langle\frac{1}{2},\frac{1}{2}|S|\frac{1}{2},\frac{1}{2}\rangle & \langle\frac{1}{2},\frac{1}{2}|S|\frac{1}{2},-\frac{1}{2}\rangle & \langle\frac{1}{2},\frac{1}{2}|S|\frac{1}{2},-\frac{1}{2}\rangle & \langle\frac{1}{2},\frac{1}{2}|S|\frac{1}{2},-\frac{1}{2}\rangle & \langle\frac{1}{2},\frac{1}{2}|S|\frac{1}{2},-\frac{1}{2}\rangle & \langle\frac{1}{2},\frac{1}{2}|S|\frac{1}{2},\frac{1}{2}\rangle & \langle\frac{1}{2},\frac{1}{2}|S|\frac{1}{2},\frac{1}{2}\rangle & \langle\frac{1}{2},\frac{1}{2}|S|\frac{1}{2},\frac{1}{2}\rangle & \langle\frac{1}{2},\frac{1}{2}|S|\frac{1}{2},\frac{1}{2}\rangle & \langle\frac{1}{2},\frac{1}{2}|S|\frac{1}{2},-\frac{1}{2}\rangle & \langle\frac{1}{2},\frac{1}{2}|S|\frac{1}{2},-\frac{1}{2}\rangle & \langle\frac{1}{2},\frac{1}{2}|S|\frac{1}{2},-\frac{1}{2}\rangle & \langle\frac{1}{2},\frac{1}{2}|S|\frac{1}{2},-\frac{1}{2}\rangle \\
\langle\frac{1}{2},-\frac{1}{2}|S|\frac{1}{2},\frac{1}{2}\rangle & \langle\frac{1}{2},-\frac{1}{2}|S|\frac{1}{2},\frac{1}{2}\rangle & \langle\frac{1}{2},-\frac{1}{2}|S|\frac{1}{2},\frac{1}{2}\rangle & \langle\frac{1}{2},-\frac{1}{2}|S|\frac{1}{2},\frac{1}{2}\rangle & \langle\frac{1}{2},-\frac{1}{2}|S|\frac{1}{2},-\frac{1}{2}\rangle & \langle\frac{1}{2},-\frac{1}{2}|S|\frac{1}{2},-\frac{1}{2}\rangle & \langle\frac{1}{2},-\frac{1}{2}|S|\frac{1}{2},-\frac{1}{2}\rangle & \langle\frac{1}{2},-\frac{1}{2}|S|\frac{1}{2},-\frac{1}{2}\rangle & \langle\frac{1}{2},-\frac{1}{2}|S|\frac{1}{2},\frac{1}{2}\rangle & \langle\frac{1}{2},-\frac{1}{2}|S|\frac{1}{2},\frac{1}{2}\rangle & \langle\frac{1}{2},-\frac{1}{2}|S|\frac{1}{2},\frac{1}{2}\rangle & \langle\frac{1}{2},-\frac{1}{2}|S|\frac{1}{2},\frac{1}{2}\rangle & \langle\frac{1}{2},-\frac{1}{2}|S|\frac{1}{2},-\frac{1}{2}\rangle & \langle\frac{1}{2},-\frac{1}{2}|S|\frac{1}{2},-\frac{1}{2}\rangle & \langle\frac{1}{2},-\frac{1}{2}|S|\frac{1}{2},-\frac{1}{2}\rangle & \langle\frac{1}{2},-\frac{1}{2}|S|\frac{1}{2},-\frac{1}{2}\rangle \\
\langle\frac{1}{2},-\frac{1}{2}|S|\frac{1}{2},\frac{1}{2}\rangle & \langle\frac{1}{2},-\frac{1}{2}|S|\frac{1}{2},\frac{1}{2}\rangle & \langle\frac{1}{2},-\frac{1}{2}|S|\frac{1}{2},\frac{1}{2}\rangle & \langle\frac{1}{2},-\frac{1}{2}|S|\frac{1}{2},\frac{1}{2}\rangle & \langle\frac{1}{2},-\frac{1}{2}|S|\frac{1}{2},-\frac{1}{2}\rangle & \langle\frac{1}{2},-\frac{1}{2}|S|\frac{1}{2},-\frac{1}{2}\rangle & \langle\frac{1}{2},-\frac{1}{2}|S|\frac{1}{2},-\frac{1}{2}\rangle & \langle\frac{1}{2},-\frac{1}{2}|S|\frac{1}{2},-\frac{1}{2}\rangle & \langle\frac{1}{2},-\frac{1}{2}|S|\frac{1}{2},\frac{1}{2}\rangle & \langle\frac{1}{2},-\frac{1}{2}|S|\frac{1}{2},\frac{1}{2}\rangle & \langle\frac{1}{2},-\frac{1}{2}|S|\frac{1}{2},\frac{1}{2}\rangle & \langle\frac{1}{2},-\frac{1}{2}|S|\frac{1}{2},\frac{1}{2}\rangle & \langle\frac{1}{2},-\frac{1}{2}|S|\frac{1}{2},-\frac{1}{2}\rangle & \langle\frac{1}{2},-\frac{1}{2}|S|\frac{1}{2},-\frac{1}{2}\rangle & \langle\frac{1}{2},-\frac{1}{2}|S|\frac{1}{2},-\frac{1}{2}\rangle & \langle\frac{1}{2},-\frac{1}{2}|S|\frac{1}{2},-\frac{1}{2}\rangle \\
\langle\frac{1}{2},-\frac{1}{2}|S|\frac{1}{2},\frac{1}{2}\rangle & \langle\frac{1}{2},-\frac{1}{2}|S|\frac{1}{2},\frac{1}{2}\rangle & \langle\frac{1}{2},-\frac{1}{2}|S|\frac{1}{2},\frac{1}{2}\rangle & \langle\frac{1}{2},-\frac{1}{2}|S|\frac{1}{2},\frac{1}{2}\rangle & \langle\frac{1}{2},-\frac{1}{2}|S|\frac{1}{2},-\frac{1}{2}\rangle & \langle\frac{1}{2},-\frac{1}{2}|S|\frac{1}{2},-\frac{1}{2}\rangle & \langle\frac{1}{2},-\frac{1}{2}|S|\frac{1}{2},-\frac{1}{2}\rangle & \langle\frac{1}{2},-\frac{1}{2}|S|\frac{1}{2},-\frac{1}{2}\rangle & \langle\frac{1}{2},-\frac{1}{2}|S|\frac{1}{2},\frac{1}{2}\rangle & \langle\frac{1}{2},-\frac{1}{2}|S|\frac{1}{2},\frac{1}{2}\rangle & \langle\frac{1}{2},-\frac{1}{2}|S|\frac{1}{2},\frac{1}{2}\rangle & \langle\frac{1}{2},-\frac{1}{2}|S|\frac{1}{2},\frac{1}{2}\rangle & \langle\frac{1}{2},-\frac{1}{2}|S|\frac{1}{2},-\frac{1}{2}\rangle & \langle\frac{1}{2},-\frac{1}{2}|S|\frac{1}{2},-\frac{1}{2}\rangle & \langle\frac{1}{2},-\frac{1}{2}|S|\frac{1}{2},-\frac{1}{2}\rangle & \langle\frac{1}{2},-\frac{1}{2}|S|\frac{1}{2},-\frac{1}{2}\rangle \\
\langle\frac{1}{2},-\frac{1}{2}|S|\frac{1}{2},\frac{1}{2}\rangle & \langle\frac{1}{2},-\frac{1}{2}|S|\frac{1}{2},\frac{1}{2}\rangle & \langle\frac{1}{2},-\frac{1}{2}|S|\frac{1}{2},\frac{1}{2}\rangle & \langle\frac{1}{2},-\frac{1}{2}|S|\frac{1}{2},\frac{1}{2}\rangle & \langle\frac{1}{2},-\frac{1}{2}|S|\frac{1}{2},-\frac{1}{2}\rangle & \langle\frac{1}{2},-\frac{1}{2}|S|\frac{1}{2},-\frac{1}{2}\rangle & \langle\frac{1}{2},-\frac{1}{2}|S|\frac{1}{2},-\frac{1}{2}\rangle & \langle\frac{1}{2},-\frac{1}{2}|S|\frac{1}{2},-\frac{1}{2}\rangle & \langle\frac{1}{2},-\frac{1}{2}|S|\frac{1}{2},\frac{1}{2}\rangle & \langle\frac{1}{2},-\frac{1}{2}|S|\frac{1}{2},\frac{1}{2}\rangle & \langle\frac{1}{2},-\frac{1}{2}|S|\frac{1}{2},\frac{1}{2}\rangle & \langle\frac{1}{2},-\frac{1}{2}|S|\frac{1}{2},\frac{1}{2}\rangle & \langle\frac{1}{2},-\frac{1}{2}|S|\frac{1}{2},-\frac{1}{2}\rangle & \langle\frac{1}{2},-\frac{1}{2}|S|\frac{1}{2},-\frac{1}{2}\rangle & \langle\frac{1}{2},-\frac{1}{2}|S|\frac{1}{2},-\frac{1}{2}\rangle & \langle\frac{1}{2},-\frac{1}{2}|S|\frac{1}{2},-\frac{1}{2}\rangle
\end{bmatrix}_{16\times 16} \tag{A.5}
$$

$$
s_3=\begin{bmatrix}
\langle\frac{1}{2},\frac{1}{2}|S|\frac{1}{2},\frac{1}{2}\rangle & \langle\frac{1}{2},\frac{1}{2}|S|\frac{1}{2},\frac{1}{2}\rangle & \langle\frac{1}{2},\frac{1}{2}|S|\frac{1}{2},-\frac{1}{2}\rangle & \langle\frac{1}{2},\frac{1}{2}|S|\frac{1}{2},-\frac{1}{2}\rangle & \langle\frac{1}{2},\frac{1}{2}|S|\frac{1}{2},\frac{1}{2}\rangle & \langle\frac{1}{2},\frac{1}{2}|S|\frac{1}{2},\frac{1}{2}\rangle & \langle\frac{1}{2},\frac{1}{2}|S|\frac{1}{2},-\frac{1}{2}\rangle & \langle\frac{1}{2},\frac{1}{2}|S|\frac{1}{2},-\frac{1}{2}\rangle & \langle\frac{1}{2},\frac{1}{2}|S|\frac{1}{2},\frac{1}{2}\rangle & \langle\frac{1}{2},\frac{1}{2}|S|\frac{1}{2},\frac{1}{2}\rangle & \langle\frac{1}{2},\frac{1}{2}|S|\frac{1}{2},-\frac{1}{2}\rangle & \langle\frac{1}{2},\frac{1}{2}|S|\frac{1}{2},-\frac{1}{2}\rangle & \langle\frac{1}{2},\frac{1}{2}|S|\frac{1}{2},\frac{1}{2}\rangle & \langle\frac{1}{2},\frac{1}{2}|S|\frac{1}{2},\frac{1}{2}\rangle & \langle\frac{1}{2},\frac{1}{2}|S|\frac{1}{2},-\frac{1}{2}\rangle & \langle\frac{1}{2},\frac{1}{2}|S|\frac{1}{2},-\frac{1}{2}\rangle \\
\langle\frac{1}{2},\frac{1}{2}|S|\frac{1}{2},\frac{1}{2}\rangle & \langle\frac{1}{2},\frac{1}{2}|S|\frac{1}{2},\frac{1}{2}\rangle & \langle\frac{1}{2},\frac{1}{2}|S|\frac{1}{2},-\frac{1}{2}\rangle & \langle\frac{1}{2},\frac{1}{2}|S|\frac{1}{2},-\frac{1}{2}\rangle & \langle\frac{1}{2},\frac{1}{2}|S|\frac{1}{2},\frac{1}{2}\rangle & \langle\frac{1}{2},\frac{1}{2}|S|\frac{1}{2},\frac{1}{2}\rangle & \langle\frac{1}{2},\frac{1}{2}|S|\frac{1}{2},-\frac{1}{2}\rangle & \langle\frac{1}{2},\frac{1}{2}|S|\frac{1}{2},-\frac{1}{2}\rangle & \langle\frac{1}{2},\frac{1}{2}|S|\frac{1}{2},\frac{1}{2}\rangle & \langle\frac{1}{2},\frac{1}{2}|S|\frac{1}{2},\frac{1}{2}\rangle & \langle\frac{1}{2},\frac{1}{2}|S|\frac{1}{2},-\frac{1}{2}\rangle & \langle\frac{1}{2},\frac{1}{2}|S|\frac{1}{2},-\frac{1}{2}\rangle & \langle\frac{1}{2},\frac{1}{2}|S|\frac{1}{2},\frac{1}{2}\rangle & \langle\frac{1}{2},\frac{1}{2}|S|\frac{1}{2},\frac{1}{2}\rangle & \langle\frac{1}{2},\frac{1}{2}|S|\frac{1}{2},-\frac{1}{2}\rangle & \langle\frac{1}{2},\frac{1}{2}|S|\frac{1}{2},-\frac{1}{2}\rangle \\
\langle\frac{1}{2},-\frac{1}{2}|S|\frac{1}{2},\frac{1}{2}\rangle & \langle\frac{1}{2},-\frac{1}{2}|S|\frac{1}{2},\frac{1}{2}\rangle & \langle\frac{1}{2},-\frac{1}{2}|S|\frac{1}{2},-\frac{1}{2}\rangle & \langle\frac{1}{2},-\frac{1}{2}|S|\frac{1}{2},-\frac{1}{2}\rangle & \langle\frac{1}{2},-\frac{1}{2}|S|\frac{1}{2},\frac{1}{2}\rangle & \langle\frac{1}{2},-\frac{1}{2}|S|\frac{1}{2},\frac{1}{2}\rangle & \langle\frac{1}{2},-\frac{1}{2}|S|\frac{1}{2},-\frac{1}{2}\rangle & \langle\frac{1}{2},-\frac{1}{2}|S|\frac{1}{2},-\frac{1}{2}\rangle & \langle\frac{1}{2},-\frac{1}{2}|S|\frac{1}{2},\frac{1}{2}\rangle & \langle\frac{1}{2},-\frac{1}{2}|S|\frac{1}{2},\frac{1}{2}\rangle & \langle\frac{1}{2},-\frac{1}{2}|S|\frac{1}{2},-\frac{1}{2}\rangle & \langle\frac{1}{2},-\frac{1}{2}|S|\frac{1}{2},-\frac{1}{2}\rangle & \langle\frac{1}{2},-\frac{1}{2}|S|\frac{1}{2},\frac{1}{2}\rangle & \langle\frac{1}{2},-\frac{1}{2}|S|\frac{1}{2},\frac{1}{2}\rangle & \langle\frac{1}{2},-\frac{1}{2}|S|\frac{1}{2},-\frac{1}{2}\rangle & \langle\frac{1}{2},-\frac{1}{2}|S|\frac{1}{2},-\frac{1}{2}\rangle \\
\langle\frac{1}{2},-\frac{1}{2}|S|\frac{1}{2},\frac{1}{2}\rangle & \langle\frac{1}{2},-\frac{1}{2}|S|\frac{1}{2},\frac{1}{2}\rangle & \langle\frac{1}{2},-\frac{1}{2}|S|\frac{1}{2},-\frac{1}{2}\rangle & \langle\frac{1}{2},-\frac{1}{2}|S|\frac{1}{2},-\frac{1}{2}\rangle & \langle\frac{1}{2},-\frac{1}{2}|S|\frac{1}{2},\frac{1}{2}\rangle & \langle\frac{1}{2},-\frac{1}{2}|S|\frac{1}{2},\frac{1}{2}\rangle & \langle\frac{1}{2},-\frac{1}{2}|S|\frac{1}{2},-\frac{1}{2}\rangle & \langle\frac{1}{2},-\frac{1}{2}|S|\frac{1}{2},-\frac{1}{2}\rangle & \langle\frac{1}{2},-\frac{1}{2}|S|\frac{1}{2},\frac{1}{2}\rangle & \langle\frac{1}{2},-\frac{1}{2}|S|\frac{1}{2},\frac{1}{2}\rangle & \langle\frac{1}{2},-\frac{1}{2}|S|\frac{1}{2},-\frac{1}{2}\rangle & \langle\frac{1}{2},-\frac{1}{2}|S|\frac{1}{2},-\frac{1}{2}\rangle & \langle\frac{1}{2},-\frac{1}{2}|S|\frac{1}{2},\frac{1}{2}\rangle & \langle\frac{1}{2},-\frac{1}{2}|S|\frac{1}{2},\frac{1}{2}\rangle & \langle\frac{1}{2},-\frac{1}{2}|S|\frac{1}{2},-\frac{1}{2}\rangle & \langle\frac{1}{2},-\frac{1}{2}|S|\frac{1}{2},-\frac{1}{2}\rangle \\
\langle\frac{1}{2},\frac{1}{2}|S|\frac{1}{2},\frac{1}{2}\rangle & \langle\frac{1}{2},\frac{1}{2}|S|\frac{1}{2},\frac{1}{2}\rangle & \langle\frac{1}{2},\frac{1}{2}|S|\frac{1}{2},-\frac{1}{2}\rangle & \langle\frac{1}{2},\frac{1}{2}|S|\frac{1}{2},-\frac{1}{2}\rangle & \langle\frac{1}{2},\frac{1}{2}|S|\frac{1}{2},\frac{1}{2}\rangle & \langle\frac{1}{2},\frac{1}{2}|S|\frac{1}{2},\frac{1}{2}\rangle & \langle\frac{1}{2},\frac{1}{2}|S|\frac{1}{2},-\frac{1}{2}\rangle & \langle\frac{1}{2},\frac{1}{2}|S|\frac{1}{2},-\frac{1}{2}\rangle & \langle\frac{1}{2},\frac{1}{2}|S|\frac{1}{2},\frac{1}{2}\rangle & \langle\frac{1}{2},\frac{1}{2}|S|\frac{1}{2},\frac{1}{2}\rangle & \langle\frac{1}{2},\frac{1}{2}|S|\frac{1}{2},-\frac{1}{2}\rangle & \langle\frac{1}{2},\frac{1}{2}|S|\frac{1}{2},-\frac{1}{2}\rangle & \langle\frac{1}{2},\frac{1}{2}|S|\frac{1}{2},\frac{1}{2}\rangle & \langle\frac{1}{2},\frac{1}{2}|S|\frac{1}{2},\frac{1}{2}\rangle & \langle\frac{1}{2},\frac{1}{2}|S|\frac{1}{2},-\frac{1}{2}\rangle & \langle\frac{1}{2},\frac{1}{2}|S|\frac{1}{2},-\frac{1}{2}\rangle \\
\langle\frac{1}{2},\frac{1}{2}|S|\frac{1}{2},\frac{1}{2}\rangle & \langle\frac{1}{2},\frac{1}{2}|S|\frac{1}{2},\frac{1}{2}\rangle & \langle\frac{1}{2},\frac{1}{2}|S|\frac{1}{2},-\frac{1}{2}\rangle & \langle\frac{1}{2},\frac{1}{2}|S|\frac{1}{2},-\frac{1}{2}\rangle & \langle\frac{1}{2},\frac{1}{2}|S|\frac{1}{2},\frac{1}{2}\rangle & \langle\frac{1}{2},\frac{1}{2}|S|\frac{1}{2},\frac{1}{2}\rangle & \langle\frac{1}{2},\frac{1}{2}|S|\frac{1}{2},-\frac{1}{2}\rangle & \langle\frac{1}{2},\frac{1}{2}|S|\frac{1}{2},-\frac{1}{2}\rangle & \langle\frac{1}{2},\frac{1}{2}|S|\frac{1}{2},\frac{1}{2}\rangle & \langle\frac{1}{2},\frac{1}{2}|S|\frac{1}{2},\frac{1}{2}\rangle & \langle\frac{1}{2},\frac{1}{2}|S|\frac{1}{2},-\frac{1}{2}\rangle & \langle\frac{1}{2},\frac{1}{2}|S|\frac{1}{2},-\frac{1}{2}\rangle & \langle\frac{1}{2},\frac{1}{2}|S|\frac{1}{2},\frac{1}{2}\rangle & \langle\frac{1}{2},\frac{1}{2}|S|\frac{1}{2},\frac{1}{2}\rangle & \langle\frac{1}{2},\frac{1}{2}|S|\frac{1}{2},-\frac{1}{2}\rangle & \langle\frac{1}{2},\frac{1}{2}|S|\frac{1}{2},-\frac{1}{2}\rangle \\
\langle\frac{1}{2},-\frac{1}{2}|S|\frac{1}{2},\frac{1}{2}\rangle & \langle\frac{1}{2},-\frac{1}{2}|S|\frac{1}{2},\frac{1}{2}\rangle & \langle\frac{1}{2},-\frac{1}{2}|S|\frac{1}{2},-\frac{1}{2}\rangle & \langle\frac{1}{2},-\frac{1}{2}|S|\frac{1}{2},-\frac{1}{2}\rangle & \langle\frac{1}{2},-\frac{1}{2}|S|\frac{1}{2},\frac{1}{2}\rangle & \langle\frac{1}{2},-\frac{1}{2}|S|\frac{1}{2},\frac{1}{2}\rangle & \langle\frac{1}{2},-\frac{1}{2}|S|\frac{1}{2},-\frac{1}{2}\rangle & \langle\frac{1}{2},-\frac{1}{2}|S|\frac{1}{2},-\frac{1}{2}\rangle & \langle\frac{1}{2},-\frac{1}{2}|S|\frac{1}{2},\frac{1}{2}\rangle & \langle\frac{1}{2},-\frac{1}{2}|S|\frac{1}{2},\frac{1}{2}\rangle & \langle\frac{1}{2},-\frac{1}{2}|S|\frac{1}{2},-\frac{1}{2}\rangle & \langle\frac{1}{2},-\frac{1}{2}|S|\frac{1}{2},-\frac{1}{2}\rangle & \langle\frac{1}{2},-\frac{1}{2}|S|\frac{1}{2},\frac{1}{2}\rangle & \langle\frac{1}{2},-\frac{1}{2}|S|\frac{1}{2},\frac{1}{2}\rangle & \langle\frac{1}{2},-\frac{1}{2}|S|\frac{1}{2},-\frac{1}{2}\rangle & \langle\frac{1}{2},-\frac{1}{2}|S|\frac{1}{2},-\frac{1}{2}\rangle \\
\langle\frac{1}{2},-\frac{1}{2}|S|\frac{1}{2},\frac{1}{2}\rangle & \langle\frac{1}{2},-\frac{1}{2}|S|\frac{1}{2},\frac{1}{2}\rangle & \langle\frac{1}{2},-\frac{1}{2}|S|\frac{1}{2},-\frac{1}{2}\rangle & \langle\frac{1}{2},-\frac{1}{2}|S|\frac{1}{2},-\frac{1}{2}\rangle & \langle\frac{1}{2},-\frac{1}{2}|S|\frac{1}{2},\frac{1}{2}\rangle & \langle\frac{1}{2},-\frac{1}{2}|S|\frac{1}{2},\frac{1}{2}\rangle & \langle\frac{1}{2},-\frac{1}{2}|S|\frac{1}{2},-\frac{1}{2}\rangle & \langle\frac{1}{2},-\frac{1}{2}|S|\frac{1}{2},-\frac{1}{2}\rangle & \langle\frac{1}{2},-\frac{1}{2}|S|\frac{1}{2},\frac{1}{2}\rangle & \langle\frac{1}{2},-\frac{1}{2}|S|\frac{1}{2},\frac{1}{2}\rangle & \langle\frac{1}{2},-\frac{1}{2}|S|\frac{1}{2},-\frac{1}{2}\rangle & \langle\frac{1}{2},-\frac{1}{2}|S|\frac{1}{2},-\frac{1}{2}\rangle & \langle\frac{1}{2},-\frac{1}{2}|S|\frac{1}{2},\frac{1}{2}\rangle & \langle\frac{1}{2},-\frac{1}{2}|S|\frac{1}{2},\frac{1}{2}\rangle & \langle\frac{1}{2},-\frac{1}{2}|S|\frac{1}{2},-\frac{1}{2}\rangle & \langle\frac{1}{2},-\frac{1}{2}|S|\frac{1}{2},-\frac{1}{2}\rangle \\
\langle\frac{1}{2},\frac{1}{2}|S|\frac{1}{2},\frac{1}{2}\rangle & \langle\frac{1}{2},\frac{1}{2}|S|\frac{1}{2},\frac{1}{2}\rangle & \langle\frac{1}{2},\frac{1}{2}|S|\frac{1}{2},-\frac{1}{2}\rangle & \langle\frac{1}{2},\frac{1}{2}|S|\frac{1}{2},-\frac{1}{2}\rangle & \langle\frac{1}{2},\frac{1}{2}|S|\frac{1}{2},\frac{1}{2}\rangle & \langle\frac{1}{2},\frac{1}{2}|S|\frac{1}{2},\frac{1}{2}\rangle & \langle\frac{1}{2},\frac{1}{2}|S|\frac{1}{2},-\frac{1}{2}\rangle & \langle\frac{1}{2},\frac{1}{2}|S|\frac{1}{2},-\frac{1}{2}\rangle & \langle\frac{1}{2},\frac{1}{2}|S|\frac{1}{2},\frac{1}{2}\rangle & \langle\frac{1}{2},\frac{1}{2}|S|\frac{1}{2},\frac{1}{2}\rangle & \langle\frac{1}{2},\frac{1}{2}|S|\frac{1}{2},-\frac{1}{2}\rangle & \langle\frac{1}{2},\frac{1}{2}|S|\frac{1}{2},-\frac{1}{2}\rangle & \langle\frac{1}{2},\frac{1}{2}|S|\frac{1}{2},\frac{1}{2}\rangle & \langle\frac{1}{2},\frac{1}{2}|S|\frac{1}{2},\frac{1}{2}\rangle & \langle\frac{1}{2},\frac{1}{2}|S|\frac{1}{2},-\frac{1}{2}\rangle & \langle\frac{1}{2},\frac{1}{2}|S|\frac{1}{2},-\frac{1}{2}\rangle \\
\langle\frac{1}{2},\frac{1}{2}|S|\frac{1}{2},\frac{1}{2}\rangle & \langle\frac{1}{2},\frac{1}{2}|S|\frac{1}{2},\frac{1}{2}\rangle & \langle\frac{1}{2},\frac{1}{2}|S|\frac{1}{2},-\frac{1}{2}\rangle & \langle\frac{1}{2},\frac{1}{2}|S|\frac{1}{2},-\frac{1}{2}\rangle & \langle\frac{1}{2},\frac{1}{2}|S|\frac{1}{2},\frac{1}{2}\rangle & \langle\frac{1}{2},\frac{1}{2}|S|\frac{1}{2},\frac{1}{2}\rangle & \langle\frac{1}{2},\frac{1}{2}|S|\frac{1}{2},-\frac{1}{2}\rangle & \langle\frac{1}{2},\frac{1}{2}|S|\frac{1}{2},-\frac{1}{2}\rangle & \langle\frac{1}{2},\frac{1}{2}|S|\frac{1}{2},\frac{1}{2}\rangle & \langle\frac{1}{2},\frac{1}{2}|S|\frac{1}{2},\frac{1}{2}\rangle & \langle\frac{1}{2},\frac{1}{2}|S|\frac{1}{2},-\frac{1}{2}\rangle & \langle\frac{1}{2},\frac{1}{2}|S|\frac{1}{2},-\frac{1}{2}\rangle & \langle\frac{1}{2},\frac{1}{2}|S|\frac{1}{2},\frac{1}{2}\rangle & \langle\frac{1}{2},\frac{1}{2}|S|\frac{1}{2},\frac{1}{2}\rangle & \langle\frac{1}{2},\frac{1}{2}|S|\frac{1}{2},-\frac{1}{2}\rangle & \langle\frac{1}{2},\frac{1}{2}|S|\frac{1}{2},-\frac{1}{2}\rangle \\
\langle\frac{1}{2},-\frac{1}{2}|S|\frac{1}{2},\frac{1}{2}\rangle & \langle\frac{1}{2},-\frac{1}{2}|S|\frac{1}{2},\frac{1}{2}\rangle & \langle\frac{1}{2},-\frac{1}{2}|S|\frac{1}{2},-\frac{1}{2}\rangle & \langle\frac{1}{2},-\frac{1}{2}|S|\frac{1}{2},-\frac{1}{2}\rangle & \langle\frac{1}{2},-\frac{1}{2}|S|\frac{1}{2},\frac{1}{2}\rangle & \langle\frac{1}{2},-\frac{1}{2}|S|\frac{1}{2},\frac{1}{2}\rangle & \langle\frac{1}{2},-\frac{1}{2}|S|\frac{1}{2},-\frac{1}{2}\rangle & \langle\frac{1}{2},-\frac{1}{2}|S|\frac{1}{2},-\frac{1}{2}\rangle & \langle\frac{1}{2},-\frac{1}{2}|S|\frac{1}{2},\frac{1}{2}\rangle & \langle\frac{1}{2},-\frac{1}{2}|S|\frac{1}{2},\frac{1}{2}\rangle & \langle\frac{1}{2},-\frac{1}{2}|S|\frac{1}{2},-\frac{1}{2}\rangle & \langle\frac{1}{2},-\frac{1}{2}|S|\frac{1}{2},-\frac{1}{2}\rangle & \langle\frac{1}{2},-\frac{1}{2}|S|\frac{1}{2},\frac{1}{2}\rangle & \langle\frac{1}{2},-\frac{1}{2}|S|\frac{1}{2},\frac{1}{2}\rangle & \langle\frac{1}{2},-\frac{1}{2}|S|\frac{1}{2},-\frac{1}{2}\rangle & \langle\frac{1}{2},-\frac{1}{2}|S|\frac{1}{2},-\frac{1}{2}\rangle \\
\langle\frac{1}{2},-\frac{1}{2}|S|\frac{1}{2},\frac{1}{2}\rangle & \langle\frac{1}{2},-\frac{1}{2}|S|\frac{1}{2},\frac{1}{2}\rangle & \langle\frac{1}{2},-\frac{1}{2}|S|\frac{1}{2},-\frac{1}{2}\rangle & \langle\frac{1}{2},-\frac{1}{2}|S|\frac{1}{2},-\frac{1}{2}\rangle & \langle\frac{1}{2},-\frac{1}{2}|S|\frac{1}{2},\frac{1}{2}\rangle & \langle\frac{1}{2},-\frac{1}{2}|S|\frac{1}{2},\frac{1}{2}\rangle & \langle\frac{1}{2},-\frac{1}{2}|S|\frac{1}{2},-\frac{1}{2}\rangle & \langle\frac{1}{2},-\frac{1}{2}|S|\frac{1}{2},-\frac{1}{2}\rangle & \langle\frac{1}{2},-\frac{1}{2}|S|\frac{1}{2},\frac{1}{2}\rangle & \langle\frac{1}{2},-\frac{1}{2}|S|\frac{1}{2},\frac{1}{2}\rangle & \langle\frac{1}{2},-\frac{1}{2}|S|\frac{1}{2},-\frac{1}{2}\rangle & \langle\frac{1}{2},-\frac{1}{2}|S|\frac{1}{2},-\frac{1}{2}\rangle & \langle\frac{1}{2},-\frac{1}{2}|S|\frac{1}{2},\frac{1}{2}\rangle & \langle\frac{1}{2},-\frac{1}{2}|S|\frac{1}{2},\frac{1}{2}\rangle & \langle\frac{1}{2},-\frac{1}{2}|S|\frac{1}{2},-\frac{1}{2}\rangle & \langle\frac{1}{2},-\frac{1}{2}|S|\frac{1}{2},-\frac{1}{2}\rangle \\
\langle\frac{1}{2},\frac{1}{2}|S|\frac{1}{2},\frac{1}{2}\rangle & \langle\frac{1}{2},\frac{1}{2}|S|\frac{1}{2},\frac{1}{2}\rangle & \langle\frac{1}{2},\frac{1}{2}|S|\frac{1}{2},-\frac{1}{2}\rangle & \langle\frac{1}{2},\frac{1}{2}|S|\frac{1}{2},-\frac{1}{2}\rangle & \langle\frac{1}{2},\frac{1}{2}|S|\frac{1}{2},\frac{1}{2}\rangle & \langle\frac{1}{2},\frac{1}{2}|S|\frac{1}{2},\frac{1}{2}\rangle & \langle\frac{1}{2},\frac{1}{2}|S|\frac{1}{2},-\frac{1}{2}\rangle & \langle\frac{1}{2},\frac{1}{2}|S|\frac{1}{2},-\frac{1}{2}\rangle & \langle\frac{1}{2},\frac{1}{2}|S|\frac{1}{2},\frac{1}{2}\rangle & \langle\frac{1}{2},\frac{1}{2}|S|\frac{1}{2},\frac{1}{2}\rangle & \langle\frac{1}{2},\frac{1}{2}|S|\frac{1}{2},-\frac{1}{2}\rangle & \langle\frac{1}{2},\frac{1}{2}|S|\frac{1}{2},-\frac{1}{2}\rangle & \langle\frac{1}{2},\frac{1}{2}|S|\frac{1}{2},\frac{1}{2}\rangle & \langle\frac{1}{2},\frac{1}{2}|S|\frac{1}{2},\frac{1}{2}\rangle & \langle\frac{1}{2},\frac{1}{2}|S|\frac{1}{2},-\frac{1}{2}\rangle & \langle\frac{1}{2},\frac{1}{2}|S|\frac{1}{2},-\frac{1}{2}\rangle \\
\langle\frac{1}{2},\frac{1}{2}|S|\frac{1}{2},\frac{1}{2}\rangle & \langle\frac{1}{2},\frac{1}{2}|S|\frac{1}{2},\frac{1}{2}\rangle & \langle\frac{1}{2},\frac{1}{2}|S|\frac{1}{2},-\frac{1}{2}\rangle & \langle\frac{1}{2},\frac{1}{2}|S|\frac{1}{2},-\frac{1}{2}\rangle & \langle\frac{1}{2},\frac{1}{2}|S|\frac{1}{2},\frac{1}{2}\rangle & \langle\frac{1}{2},\frac{1}{2}|S|\frac{1}{2},\frac{1}{2}\rangle & \langle\frac{1}{2},\frac{1}{2}|S|\frac{1}{2},-\frac{1}{2}\rangle & \langle\frac{1}{2},\frac{1}{2}|S|\frac{1}{2},-\frac{1}{2}\rangle & \langle\frac{1}{2},\frac{1}{2}|S|\frac{1}{2},\frac{1}{2}\rangle & \langle\frac{1}{2},\frac{1}{2}|S|\frac{1}{2},\frac{1}{2}\rangle & \langle\frac{1}{2},\frac{1}{2}|S|\frac{1}{2},-\frac{1}{2}\rangle & \langle\frac{1}{2},\frac{1}{2}|S|\frac{1}{2},-\frac{1}{2}\rangle & \langle\frac{1}{2},\frac{1}{2}|S|\frac{1}{2},\frac{1}{2}\rangle & \langle\frac{1}{2},\frac{1}{2}|S|\frac{1}{2},\frac{1}{2}\rangle & \langle\frac{1}{2},\frac{1}{2}|S|\frac{1}{2},-\frac{1}{2}\rangle & \langle\frac{1}{2},\frac{1}{2}|S|\frac{1}{2},-\frac{1}{2}\rangle \\
\langle\frac{1}{2},-\frac{1}{2}|S|\frac{1}{2},\frac{1}{2}\rangle & \langle\frac{1}{2},-\frac{1}{2}|S|\frac{1}{2},\frac{1}{2}\rangle & \langle\frac{1}{2},-\frac{1}{2}|S|\frac{1}{2},-\frac{1}{2}\rangle & \langle\frac{1}{2},-\frac{1}{2}|S|\frac{1}{2},-\frac{1}{2}\rangle & \langle\frac{1}{2},-\frac{1}{2}|S|\frac{1}{2},\frac{1}{2}\rangle & \langle\frac{1}{2},-\frac{1}{2}|S|\frac{1}{2},\frac{1}{2}\rangle & \langle\frac{1}{2},-\frac{1}{2}|S|\frac{1}{2},-\frac{1}{2}\rangle & \langle\frac{1}{2},-\frac{1}{2}|S|\frac{1}{2},-\frac{1}{2}\rangle & \langle\frac{1}{2},-\frac{1}{2}|S|\frac{1}{2},\frac{1}{2}\rangle & \langle\frac{1}{2},-\frac{1}{2}|S|\frac{1}{2},\frac{1}{2}\rangle & \langle\frac{1}{2},-\frac{1}{2}|S|\frac{1}{2},-\frac{1}{2}\rangle & \langle\frac{1}{2},-\frac{1}{2}|S|\frac{1}{2},-\frac{1}{2}\rangle & \langle\frac{1}{2},-\frac{1}{2}|S|\frac{1}{2},\frac{1}{2}\rangle & \langle\frac{1}{2},-\frac{1}{2}|S|\frac{1}{2},\frac{1}{2}\rangle & \langle\frac{1}{2},-\frac{1}{2}|S|\frac{1}{2},-\frac{1}{2}\rangle & \langle\frac{1}{2},-\frac{1}{2}|S|\frac{1}{2},-\frac{1}{2}\rangle \\
\langle\frac{1}{2},-\frac{1}{2}|S|\frac{1}{2},\frac{1}{2}\rangle & \langle\frac{1}{2},-\frac{1}{2}|S|\frac{1}{2},\frac{1}{2}\rangle & \langle\frac{1}{2},-\frac{1}{2}|S|\frac{1}{2},-\frac{1}{2}\rangle & \langle\frac{1}{2},-\frac{1}{2}|S|\frac{1}{2},-\frac{1}{2}\rangle & \langle\frac{1}{2},-\frac{1}{2}|S|\frac{1}{2},\frac{1}{2}\rangle & \langle\frac{1}{2},-\frac{1}{2}|S|\frac{1}{2},\frac{1}{2}\rangle & \langle\frac{1}{2},-\frac{1}{2}|S|\frac{1}{2},-\frac{1}{2}\rangle & \langle\frac{1}{2},-\frac{1}{2}|S|\frac{1}{2},-\frac{1}{2}\rangle & \langle\frac{1}{2},-\frac{1}{2}|S|\frac{1}{2},\frac{1}{2}\rangle & \langle\frac{1}{2},-\frac{1}{2}|S|\frac{1}{2},\frac{1}{2}\rangle & \langle\frac{1}{2},-\frac{1}{2}|S|\frac{1}{2},-\frac{1}{2}\rangle & \langle\frac{1}{2},-\frac{1}{2}|S|\frac{1}{2},-\frac{1}{2}\rangle & \langle\frac{1}{2},-\frac{1}{2}|S|\frac{1}{2},\frac{1}{2}\rangle & \langle\frac{1}{2},-\frac{1}{2}|S|\frac{1}{2},\frac{1}{2}\rangle & \langle\frac{1}{2},-\frac{1}{2}|S|\frac{1}{2},-\frac{1}{2}\rangle & \langle\frac{1}{2},-\frac{1}{2}|S|\frac{1}{2},-\frac{1}{2}\rangle
\end{bmatrix}_{16\times16} \tag{A.6}
$$

$$S_4=\begin{bmatrix}
\langle\frac{1}{2},\frac{1}{2}|S|\frac{1}{2},\frac{1}{2}\rangle & \langle\frac{1}{2},\frac{1}{2}|S|\frac{1}{2},-\frac{1}{2}\rangle & \langle\frac{1}{2},\frac{1}{2}|S|\frac{1}{2},\frac{1}{2}\rangle & \langle\frac{1}{2},\frac{1}{2}|S|\frac{1}{2},-\frac{1}{2}\rangle & \langle\frac{1}{2},\frac{1}{2}|S|\frac{1}{2},\frac{1}{2}\rangle & \langle\frac{1}{2},\frac{1}{2}|S|\frac{1}{2},-\frac{1}{2}\rangle & \langle\frac{1}{2},\frac{1}{2}|S|\frac{1}{2},\frac{1}{2}\rangle & \langle\frac{1}{2},\frac{1}{2}|S|\frac{1}{2},-\frac{1}{2}\rangle & \langle\frac{1}{2},\frac{1}{2}|S|\frac{1}{2},\frac{1}{2}\rangle & \langle\frac{1}{2},\frac{1}{2}|S|\frac{1}{2},-\frac{1}{2}\rangle & \langle\frac{1}{2},\frac{1}{2}|S|\frac{1}{2},\frac{1}{2}\rangle & \langle\frac{1}{2},\frac{1}{2}|S|\frac{1}{2},-\frac{1}{2}\rangle & \langle\frac{1}{2},\frac{1}{2}|S|\frac{1}{2},\frac{1}{2}\rangle & \langle\frac{1}{2},\frac{1}{2}|S|\frac{1}{2},-\frac{1}{2}\rangle & \langle\frac{1}{2},\frac{1}{2}|S|\frac{1}{2},\frac{1}{2}\rangle & \langle\frac{1}{2},\frac{1}{2}|S|\frac{1}{2},-\frac{1}{2}\rangle \\
\langle\frac{1}{2},-\frac{1}{2}|S|\frac{1}{2},\frac{1}{2}\rangle & \langle\frac{1}{2},-\frac{1}{2}|S|\frac{1}{2},-\frac{1}{2}\rangle & \langle\frac{1}{2},-\frac{1}{2}|S|\frac{1}{2},\frac{1}{2}\rangle & \langle\frac{1}{2},-\frac{1}{2}|S|\frac{1}{2},-\frac{1}{2}\rangle & \langle\frac{1}{2},-\frac{1}{2}|S|\frac{1}{2},\frac{1}{2}\rangle & \langle\frac{1}{2},-\frac{1}{2}|S|\frac{1}{2},-\frac{1}{2}\rangle & \langle\frac{1}{2},-\frac{1}{2}|S|\frac{1}{2},\frac{1}{2}\rangle & \langle\frac{1}{2},-\frac{1}{2}|S|\frac{1}{2},-\frac{1}{2}\rangle & \langle\frac{1}{2},-\frac{1}{2}|S|\frac{1}{2},\frac{1}{2}\rangle & \langle\frac{1}{2},-\frac{1}{2}|S|\frac{1}{2},-\frac{1}{2}\rangle & \langle\frac{1}{2},-\frac{1}{2}|S|\frac{1}{2},\frac{1}{2}\rangle & \langle\frac{1}{2},-\frac{1}{2}|S|\frac{1}{2},-\frac{1}{2}\rangle & \langle\frac{1}{2},-\frac{1}{2}|S|\frac{1}{2},\frac{1}{2}\rangle & \langle\frac{1}{2},-\frac{1}{2}|S|\frac{1}{2},-\frac{1}{2}\rangle & \langle\frac{1}{2},-\frac{1}{2}|S|\frac{1}{2},\frac{1}{2}\rangle & \langle\frac{1}{2},-\frac{1}{2}|S|\frac{1}{2},-\frac{1}{2}\rangle \\
\langle\frac{1}{2},\frac{1}{2}|S|\frac{1}{2},\frac{1}{2}\rangle & \langle\frac{1}{2},\frac{1}{2}|S|\frac{1}{2},-\frac{1}{2}\rangle & \langle\frac{1}{2},\frac{1}{2}|S|\frac{1}{2},\frac{1}{2}\rangle & \langle\frac{1}{2},\frac{1}{2}|S|\frac{1}{2},-\frac{1}{2}\rangle & \langle\frac{1}{2},\frac{1}{2}|S|\frac{1}{2},\frac{1}{2}\rangle & \langle\frac{1}{2},\frac{1}{2}|S|\frac{1}{2},-\frac{1}{2}\rangle & \langle\frac{1}{2},\frac{1}{2}|S|\frac{1}{2},\frac{1}{2}\rangle & \langle\frac{1}{2},\frac{1}{2}|S|\frac{1}{2},-\frac{1}{2}\rangle & \langle\frac{1}{2},\frac{1}{2}|S|\frac{1}{2},\frac{1}{2}\rangle & \langle\frac{1}{2},\frac{1}{2}|S|\frac{1}{2},-\frac{1}{2}\rangle & \langle\frac{1}{2},\frac{1}{2}|S|\frac{1}{2},\frac{1}{2}\rangle & \langle\frac{1}{2},\frac{1}{2}|S|\frac{1}{2},-\frac{1}{2}\rangle & \langle\frac{1}{2},\frac{1}{2}|S|\frac{1}{2},\frac{1}{2}\rangle & \langle\frac{1}{2},\frac{1}{2}|S|\frac{1}{2},-\frac{1}{2}\rangle & \langle\frac{1}{2},\frac{1}{2}|S|\frac{1}{2},\frac{1}{2}\rangle & \langle\frac{1}{2},\frac{1}{2}|S|\frac{1}{2},-\frac{1}{2}\rangle \\
\langle\frac{1}{2},-\frac{1}{2}|S|\frac{1}{2},\frac{1}{2}\rangle & \langle\frac{1}{2},-\frac{1}{2}|S|\frac{1}{2},-\frac{1}{2}\rangle & \langle\frac{1}{2},-\frac{1}{2}|S|\frac{1}{2},\frac{1}{2}\rangle & \langle\frac{1}{2},-\frac{1}{2}|S|\frac{1}{2},-\frac{1}{2}\rangle & \langle\frac{1}{2},-\frac{1}{2}|S|\frac{1}{2},\frac{1}{2}\rangle & \langle\frac{1}{2},-\frac{1}{2}|S|\frac{1}{2},-\frac{1}{2}\rangle & \langle\frac{1}{2},-\frac{1}{2}|S|\frac{1}{2},\frac{1}{2}\rangle & \langle\frac{1}{2},-\frac{1}{2}|S|\frac{1}{2},-\frac{1}{2}\rangle & \langle\frac{1}{2},-\frac{1}{2}|S|\frac{1}{2},\frac{1}{2}\rangle & \langle\frac{1}{2},-\frac{1}{2}|S|\frac{1}{2},-\frac{1}{2}\rangle & \langle\frac{1}{2},-\frac{1}{2}|S|\frac{1}{2},\frac{1}{2}\rangle & \langle\frac{1}{2},-\frac{1}{2}|S|\frac{1}{2},-\frac{1}{2}\rangle & \langle\frac{1}{2},-\frac{1}{2}|S|\frac{1}{2},\frac{1}{2}\rangle & \langle\frac{1}{2},-\frac{1}{2}|S|\frac{1}{2},-\frac{1}{2}\rangle & \langle\frac{1}{2},-\frac{1}{2}|S|\frac{1}{2},\frac{1}{2}\rangle & \langle\frac{1}{2},-\frac{1}{2}|S|\frac{1}{2},-\frac{1}{2}\rangle \\
\langle\frac{1}{2},\frac{1}{2}|S|\frac{1}{2},\frac{1}{2}\rangle & \langle\frac{1}{2},\frac{1}{2}|S|\frac{1}{2},-\frac{1}{2}\rangle & \langle\frac{1}{2},\frac{1}{2}|S|\frac{1}{2},\frac{1}{2}\rangle & \langle\frac{1}{2},\frac{1}{2}|S|\frac{1}{2},-\frac{1}{2}\rangle & \langle\frac{1}{2},\frac{1}{2}|S|\frac{1}{2},\frac{1}{2}\rangle & \langle\frac{1}{2},\frac{1}{2}|S|\frac{1}{2},-\frac{1}{2}\rangle & \langle\frac{1}{2},\frac{1}{2}|S|\frac{1}{2},\frac{1}{2}\rangle & \langle\frac{1}{2},\frac{1}{2}|S|\frac{1}{2},-\frac{1}{2}\rangle & \langle\frac{1}{2},\frac{1}{2}|S|\frac{1}{2},\frac{1}{2}\rangle & \langle\frac{1}{2},\frac{1}{2}|S|\frac{1}{2},-\frac{1}{2}\rangle & \langle\frac{1}{2},\frac{1}{2}|S|\frac{1}{2},\frac{1}{2}\rangle & \langle\frac{1}{2},\frac{1}{2}|S|\frac{1}{2},-\frac{1}{2}\rangle & \langle\frac{1}{2},\frac{1}{2}|S|\frac{1}{2},\frac{1}{2}\rangle & \langle\frac{1}{2},\frac{1}{2}|S|\frac{1}{2},-\frac{1}{2}\rangle & \langle\frac{1}{2},\frac{1}{2}|S|\frac{1}{2},\frac{1}{2}\rangle & \langle\frac{1}{2},\frac{1}{2}|S|\frac{1}{2},-\frac{1}{2}\rangle \\
\langle\frac{1}{2},-\frac{1}{2}|S|\frac{1}{2},\frac{1}{2}\rangle & \langle\frac{1}{2},-\frac{1}{2}|S|\frac{1}{2},-\frac{1}{2}\rangle & \langle\frac{1}{2},-\frac{1}{2}|S|\frac{1}{2},\frac{1}{2}\rangle & \langle\frac{1}{2},-\frac{1}{2}|S|\frac{1}{2},-\frac{1}{2}\rangle & \langle\frac{1}{2},-\frac{1}{2}|S|\frac{1}{2},\frac{1}{2}\rangle & \langle\frac{1}{2},-\frac{1}{2}|S|\frac{1}{2},-\frac{1}{2}\rangle & \langle\frac{1}{2},-\frac{1}{2}|S|\frac{1}{2},\frac{1}{2}\rangle & \langle\frac{1}{2},-\frac{1}{2}|S|\frac{1}{2},-\frac{1}{2}\rangle & \langle\frac{1}{2},-\frac{1}{2}|S|\frac{1}{2},\frac{1}{2}\rangle & \langle\frac{1}{2},-\frac{1}{2}|S|\frac{1}{2},-\frac{1}{2}\rangle & \langle\frac{1}{2},-\frac{1}{2}|S|\frac{1}{2},\frac{1}{2}\rangle & \langle\frac{1}{2},-\frac{1}{2}|S|\frac{1}{2},-\frac{1}{2}\rangle & \langle\frac{1}{2},-\frac{1}{2}|S|\frac{1}{2},\frac{1}{2}\rangle & \langle\frac{1}{2},-\frac{1}{2}|S|\frac{1}{2},-\frac{1}{2}\rangle & \langle\frac{1}{2},-\frac{1}{2}|S|\frac{1}{2},\frac{1}{2}\rangle & \langle\frac{1}{2},-\frac{1}{2}|S|\frac{1}{2},-\frac{1}{2}\rangle \\
\langle\frac{1}{2},\frac{1}{2}|S|\frac{1}{2},\frac{1}{2}\rangle & \langle\frac{1}{2},\frac{1}{2}|S|\frac{1}{2},-\frac{1}{2}\rangle & \langle\frac{1}{2},\frac{1}{2}|S|\frac{1}{2},\frac{1}{2}\rangle & \langle\frac{1}{2},\frac{1}{2}|S|\frac{1}{2},-\frac{1}{2}\rangle & \langle\frac{1}{2},\frac{1}{2}|S|\frac{1}{2},\frac{1}{2}\rangle & \langle\frac{1}{2},\frac{1}{2}|S|\frac{1}{2},-\frac{1}{2}\rangle & \langle\frac{1}{2},\frac{1}{2}|S|\frac{1}{2},\frac{1}{2}\rangle & \langle\frac{1}{2},\frac{1}{2}|S|\frac{1}{2},-\frac{1}{2}\rangle & \langle\frac{1}{2},\frac{1}{2}|S|\frac{1}{2},\frac{1}{2}\rangle & \langle\frac{1}{2},\frac{1}{2}|S|\frac{1}{2},-\frac{1}{2}\rangle & \langle\frac{1}{2},\frac{1}{2}|S|\frac{1}{2},\frac{1}{2}\rangle & \langle\frac{1}{2},\frac{1}{2}|S|\frac{1}{2},-\frac{1}{2}\rangle & \langle\frac{1}{2},\frac{1}{2}|S|\frac{1}{2},\frac{1}{2}\rangle & \langle\frac{1}{2},\frac{1}{2}|S|\frac{1}{2},-\frac{1}{2}\rangle & \langle\frac{1}{2},\frac{1}{2}|S|\frac{1}{2},\frac{1}{2}\rangle & \langle\frac{1}{2},\frac{1}{2}|S|\frac{1}{2},-\frac{1}{2}\rangle \\
\langle\frac{1}{2},-\frac{1}{2}|S|\frac{1}{2},\frac{1}{2}\rangle & \langle\frac{1}{2},-\frac{1}{2}|S|\frac{1}{2},-\frac{1}{2}\rangle & \langle\frac{1}{2},-\frac{1}{2}|S|\frac{1}{2},\frac{1}{2}\rangle & \langle\frac{1}{2},-\frac{1}{2}|S|\frac{1}{2},-\frac{1}{2}\rangle & \langle\frac{1}{2},-\frac{1}{2}|S|\frac{1}{2},\frac{1}{2}\rangle & \langle\frac{1}{2},-\frac{1}{2}|S|\frac{1}{2},-\frac{1}{2}\rangle & \langle\frac{1}{2},-\frac{1}{2}|S|\frac{1}{2},\frac{1}{2}\rangle & \langle\frac{1}{2},-\frac{1}{2}|S|\frac{1}{2},-\frac{1}{2}\rangle & \langle\frac{1}{2},-\frac{1}{2}|S|\frac{1}{2},\frac{1}{2}\rangle & \langle\frac{1}{2},-\frac{1}{2}|S|\frac{1}{2},-\frac{1}{2}\rangle & \langle\frac{1}{2},-\frac{1}{2}|S|\frac{1}{2},\frac{1}{2}\rangle & \langle\frac{1}{2},-\frac{1}{2}|S|\frac{1}{2},-\frac{1}{2}\rangle & \langle\frac{1}{2},-\frac{1}{2}|S|\frac{1}{2},\frac{1}{2}\rangle & \langle\frac{1}{2},-\frac{1}{2}|S|\frac{1}{2},-\frac{1}{2}\rangle & \langle\frac{1}{2},-\frac{1}{2}|S|\frac{1}{2},\frac{1}{2}\rangle & \langle\frac{1}{2},-\frac{1}{2}|S|\frac{1}{2},-\frac{1}{2}\rangle \\
\langle\frac{1}{2},\frac{1}{2}|S|\frac{1}{2},\frac{1}{2}\rangle & \langle\frac{1}{2},\frac{1}{2}|S|\frac{1}{2},-\frac{1}{2}\rangle & \langle\frac{1}{2},\frac{1}{2}|S|\frac{1}{2},\frac{1}{2}\rangle & \langle\frac{1}{2},\frac{1}{2}|S|\frac{1}{2},-\frac{1}{2}\rangle & \langle\frac{1}{2},\frac{1}{2}|S|\frac{1}{2},\frac{1}{2}\rangle & \langle\frac{1}{2},\frac{1}{2}|S|\frac{1}{2},-\frac{1}{2}\rangle & \langle\frac{1}{2},\frac{1}{2}|S|\frac{1}{2},\frac{1}{2}\rangle & \langle\frac{1}{2},\frac{1}{2}|S|\frac{1}{2},-\frac{1}{2}\rangle & \langle\frac{1}{2},\frac{1}{2}|S|\frac{1}{2},\frac{1}{2}\rangle & \langle\frac{1}{2},\frac{1}{2}|S|\frac{1}{2},-\frac{1}{2}\rangle & \langle\frac{1}{2},\frac{1}{2}|S|\frac{1}{2},\frac{1}{2}\rangle & \langle\frac{1}{2},\frac{1}{2}|S|\frac{1}{2},-\frac{1}{2}\rangle & \langle\frac{1}{2},\frac{1}{2}|S|\frac{1}{2},\frac{1}{2}\rangle & \langle\frac{1}{2},\frac{1}{2}|S|\frac{1}{2},-\frac{1}{2}\rangle & \langle\frac{1}{2},\frac{1}{2}|S|\frac{1}{2},\frac{1}{2}\rangle & \langle\frac{1}{2},\frac{1}{2}|S|\frac{1}{2},-\frac{1}{2}\rangle \\
\langle\frac{1}{2},-\frac{1}{2}|S|\frac{1}{2},\frac{1}{2}\rangle & \langle\frac{1}{2},-\frac{1}{2}|S|\frac{1}{2},-\frac{1}{2}\rangle & \langle\frac{1}{2},-\frac{1}{2}|S|\frac{1}{2},\frac{1}{2}\rangle & \langle\frac{1}{2},-\frac{1}{2}|S|\frac{1}{2},-\frac{1}{2}\rangle & \langle\frac{1}{2},-\frac{1}{2}|S|\frac{1}{2},\frac{1}{2}\rangle & \langle\frac{1}{2},-\frac{1}{2}|S|\frac{1}{2},-\frac{1}{2}\rangle & \langle\frac{1}{2},-\frac{1}{2}|S|\frac{1}{2},\frac{1}{2}\rangle & \langle\frac{1}{2},-\frac{1}{2}|S|\frac{1}{2},-\frac{1}{2}\rangle & \langle\frac{1}{2},-\frac{1}{2}|S|\frac{1}{2},\frac{1}{2}\rangle & \langle\frac{1}{2},-\frac{1}{2}|S|\frac{1}{2},-\frac{1}{2}\rangle & \langle\frac{1}{2},-\frac{1}{2}|S|\frac{1}{2},\frac{1}{2}\rangle & \langle\frac{1}{2},-\frac{1}{2}|S|\frac{1}{2},-\frac{1}{2}\rangle & \langle\frac{1}{2},-\frac{1}{2}|S|\frac{1}{2},\frac{1}{2}\rangle & \langle\frac{1}{2},-\frac{1}{2}|S|\frac{1}{2},-\frac{1}{2}\rangle & \langle\frac{1}{2},-\frac{1}{2}|S|\frac{1}{2},\frac{1}{2}\rangle & \langle\frac{1}{2},-\frac{1}{2}|S|\frac{1}{2},-\frac{1}{2}\rangle \\
\langle\frac{1}{2},\frac{1}{2}|S|\frac{1}{2},\frac{1}{2}\rangle & \langle\frac{1}{2},\frac{1}{2}|S|\frac{1}{2},-\frac{1}{2}\rangle & \langle\frac{1}{2},\frac{1}{2}|S|\frac{1}{2},\frac{1}{2}\rangle & \langle\frac{1}{2},\frac{1}{2}|S|\frac{1}{2},-\frac{1}{2}\rangle & \langle\frac{1}{2},\frac{1}{2}|S|\frac{1}{2},\frac{1}{2}\rangle & \langle\frac{1}{2},\frac{1}{2}|S|\frac{1}{2},-\frac{1}{2}\rangle & \langle\frac{1}{2},\frac{1}{2}|S|\frac{1}{2},\frac{1}{2}\rangle & \langle\frac{1}{2},\frac{1}{2}|S|\frac{1}{2},-\frac{1}{2}\rangle & \langle\frac{1}{2},\frac{1}{2}|S|\frac{1}{2},\frac{1}{2}\rangle & \langle\frac{1}{2},\frac{1}{2}|S|\frac{1}{2},-\frac{1}{2}\rangle & \langle\frac{1}{2},\frac{1}{2}|S|\frac{1}{2},\frac{1}{2}\rangle & \langle\frac{1}{2},\frac{1}{2}|S|\frac{1}{2},-\frac{1}{2}\rangle & \langle\frac{1}{2},\frac{1}{2}|S|\frac{1}{2},\frac{1}{2}\rangle & \langle\frac{1}{2},\frac{1}{2}|S|\frac{1}{2},-\frac{1}{2}\rangle & \langle\frac{1}{2},\frac{1}{2}|S|\frac{1}{2},\frac{1}{2}\rangle & \langle\frac{1}{2},\frac{1}{2}|S|\frac{1}{2},-\frac{1}{2}\rangle \\
\langle\frac{1}{2},-\frac{1}{2}|S|\frac{1}{2},\frac{1}{2}\rangle & \langle\frac{1}{2},-\frac{1}{2}|S|\frac{1}{2},-\frac{1}{2}\rangle & \langle\frac{1}{2},-\frac{1}{2}|S|\frac{1}{2},\frac{1}{2}\rangle & \langle\frac{1}{2},-\frac{1}{2}|S|\frac{1}{2},-\frac{1}{2}\rangle & \langle\frac{1}{2},-\frac{1}{2}|S|\frac{1}{2},\frac{1}{2}\rangle & \langle\frac{1}{2},-\frac{1}{2}|S|\frac{1}{2},-\frac{1}{2}\rangle & \langle\frac{1}{2},-\frac{1}{2}|S|\frac{1}{2},\frac{1}{2}\rangle & \langle\frac{1}{2},-\frac{1}{2}|S|\frac{1}{2},-\frac{1}{2}\rangle & \langle\frac{1}{2},-\frac{1}{2}|S|\frac{1}{2},\frac{1}{2}\rangle & \langle\frac{1}{2},-\frac{1}{2}|S|\frac{1}{2},-\frac{1}{2}\rangle & \langle\frac{1}{2},-\frac{1}{2}|S|\frac{1}{2},\frac{1}{2}\rangle & \langle\frac{1}{2},-\frac{1}{2}|S|\frac{1}{2},-\frac{1}{2}\rangle & \langle\frac{1}{2},-\frac{1}{2}|S|\frac{1}{2},\frac{1}{2}\rangle & \langle\frac{1}{2},-\frac{1}{2}|S|\frac{1}{2},-\frac{1}{2}\rangle & \langle\frac{1}{2},-\frac{1}{2}|S|\frac{1}{2},\frac{1}{2}\rangle & \langle\frac{1}{2},-\frac{1}{2}|S|\frac{1}{2},-\frac{1}{2}\rangle \\
\langle\frac{1}{2},\frac{1}{2}|S|\frac{1}{2},\frac{1}{2}\rangle & \langle\frac{1}{2},\frac{1}{2}|S|\frac{1}{2},-\frac{1}{2}\rangle & \langle\frac{1}{2},\frac{1}{2}|S|\frac{1}{2},\frac{1}{2}\rangle & \langle\frac{1}{2},\frac{1}{2}|S|\frac{1}{2},-\frac{1}{2}\rangle & \langle\frac{1}{2},\frac{1}{2}|S|\frac{1}{2},\frac{1}{2}\rangle & \langle\frac{1}{2},\frac{1}{2}|S|\frac{1}{2},-\frac{1}{2}\rangle & \langle\frac{1}{2},\frac{1}{2}|S|\frac{1}{2},\frac{1}{2}\rangle & \langle\frac{1}{2},\frac{1}{2}|S|\frac{1}{2},-\frac{1}{2}\rangle & \langle\frac{1}{2},\frac{1}{2}|S|\frac{1}{2},\frac{1}{2}\rangle & \langle\frac{1}{2},\frac{1}{2}|S|\frac{1}{2},-\frac{1}{2}\rangle & \langle\frac{1}{2},\frac{1}{2}|S|\frac{1}{2},\frac{1}{2}\rangle & \langle\frac{1}{2},\frac{1}{2}|S|\frac{1}{2},-\frac{1}{2}\rangle & \langle\frac{1}{2},\frac{1}{2}|S|\frac{1}{2},\frac{1}{2}\rangle & \langle\frac{1}{2},\frac{1}{2}|S|\frac{1}{2},-\frac{1}{2}\rangle & \langle\frac{1}{2},\frac{1}{2}|S|\frac{1}{2},\frac{1}{2}\rangle & \langle\frac{1}{2},\frac{1}{2}|S|\frac{1}{2},-\frac{1}{2}\rangle \\
\langle\frac{1}{2},-\frac{1}{2}|S|\frac{1}{2},\frac{1}{2}\rangle & \langle\frac{1}{2},-\frac{1}{2}|S|\frac{1}{2},-\frac{1}{2}\rangle & \langle\frac{1}{2},-\frac{1}{2}|S|\frac{1}{2},\frac{1}{2}\rangle & \langle\frac{1}{2},-\frac{1}{2}|S|\frac{1}{2},-\frac{1}{2}\rangle & \langle\frac{1}{2},-\frac{1}{2}|S|\frac{1}{2},\frac{1}{2}\rangle & \langle\frac{1}{2},-\frac{1}{2}|S|\frac{1}{2},-\frac{1}{2}\rangle & \langle\frac{1}{2},-\frac{1}{2}|S|\frac{1}{2},\frac{1}{2}\rangle & \langle\frac{1}{2},-\frac{1}{2}|S|\frac{1}{2},-\frac{1}{2}\rangle & \langle\frac{1}{2},-\frac{1}{2}|S|\frac{1}{2},\frac{1}{2}\rangle & \langle\frac{1}{2},-\frac{1}{2}|S|\frac{1}{2},-\frac{1}{2}\rangle & \langle\frac{1}{2},-\frac{1}{2}|S|\frac{1}{2},\frac{1}{2}\rangle & \langle\frac{1}{2},-\frac{1}{2}|S|\frac{1}{2},-\frac{1}{2}\rangle & \langle\frac{1}{2},-\frac{1}{2}|S|\frac{1}{2},\frac{1}{2}\rangle & \langle\frac{1}{2},-\frac{1}{2}|S|\frac{1}{2},-\frac{1}{2}\rangle & \langle\frac{1}{2},-\frac{1}{2}|S|\frac{1}{2},\frac{1}{2}\rangle & \langle\frac{1}{2},-\frac{1}{2}|S|\frac{1}{2},-\frac{1}{2}\rangle \\
\langle\frac{1}{2},\frac{1}{2}|S|\frac{1}{2},\frac{1}{2}\rangle & \langle\frac{1}{2},\frac{1}{2}|S|\frac{1}{2},-\frac{1}{2}\rangle & \langle\frac{1}{2},\frac{1}{2}|S|\frac{1}{2},\frac{1}{2}\rangle & \langle\frac{1}{2},\frac{1}{2}|S|\frac{1}{2},-\frac{1}{2}\rangle & \langle\frac{1}{2},\frac{1}{2}|S|\frac{1}{2},\frac{1}{2}\rangle & \langle\frac{1}{2},\frac{1}{2}|S|\frac{1}{2},-\frac{1}{2}\rangle & \langle\frac{1}{2},\frac{1}{2}|S|\frac{1}{2},\frac{1}{2}\rangle & \langle\frac{1}{2},\frac{1}{2}|S|\frac{1}{2},-\frac{1}{2}\rangle & \langle\frac{1}{2},\frac{1}{2}|S|\frac{1}{2},\frac{1}{2}\rangle & \langle\frac{1}{2},\frac{1}{2}|S|\frac{1}{2},-\frac{1}{2}\rangle & \langle\frac{1}{2},\frac{1}{2}|S|\frac{1}{2},\frac{1}{2}\rangle & \langle\frac{1}{2},\frac{1}{2}|S|\frac{1}{2},-\frac{1}{2}\rangle & \langle\frac{1}{2},\frac{1}{2}|S|\frac{1}{2},\frac{1}{2}\rangle & \langle\frac{1}{2},\frac{1}{2}|S|\frac{1}{2},-\frac{1}{2}\rangle & \langle\frac{1}{2},\frac{1}{2}|S|\frac{1}{2},\frac{1}{2}\rangle & \langle\frac{1}{2},\frac{1}{2}|S|\frac{1}{2},-\frac{1}{2}\rangle \\
\langle\frac{1}{2},-\frac{1}{2}|S|\frac{1}{2},\frac{1}{2}\rangle & \langle\frac{1}{2},-\frac{1}{2}|S|\frac{1}{2},-\frac{1}{2}\rangle & \langle\frac{1}{2},-\frac{1}{2}|S|\frac{1}{2},\frac{1}{2}\rangle & \langle\frac{1}{2},-\frac{1}{2}|S|\frac{1}{2},-\frac{1}{2}\rangle & \langle\frac{1}{2},-\frac{1}{2}|S|\frac{1}{2},\frac{1}{2}\rangle & \langle\frac{1}{2},-\frac{1}{2}|S|\frac{1}{2},-\frac{1}{2}\rangle & \langle\frac{1}{2},-\frac{1}{2}|S|\frac{1}{2},\frac{1}{2}\rangle & \langle\frac{1}{2},-\frac{1}{2}|S|\frac{1}{2},-\frac{1}{2}\rangle & \langle\frac{1}{2},-\frac{1}{2}|S|\frac{1}{2},\frac{1}{2}\rangle & \langle\frac{1}{2},-\frac{1}{2}|S|\frac{1}{2},-\frac{1}{2}\rangle & \langle\frac{1}{2},-\frac{1}{2}|S|\frac{1}{2},\frac{1}{2}\rangle & \langle\frac{1}{2},-\frac{1}{2}|S|\frac{1}{2},-\frac{1}{2}\rangle & \langle\frac{1}{2},-\frac{1}{2}|S|\frac{1}{2},\frac{1}{2}\rangle & \langle\frac{1}{2},-\frac{1}{2}|S|\frac{1}{2},-\frac{1}{2}\rangle & \langle\frac{1}{2},-\frac{1}{2}|S|\frac{1}{2},\frac{1}{2}\rangle & \langle\frac{1}{2},-\frac{1}{2}|S|\frac{1}{2},-\frac{1}{2}\rangle
\end{bmatrix}_{16\times16} \quad (A.7)$$

B

Electrostatic Energy Calculation

Computing with Multi-Value Logic in Quantum Dot Cellular Automata, First Edition.
Reza Sabbaghi-Nadooshan, Reza Akbari-Hasanjani, Leila Dehbozorgi, Majid Haghparast, and Hamid Reza Akbari-Hasanjani.

Table B.1 Output results of the ternary AND gate in all cases.

	Output cell in A state	Output cell in C state	Output cell in D state	Output cell in B state	Final result
In1 = A	$E_t(J) = 0.1343*10^{-19}$	$E_t(J) = 0.1500*10^{-19}$	$E_t(J) = 0.2378*10^{-19}$	$E_t(J) = 0.1731*10^{-19}$	$E_t(J) = 0.1343*10^{-19}$
In2 = A					
In3 = A					
In1 = A	$E_t(J) = 0.1289*10^{-19}$	$E_t(J) = 0.1378*10^{-19}$	$E_t(J) = 0.2658*10^{-19}$	$E_t(J) = 0.2229*10^{-19}$	$E_t(J) = 0.1289*10^{-19}$
In2 = A					
In3 = C					
In1 = A	$E_t(J) = 0.1049*10^{-19}$	$E_t(J) = 0.1759*10^{-19}$	$E_t(J) = 0.2637*10^{-19}$	$E_t(J) = 0.2534*10^{-19}$	$E_t(J) = 0.1049*10^{-19}$
In2 = A					
In3 = B					
In1 = A	$E_t(J) = 0.1196*10^{-19}$	$E_t(J) = 0.1272*10^{-19}$	$E_t(J) = 0.2163*10^{-19}$	$E_t(J) = 0.1797*10^{-19}$	$E_t(J) = 0.1196*10^{-19}$
In2 = C					
In3 = A					
In1 = A	$E_t(J) = 0.2415*10^{-19}$	$E_t(J) = 0.0471*10^{-19}$	$E_t(J) = 0.3845*10^{-19}$	$E_t(J) = 0.2213*10^{-19}$	$E_t(J) = 0.0471*10^{-19}$
In2 = C					
In3 = C					
In1 = A	$E_t(J) = 0.2175*10^{-19}$	$E_t(J) = 0.0852*10^{-19}$	$E_t(J) = 0.3824*10^{-19}$	$E_t(J) = 0.2517*10^{-19}$	$E_t(J) = 0.0852*10^{-19}$
In2 = C					
In3 = B					
In1 = A	$E_t(J) = 0.0541*10^{-19}$	$E_t(J) = 0.1242*10^{-19}$	$E_t(J) = 0.2119*10^{-19}$	$E_t(J) = 0.2025*10^{-19}$	$E_t(J) = 0.0541*10^{-19}$
In2 = B					
In3 = A					
In1 = A	$E_t(J) = 0.1759*10^{-19}$	$E_t(J) = 0.0441*10^{-19}$	$E_t(J) = 0.3802*10^{-19}$	$E_t(J) = 0.2441*10^{-19}$	$E_t(J) = 0.0441*10^{-19}$
In2 = B					
In3 = C					
In1 = A	$E_t(J) = 0.1602*10^{-19}$	$E_t(J) = 0.1500*10^{-19}$	$E_t(J) = 0.2378*10^{-19}$	$E_t(J) = 0.1473*10^{-19}$	$E_t(J) = 0.1473*10^{-19}$
In2 = B					
In3 = B					

Table B.2 Output of ternary OR gates in all cases.

	Output cell in A state	Output cell in C state	Output cell in D state	Output cell in B state	Final result
In1 = B In2 = A In3 = A	$E_t(J) = 0.1473*10^{-19}$	$E_t(J) = 0.1500*10^{-19}$	$E_t(J) = 0.2378*10^{-19}$	$E_t(J) = 0.1602*10^{-19}$	$E_t(J) = 0.1473*10^{-19}$
In1 = B In2 = A In3 = C	$E_t(J) = 0.2691*10^{-19}$	$E_t(J) = 0.0700*10^{-19}$	$E_t(J) = 0.4060*10^{-19}$	$E_t(J) = 0.2018*10^{-19}$	$E_t(J) = 0.0700*10^{-19}$
In1 = B In2 = A In3 = B	$E_t(J) = 0.2534*10^{-19}$	$E_t(J) = 0.1759*10^{-19}$	$E_t(J) = 0.2637*10^{-19}$	$E_t(J) = 0.1049*10^{-19}$	$E_t(J) = 0.1049*10^{-19}$
In1 = B In2 = C In3 = A	$E_t(J) = 0.2599*10^{-19}$	$E_t(J) = 0.0593*10^{-19}$	$E_t(J) = 0.3566*10^{-19}$	$E_t(J) = 0.1585 *10^{-19}$	$E_t(J) = 0.0593*10^{-19}$
In1 = B In2 = C In3 = C	$E_t(J) = 0.2544*10^{-19}$	$E_t(J) = 0.0471*10^{-19}$	$E_t(J) = 0.3845*10^{-19}$	$E_t(J) = 0.2083*10^{-19}$	$E_t(J) = 0.0471*10^{-19}$
In1 = B In2 = C In3 = B	$E_t(J) = 0.2387*10^{-19}$	$E_t(J) = 0.1531*10^{-19}$	$E_t(J) = 0.2422*10^{-19}$	$E_t(J) = 0.1115 *10^{-19}$	$E_t(J) = 0.1115 *10^{-19}$
In1 = B In2 = B In3 = A	$E_t(J) = 0.2025*10^{-19}$	$E_t(J) = 0.1242*10^{-19}$	$E_t(J) = 0.2119*10^{-19}$	$E_t(J) = 0.0541*10^{-19}$	$E_t(J) = 0.0541*10^{-19}$
In1 = B In2 = B In3 = C	$E_t(J) = 0.1971*10^{-19}$	$E_t(J) = 0.1120*10^{-19}$	$E_t(J) = 0.2399*10^{-19}$	$E_t(J) = 0.1039*10^{-19}$	$E_t(J) = 0.1039*10^{-19}$
In1 = B In2 = B In3 = B	$E_t(J) = 0.1731*10^{-19}$	$E_t(J) = 0.1500*10^{-19}$	$E_t(J) = 0.2378*10^{-19}$	$E_t(J) = 0.1343*10^{-19}$	$E_t(J) = 0.1343*10^{-19}$

Table B.3 Outputs of the quaternary AND gates in all cases.

	Output cell in A state	Output cell in B state	Output cell in C state	Output cell in D state	Final result
$In1 = A$	$E_{t_{bQCA}} = 0.919*10^{-20}$	$E_{t_{bQCA}} = 1.731*10^{-20}$	$E_{t_{bQCA}} = 0.919*10^{-20}$	$E_{t_{bQCA}} = 1.731*10^{-20}$	$E_{t_{bQCA}} = 0.919*10^{-20}$
$In2 = A$	$E_{t_{TQCA}} = 1.343*10^{-20}$	$E_{t_{TQCA}} = 1.343*10^{-20}$	$E_{t_{TQCA}} = 1.731*10^{-20}$	$E_{t_{TQCA}} = 1.731*10^{-20}$	$E_{t_{TQCA}} = 1.343*10^{-20}$
$In3 = A$	$E_{t_{QQCA}} = 2.262*10^{-20}$	$E_{t_{QQCA}} = 3.074*10^{-20}$	$E_{t_{QQCA}} = 2.65*10^{-20}$	$E_{t_{QQCA}} = 3.462*10^{-20}$	$E_{t_{QQCA}} = 2.262*10^{-20}$
$In1 = A$	$E_{t_{bQCA}} = 0.625*10^{-20}$	$E_{t_{bQCA}} = 2.534*10^{-20}$	$E_{t_{bQCA}} = 0.625*10^{-20}$	$E_{t_{bQCA}} = 2.534*10^{-20}$	$E_{t_{bQCA}} = 0.625*10^{-20}$
$In2 = A$	$E_{t_{TQCA}} = 1.343*10^{-20}$	$E_{t_{TQCA}} = 1.343*10^{-20}$	$E_{t_{TQCA}} = 1.731*10^{-20}$	$E_{t_{TQCA}} = 1.731*10^{-20}$	$E_{t_{TQCA}} = 1.343*10^{-20}$
$In3 = B$	$E_{t_{QQCA}} = 1.968*10^{-20}$	$E_{t_{QQCA}} = 3.887*10^{-20}$	$E_{t_{QQCA}} = 2.356*10^{-20}$	$E_{t_{QQCA}} = 4.265*10^{-20}$	$E_{t_{QQCA}} = 1.968*10^{-20}$
$In1 = A$	$E_{t_{bQCA}} = 0.919*10^{-20}$	$E_{t_{bQCA}} = 1.731*10^{-20}$	$E_{t_{bQCA}} = 0.919*10^{-20}$	$E_{t_{bQCA}} = 1.731*10^{-20}$	$E_{t_{bQCA}} = 0.919*10^{-20}$
$In2 = A$	$E_{t_{TQCA}} = 1.049*10^{-20}$	$E_{t_{TQCA}} = 1.049*10^{-20}$	$E_{t_{TQCA}} = 2.534*10^{-20}$	$E_{t_{TQCA}} = 2.534*10^{-20}$	$E_{t_{TQCA}} = 1.049*10^{-20}$
$In3 = C$	$E_{t_{QQCA}} = 1.968*10^{-20}$	$E_{t_{QQCA}} = 2.78*10^{-20}$	$E_{t_{QQCA}} = 3.453*10^{-20}$	$E_{t_{QQCA}} = 4.265*10^{-20}$	$E_{t_{QQCA}} = 1.968*10^{-20}$
$In1 = A$	$E_{t_{bQCA}} = 0.625*10^{-20}$	$E_{t_{bQCA}} = 2.534*10^{-20}$	$E_{t_{bQCA}} = 0.625*10^{-20}$	$E_{t_{bQCA}} = 2.534*10^{-20}$	$E_{t_{bQCA}} = 0.625*10^{-20}$
$In2 = A$	$E_{t_{TQCA}} = 1.049*10^{-20}$	$E_{t_{TQCA}} = 1.049*10^{-20}$	$E_{t_{TQCA}} = 2.534*10^{-20}$	$E_{t_{TQCA}} = 2.534*10^{-20}$	$E_{t_{TQCA}} = 1.049*10^{-20}$
$In3 = D$	$E_{t_{QQCA}} = 1.674*10^{-20}$	$E_{t_{QQCA}} = 3.583*10^{-20}$	$E_{t_{QQCA}} = 3.159*10^{-20}$	$E_{t_{QQCA}} = 5.068*10^{-20}$	$E_{t_{QQCA}} = 1.674*10^{-20}$
$In1 = A$	$E_{t_{bQCA}} = 0.541*10^{-20}$	$E_{t_{bQCA}} = 2.025*10^{-20}$	$E_{t_{bQCA}} = 0.541*10^{-20}$	$E_{t_{bQCA}} = 2.025*10^{-20}$	$E_{t_{bQCA}} = 0.541*10^{-20}$
$In2 = B$	$E_{t_{TQCA}} = 1.343*10^{-20}$	$E_{t_{TQCA}} = 1.343*10^{-20}$	$E_{t_{TQCA}} = 1.731*10^{-20}$	$E_{t_{TQCA}} = 1.731*10^{-20}$	$E_{t_{TQCA}} = 1.343*10^{-20}$
$In3 = A$	$E_{t_{QQCA}} = 1.884*10^{-20}$	$E_{t_{QQCA}} = 3.368*10^{-20}$	$E_{t_{QQCA}} = 2.272*10^{-20}$	$E_{t_{QQCA}} = 3.756*10^{-20}$	$E_{t_{QQCA}} = 1.884*10^{-20}$
$In1 = A$	$E_{t_{bQCA}} = 1.602*10^{-20}$	$E_{t_{bQCA}} = 1.473*10^{-20}$	$E_{t_{bQCA}} = 1.602*10^{-20}$	$E_{t_{bQCA}} = 1.473*10^{-20}$	$E_{t_{bQCA}} = 1.473*10^{-20}$
$In2 = B$	$E_{t_{TQCA}} = 1.343*10^{-20}$	$E_{t_{TQCA}} = 1.343*10^{-20}$	$E_{t_{TQCA}} = 1.731*10^{-20}$	$E_{t_{TQCA}} = 1.731*10^{-20}$	$E_{t_{TQCA}} = 1.343*10^{-20}$
$In3 = B$	$E_{t_{QQCA}} = 2.945*10^{-20}$	$E_{t_{QQCA}} = 2.816*10^{-20}$	$E_{t_{QQCA}} = 3.333*10^{-20}$	$E_{t_{QQCA}} = 3.204*10^{-20}$	$E_{t_{QQCA}} = 2.816*10^{-20}$
$In1 = A$	$E_{t_{bQCA}} = 0.541*10^{-20}$	$E_{t_{bQCA}} = 2.025*10^{-20}$	$E_{t_{bQCA}} = 0.541*10^{-20}$	$E_{t_{bQCA}} = 2.025*10^{-20}$	$E_{t_{bQCA}} = 0.541*10^{-20}$
$In2 = B$	$E_{t_{TQCA}} = 1.049*10^{-20}$	$E_{t_{TQCA}} = 1.049*10^{-20}$	$E_{t_{TQCA}} = 2.534*10^{-20}$	$E_{t_{TQCA}} = 2.534*10^{-20}$	$E_{t_{TQCA}} = 1.049*10^{-20}$
$In3 = C$	$E_{t_{QQCA}} = 1.59*10^{-20}$	$E_{t_{QQCA}} = 3.074*10^{-20}$	$E_{t_{QQCA}} = 3.075*10^{-20}$	$E_{t_{QQCA}} = 4.559*10^{-20}$	$E_{t_{QQCA}} = 1.59*10^{-20}$
$In1 = A$	$E_{t_{bQCA}} = 1.602*10^{-20}$	$E_{t_{bQCA}} = 1.473*10^{-20}$	$E_{t_{bQCA}} = 1.602*10^{-20}$	$E_{t_{bQCA}} = 1.473*10^{-20}$	$E_{t_{bQCA}} = 1.473*10^{-20}$
$In2 = B$	$E_{t_{TQCA}} = 1.049*10^{-20}$	$E_{t_{TQCA}} = 1.049*10^{-20}$	$E_{t_{TQCA}} = 2.534*10^{-20}$	$E_{t_{TQCA}} = 2.534*10^{-20}$	$E_{t_{TQCA}} = 1.049*10^{-20}$
$In3 = D$	$E_{t_{QQCA}} = 2.651*10^{-20}$	$E_{t_{QQCA}} = 2.522*10^{-20}$	$E_{t_{QQCA}} = 4.136*10^{-20}$	$E_{t_{QQCA}} = 4.007*10^{-20}$	$E_{t_{QQCA}} = 2.522*10^{-20}$

In1 = A	$E_{t_{bQCA}} = 1.343*10^{-20}$	$E_{t_{bQCA}} = 1.731*10^{-20}$	$E_{t_{bQCA}} = 1.343*10^{-20}$	$E_{t_{bQCA}} = 1.731*10^{-20}$	$E_{t_{bQCA}} = 1.343*10^{-20}$
In2 = C	$E_{t_{TQCA}} = 0.541*10^{-20}$	$E_{t_{TQCA}} = 0.541*10^{-20}$	$E_{t_{TQCA}} = 2.025*10^{-20}$	$E_{t_{TQCA}} = 2.025*10^{-20}$	$E_{t_{TQCA}} = 0.541*10^{-20}$
In3 = A	$E_{t_{QQCA}} = 1.884*10^{-20}$	$E_{t_{QQCA}} = 2.272*10^{-20}$	$E_{t_{QQCA}} = 2.368*10^{-20}$	$E_{t_{QQCA}} = 3.756*10^{-20}$	$E_{t_{QQCA}} = 1.884*10^{-20}$
In1 = A	$E_{t_{bQCA}} = 1.049*10^{-20}$	$E_{t_{bQCA}} = 2.534*10^{-20}$	$E_{t_{bQCA}} = 1.049*10^{-20}$	$E_{t_{bQCA}} = 2.534*10^{-20}$	$E_{t_{bQCA}} = 1.049*10^{-20}$
In2 = C	$E_{t_{TQCA}} = 0.541*10^{-20}$	$E_{t_{TQCA}} = 0.541*10^{-20}$	$E_{t_{TQCA}} = 2.025*10^{-20}$	$E_{t_{TQCA}} = 2.025*10^{-20}$	$E_{t_{TQCA}} = 0.541*10^{-20}$
In3 = B	$E_{t_{QQCA}} = 1.59*10^{-20}$	$E_{t_{QQCA}} = 3.075*10^{-20}$	$E_{t_{QQCA}} = 3.074*10^{-20}$	$E_{t_{QQCA}} = 4.559*10^{-20}$	$E_{t_{QQCA}} = 1.59*10^{-20}$
In1 = A	$E_{t_{bQCA}} = 1.343*10^{-20}$	$E_{t_{bQCA}} = 1.731*10^{-20}$	$E_{t_{bQCA}} = 1.343*10^{-20}$	$E_{t_{bQCA}} = 1.731*10^{-20}$	$E_{t_{bQCA}} = 1.343*10^{-20}$
In2 = C	$E_{t_{TQCA}} = 1.602*10^{-20}$	$E_{t_{TQCA}} = 1.602*10^{-20}$	$E_{t_{TQCA}} = 1.473*10^{-20}$	$E_{t_{TQCA}} = 1.473*10^{-20}$	$E_{t_{TQCA}} = 1.473*10^{-20}$
In3 = C	$E_{t_{QQCA}} = 2.945*10^{-20}$	$E_{t_{QQCA}} = 3.333*10^{-20}$	$E_{t_{QQCA}} = 2.816*10^{-20}$	$E_{t_{QQCA}} = 3.204*10^{-20}$	$E_{t_{QQCA}} = 2.816*10^{-20}$
In1 = A	$E_{t_{bQCA}} = 1.049*10^{-20}$	$E_{t_{bQCA}} = 2.534*10^{-20}$	$E_{t_{bQCA}} = 1.049*10^{-20}$	$E_{t_{bQCA}} = 2.534*10^{-20}$	$E_{t_{bQCA}} = 1.049*10^{-20}$
In2 = C	$E_{t_{TQCA}} = 1.602*10^{-20}$	$E_{t_{TQCA}} = 1.602*10^{-20}$	$E_{t_{TQCA}} = 1.473*10^{-20}$	$E_{t_{TQCA}} = 1.473*10^{-20}$	$E_{t_{TQCA}} = 1.473*10^{-20}$
In3 = D	$E_{t_{QQCA}} = 2.651*10^{-20}$	$E_{t_{QQCA}} = 4.136*10^{-20}$	$E_{t_{QQCA}} = 2.522*10^{-20}$	$E_{t_{QQCA}} = 4.007*10^{-20}$	$E_{t_{QQCA}} = 2.522*10^{-20}$
In1 = A	$E_{t_{bQCA}} = 0.541*10^{-20}$	$E_{t_{bQCA}} = 2.025*10^{-20}$	$E_{t_{bQCA}} = 0.541*10^{-20}$	$E_{t_{bQCA}} = 2.025*10^{-20}$	$E_{t_{bQCA}} = 0.541*10^{-20}$
In2 = D	$E_{t_{TQCA}} = 0.541*10^{-20}$	$E_{t_{TQCA}} = 0.541*10^{-20}$	$E_{t_{TQCA}} = 2.025*10^{-20}$	$E_{t_{TQCA}} = 2.025*10^{-20}$	$E_{t_{TQCA}} = 0.541*10^{-20}$
In3 = A	$E_{t_{QQCA}} = 1.082*10^{-20}$	$E_{t_{QQCA}} = 2.566*10^{-20}$	$E_{t_{QQCA}} = 2.566*10^{-20}$	$E_{t_{QQCA}} = 4.05*10^{-20}$	$E_{t_{QQCA}} = 1.082*10^{-20}$
In1 = A	$E_{t_{bQCA}} = 1.602*10^{-20}$	$E_{t_{bQCA}} = 1.473*10^{-20}$	$E_{t_{bQCA}} = 1.602*10^{-20}$	$E_{t_{bQCA}} = 1.473*10^{-20}$	$E_{t_{bQCA}} = 1.473*10^{-20}$
In2 = D	$E_{t_{TQCA}} = 0.541*10^{-20}$	$E_{t_{TQCA}} = 0.541*10^{-20}$	$E_{t_{TQCA}} = 2.025*10^{-20}$	$E_{t_{TQCA}} = 2.025*10^{-20}$	$E_{t_{TQCA}} = 0.541*10^{-20}$
In3 = B	$E_{t_{QQCA}} = 2.143*10^{-20}$	$E_{t_{QQCA}} = 2.014*10^{-20}$	$E_{t_{QQCA}} = 3.627*10^{-20}$	$E_{t_{QQCA}} = 3.498*10^{-20}$	$E_{t_{QQCA}} = 2.014*10^{-20}$
In1 = A	$E_{t_{bQCA}} = 0.541*10^{-20}$	$E_{t_{bQCA}} = 2.025*10^{-20}$	$E_{t_{bQCA}} = 0.541*10^{-20}$	$E_{t_{bQCA}} = 2.025*10^{-20}$	$E_{t_{bQCA}} = 0.541*10^{-20}$
In2 = D	$E_{t_{TQCA}} = 1.602*10^{-20}$	$E_{t_{TQCA}} = 1.602*10^{-20}$	$E_{t_{TQCA}} = 1.473*10^{-20}$	$E_{t_{TQCA}} = 1.473*10^{-20}$	$E_{t_{TQCA}} = 1.473*10^{-20}$
In3 = C	$E_{t_{QQCA}} = 2.143*10^{-20}$	$E_{t_{QQCA}} = 3.627*10^{-20}$	$E_{t_{QQCA}} = 2.014*10^{-20}$	$E_{t_{QQCA}} = 3.498*10^{-20}$	$E_{t_{QQCA}} = 2.014*10^{-20}$
In1 = A	$E_{t_{bQCA}} = 1.602*10^{-20}$	$E_{t_{bQCA}} = 1.473*10^{-20}$	$E_{t_{bQCA}} = 1.602 *10^{-20}$	$E_{t_{bQCA}} = 1.473*10^{-20}$	$E_{t_{bQCA}} = 1.473*10^{-20}$
In2 = D	$E_{t_{TQCA}} = 1.602*10^{-20}$	$E_{t_{TQCA}} = 1.602*10^{-20}$	$E_{t_{TQCA}} = 1.473*10^{-20}$	$E_{t_{TQCA}} = 1.473*10^{-20}$	$E_{t_{TQCA}} = 1.473*10^{-20}$
In3 = D	$E_{t_{QQCA}} = 3.204*10^{-20}$	$E_{t_{QQCA}} = 3.075*10^{-20}$	$E_{t_{QQCA}} = 3.075*10^{-20}$	$E_{t_{QQCA}} = 2.946*10^{-20}$	$E_{t_{QQCA}} = 2.946*10^{-20}$

Table B.4 Output results of quaternary OR gate.

	Output cell in A state	Output cell in B state	Output cell in C state	Output cell in D state	Final result
In1 = D	$E_{t_{bQCA}} = 1.048*10^{-20}$	$E_{t_{bQCA}} = 1.602*10^{-20}$	$E_{t_{bQCA}} = 1.048*10^{-20}$	$E_{t_{bQCA}} = 1.602*10^{-20}$	$E_{t_{bQCA}} = 1.048*10-20$
In2 = A	$E_{t_{TQCA}} = 1.473*10^{-20}$	$E_{t_{TQCA}} = 1.473*10^{-20}$	$E_{t_{TQCA}} = 1.602*10^{-20}$	$E_{t_{TQCA}} = 1.602*10^{-20}$	$E_{t_{TQCA}} = 1.473*10^{-20}$
In3 = A	$E_{t_{QQCA}} = 2.521*10^{-20}$	$E_{t_{QQCA}} = 3.075*10^{-20}$	$E_{t_{QQCA}} = 2.65*10^{-20}$	$E_{t_{QQCA}} = 3.204*10^{-20}$	$E_{t_{QQCA}} = 2.521*10^{-20}$
In1 = D	$E_{t_{bQCA}} = 2.110*10^{-20}$	$E_{t_{bQCA}} = 1.049*10^{-20}$	$E_{t_{bQCA}} = 2.110*10^{-20}$	$E_{t_{bQCA}} = 1.049*10^{-20}$	$E_{t_{bQCA}} = 1.049*10^{-20}$
In2 = A	$E_{t_{TQCA}} = 1.473*10^{-20}$	$E_{t_{TQCA}} = 1.473*10^{-20}$	$E_{t_{TQCA}} = 1.602*10^{-20}$	$E_{t_{TQCA}} = 1.602*10^{-20}$	$E_{t_{TQCA}} = 1.473*10^{-20}$
In3 = B	$E_{t_{QQCA}} = 3.583*10^{-20}$	$E_{t_{QQCA}} = 2.522*10^{-20}$	$E_{t_{QQCA}} = 3.712*10^{-20}$	$E_{t_{QQCA}} = 2.651*10^{-20}$	$E_{t_{QQCA}} = 2.522*10^{-20}$
In1 = D	$E_{t_{bQCA}} = 1.048*10^{-20}$	$E_{t_{bQCA}} = 1.602*10^{-20}$	$E_{t_{bQCA}} = 1.048*10^{-20}$	$E_{t_{bQCA}} = 1.602*10^{-20}$	$E_{t_{bQCA}} = 1.048*10^{-20}$
In2 = A	$E_{t_{TQCA}} = 2.534*10^{-20}$	$E_{t_{TQCA}} = 2.534*10^{-20}$	$E_{t_{TQCA}} = 1.049*10^{-20}$	$E_{t_{TQCA}} = 1.049*10^{-20}$	$E_{t_{TQCA}} = 1.049*10^{-20}$
In3 = C	$E_{t_{QQCA}} = 3.582*10^{-20}$	$E_{t_{QQCA}} = 4.136*10^{-20}$	$E_{t_{QQCA}} = 2.097*10^{-20}$	$E_{t_{QQCA}} = 2.651*10^{-20}$	$E_{t_{QQCA}} = 2.097*10^{-20}$
In1 = D	$E_{t_{bQCA}} = 2.110*10^{-20}$	$E_{t_{bQCA}} = 1.049*10^{-20}$	$E_{t_{bQCA}} = 2.110*10^{-20}$	$E_{t_{bQCA}} = 1.049*10^{-20}$	$E_{t_{bQCA}} = 1.049*10^{-20}$
In2 = A	$E_{t_{TQCA}} = 2.534*10^{-20}$	$E_{t_{TQCA}} = 2.534*10^{-20}$	$E_{t_{TQCA}} = 1.049*10^{-20}$	$E_{t_{TQCA}} = 1.049*10^{-20}$	$E_{t_{TQCA}} = 1.049*10^{-20}$
In3 = D	$E_{t_{QQCA}} = 4.644*10^{-20}$	$E_{t_{QQCA}} = 3.583*10^{-20}$	$E_{t_{QQCA}} = 3.159*10^{-20}$	$E_{t_{QQCA}} = 2.098*10^{-20}$	$E_{t_{QQCA}} = 2.098*10^{-20}$
In1 = D	$E_{t_{bQCA}} = 2.025*10^{-20}$	$E_{t_{bQCA}} = 0.541*10^{-20}$	$E_{t_{bQCA}} = 2.025*10^{-20}$	$E_{t_{bQCA}} = 0.541*10^{-20}$	$E_{t_{bQCA}} = 0.541*10^{-20}$
In2 = B	$E_{t_{TQCA}} = 1.473*10^{-20}$	$E_{t_{TQCA}} = 1.473*10^{-20}$	$E_{t_{TQCA}} = 1.602*10^{-20}$	$E_{t_{TQCA}} = 1.602*10^{-20}$	$E_{t_{TQCA}} = 1.473*10^{-20}$
In3 = A	$E_{t_{QQCA}} = 3.498*10^{-20}$	$E_{t_{QQCA}} = 2.014*10^{-20}$	$E_{t_{QQCA}} = 3.627*10^{-20}$	$E_{t_{QQCA}} = 2.143*10^{-20}$	$E_{t_{QQCA}} = 2.014*10^{-20}$
In1 = D	$E_{t_{bQCA}} = 1.731*10^{-20}$	$E_{t_{bQCA}} = 1.343*10^{-20}$	$E_{t_{bQCA}} = 1.731*10^{-20}$	$E_{t_{bQCA}} = 1.343*10^{-20}$	$E_{t_{bQCA}} = 1.343*10^{-20}$
In2 = B	$E_{t_{TQCA}} = 1.473*10^{-20}$	$E_{t_{TQCA}} = 1.473*10^{-20}$	$E_{t_{TQCA}} = 1.602*10^{-20}$	$E_{t_{TQCA}} = 1.602*10^{-20}$	$E_{t_{TQCA}} = 1.473*10^{-20}$
In3 = B	$E_{t_{QQCA}} = 3.204*10^{-20}$	$E_{t_{QQCA}} = 2.816*10^{-20}$	$E_{t_{QQCA}} = 3.333*10^{-20}$	$E_{t_{QQCA}} = 2.945*10^{-20}$	$E_{t_{QQCA}} = 2.816*10^{-20}$
In1 = D	$E_{t_{bQCA}} = 2.025*10^{-20}$	$E_{t_{bQCA}} = 0.541*10^{-20}$	$E_{t_{bQCA}} = 2.025*10^{-20}$	$E_{t_{bQCA}} = 0.541*10^{-20}$	$E_{t_{bQCA}} = 0.541*10^{-20}$
In2 = B	$E_{t_{TQCA}} = 2.534*10^{-20}$	$E_{t_{TQCA}} = 2.534*10^{-20}$	$E_{t_{TQCA}} = 1.049*10^{-20}$	$E_{t_{TQCA}} = 1.049*10^{-20}$	$E_{t_{TQCA}} = 1.049*10^{-20}$
In3 = C	$E_{t_{QQCA}} = 4.559*10^{-20}$	$E_{t_{QQCA}} = 3.075*10^{-20}$	$E_{t_{QQCA}} = 3.074*10^{-20}$	$E_{t_{QQCA}} = 1.59*10^{-20}$	$E_{t_{QQCA}} = 1.59*10^{-20}$
In1 = D	$E_{t_{bQCA}} = 1.731*10^{-20}$	$E_{t_{bQCA}} = 1.343*10^{-20}$	$E_{t_{bQCA}} = 1.731*10^{-20}$	$E_{t_{bQCA}} = 1.343*10^{-20}$	$E_{t_{bQCA}} = 1.343*10^{-20}$
In2 = B	$E_{t_{TQCA}} = 2.534*10^{-20}$	$E_{t_{TQCA}} = 2.534*10^{-20}$	$E_{t_{TQCA}} = 1.049*10^{-20}$	$E_{t_{TQCA}} = 1.049*10^{-20}$	$E_{t_{TQCA}} = 1.049*10^{-20}$
In3 = D	$E_{t_{QQCA}} = 4.265*10^{-20}$	$E_{t_{QQCA}} = 3.877*10^{-20}$	$E_{t_{QQCA}} = 2.78*10^{-20}$	$E_{t_{QQCA}} = 2.392*10^{-20}$	$E_{t_{QQCA}} = 2.392*10^{-20}$

In1 = D	$E_{t_{bQCA}} = 1.473*10^{-20}$	$E_{t_{bQCA}} = 1.602*10^{-20}$	$E_{t_{bQCA}} = 1.473*10^{-20}$	$E_{t_{bQCA}} = 1.602*10^{-20}$	$E_{t_{bQCA}} = 1.473*10^{-20}$
In2 = C	$E_{t_{TQCA}} = 2.025*10^{-20}$	$E_{t_{TQCA}} = 2.025*10^{-20}$	$E_{t_{TQCA}} = 0.541*10^{-20}$	$E_{t_{TQCA}} = 0.541*10^{-20}$	$E_{t_{TQCA}} = 0.541*10^{-20}$
In3 = A	$E_{t_{QQCA}} = 3.498*10^{-20}$	$E_{t_{QQCA}} = 3.627*10^{-20}$	$E_{t_{QQCA}} = 2.014*10^{-20}$	$E_{t_{QQCA}} = 2.143*10^{-20}$	$E_{t_{QQCA}} = 2.014*10^{-20}$
In1 = D	$E_{t_{bQCA}} = 2.534*10^{-20}$	$E_{t_{bQCA}} = 1.049*10^{-20}$	$E_{t_{bQCA}} = 2.534*10^{-20}$	$E_{t_{bQCA}} = 1.049*10^{-20}$	$E_{t_{bQCA}} = 1.049*10^{-20}$
In2 = C	$E_{t_{TQCA}} = 2.025*10^{-20}$	$E_{t_{TQCA}} = 2.025*10^{-20}$	$E_{t_{TQCA}} = 0.541*10^{-20}$	$E_{t_{TQCA}} = 0.541*10^{-20}$	$E_{t_{TQCA}} = 0.541*10^{-20}$
In3 = B	$E_{t_{QQCA}} = 4.559*10^{-20}$	$E_{t_{QQCA}} = 3.074*10^{-20}$	$E_{t_{QQCA}} = 3.075*10^{-20}$	$E_{t_{QQCA}} = 1.59*10^{-20}$	$E_{t_{QQCA}} = 1.59*10^{-20}$
In1 = D	$E_{t_{bQCA}} = 1.473*10^{-20}$	$E_{t_{bQCA}} = 1.602*10^{-20}$	$E_{t_{bQCA}} = 1.473*10^{-20}$	$E_{t_{bQCA}} = 1.602*10^{-20}$	$E_{t_{bQCA}} = 1.473*10^{-20}$
In2 = C	$E_{t_{TQCA}} = 1.731*10^{-20}$	$E_{t_{TQCA}} = 1.731*10^{-20}$	$E_{t_{TQCA}} = 1.343*10^{-20}$	$E_{t_{TQCA}} = 1.343*10^{-20}$	$E_{t_{TQCA}} = 1.343*10^{-20}$
In3 = C	$E_{t_{QQCA}} = 3.204*10^{-20}$	$E_{t_{QQCA}} = 3.333*10^{-20}$	$E_{t_{QQCA}} = 2.816*10^{-20}$	$E_{t_{QQCA}} = 2.945*10^{-20}$	$E_{t_{QQCA}} = 2.816*10^{-20}$
In1 = D	$E_{t_{bQCA}} = 2.534*10^{-20}$	$E_{t_{bQCA}} = 1.049*10^{-20}$	$E_{t_{bQCA}} = 2.534*10^{-20}$	$E_{t_{bQCA}} = 1.049*10^{-20}$	$E_{t_{bQCA}} = 1.049*10^{-20}$
In2 = C	$E_{t_{TQCA}} = 1.731*10^{-20}$	$E_{t_{TQCA}} = 1.731*10^{-20}$	$E_{t_{TQCA}} = 1.343*10^{-20}$	$E_{t_{TQCA}} = 1.343*10^{-20}$	$E_{t_{TQCA}} = 1.343*10^{-20}$
In3 = D	$E_{t_{QQCA}} = 4.265*10^{-20}$	$E_{t_{QQCA}} = 2.78*10^{-20}$	$E_{t_{QQCA}} = 3.877*10^{-20}$	$E_{t_{QQCA}} = 2.392*10^{-20}$	$E_{t_{QQCA}} = 2.392*10^{-20}$
In1 = D	$E_{t_{bQCA}} = 2.025*10^{-20}$	$E_{t_{bQCA}} = 0.541*10^{-20}$	$E_{t_{bQCA}} = 2.025*10^{-20}$	$E_{t_{bQCA}} = 0.541*10^{-20}$	$E_{t_{bQCA}} = 0.541*10^{-20}$
In2 = D	$E_{t_{TQCA}} = 2.025*10^{-20}$	$E_{t_{TQCA}} = 2.025*10^{-20}$	$E_{t_{TQCA}} = 0.541*10^{-20}$	$E_{t_{TQCA}} = 0.541*10^{-20}$	$E_{t_{TQCA}} = 0.541*10^{-20}$
In3 = A	$E_{t_{QQCA}} = 4.05*10^{-20}$	$E_{t_{QQCA}} = 2.566*10^{-20}$	$E_{t_{QQCA}} = 2.566*10^{-20}$	$E_{t_{QQCA}} = 1.082*10^{-20}$	$E_{t_{QQCA}} = 1.082*10^{-20}$
In1 = D	$E_{t_{bQCA}} = 1.731*10^{-20}$	$E_{t_{bQCA}} = 1.343*10^{-20}$	$E_{t_{bQCA}} = 1.731*10^{-20}$	$E_{t_{bQCA}} = 1.343*10^{-20}$	$E_{t_{bQCA}} = 1.343*10^{-20}$
In2 = D	$E_{t_{TQCA}} = 2.025*10^{-20}$	$E_{t_{TQCA}} = 2.025*10^{-20}$	$E_{t_{TQCA}} = 0.541*10^{-20}$	$E_{t_{TQCA}} = 0.541*10^{-20}$	$E_{t_{TQCA}} = 0.541*10^{-20}$
In3 = B	$E_{t_{QQCA}} = 3.756*10^{-20}$	$E_{t_{QQCA}} = 3.368*10^{-20}$	$E_{t_{QQCA}} = 2.272*10^{-20}$	$E_{t_{QQCA}} = 1.884*10^{-20}$	$E_{t_{QQCA}} = 1.884*10^{-20}$
In1 = D	$E_{t_{bQCA}} = 2.025*10^{-20}$	$E_{t_{bQCA}} = 0.541*10^{-20}$	$E_{t_{bQCA}} = 2.025*10^{-20}$	$E_{t_{bQCA}} = 0.541*10^{-20}$	$E_{t_{bQCA}} = 0.541*10^{-20}$
In2 = D	$E_{t_{TQCA}} = 1.731*10^{-20}$	$E_{t_{TQCA}} = 1.731*10^{-20}$	$E_{t_{TQCA}} = 1.343*10^{-20}$	$E_{t_{TQCA}} = 1.343*10^{-20}$	$E_{t_{TQCA}} = 1.343*10^{-20}$
In3 = C	$E_{t_{QQCA}} = 3.756*10^{-20}$	$E_{t_{QQCA}} = 2.272*10^{-20}$	$E_{t_{QQCA}} = 3.368*10^{-20}$	$E_{t_{QQCA}} = 1.884*10^{-20}$	$E_{t_{QQCA}} = 1.884*10^{-20}$
In1 = D	$E_{t_{bQCA}} = 1.731*10^{-20}$	$E_{t_{bQCA}} = 1.343*10^{-20}$	$E_{t_{bQCA}} = 1.731*10^{-20}$	$E_{t_{bQCA}} = 1.343*10^{-20}$	$E_{t_{bQCA}} = 1.343*10^{-20}$
In2 = D	$E_{t_{TQCA}} = 1.731*10^{-20}$	$E_{t_{TQCA}} = 1.731*10^{-20}$	$E_{t_{TQCA}} = 1.343*10^{-20}$	$E_{t_{TQCA}} = 1.343*10^{-20}$	$E_{t_{TQCA}} = 1.343*10^{-20}$
In3 = D	$E_{t_{QQCA}} = 3.462*10^{-20}$	$E_{t_{QQCA}} = 3.074*10^{-20}$	$E_{t_{QQCA}} = 3.074*10^{-20}$	$E_{t_{QQCA}} = 2.686*10^{-20}$	$E_{t_{QQCA}} = 2.686*10^{-20}$

Table B.5 Output results of Coulomb energy values in the quaternary AND gate for all states (× 1e-20 (J)).

Out In	Output cell in A state	Output cell in B state	Output cell in C state	Output cell in D state	Output cell in E state	Select final result
In1 = A	$E_{t_{TQCA1}} = 1.343$	$E_{t_{TQCA1}} = 1.343$	$E_{t_{TQCA1}} = 1.500$	$E_{t_{TQCA1}} = 1.731$	$E_{t_{TQCA1}} = 1.731$	$E_{t_{PQCA}} = 2.686$
In2 = A	$E_{t_{TQCA2}} = 1.343$	$E_{t_{TQCA2}} = 1.500$	$E_{t_{TQCA2}} = 1.500$	$E_{t_{TQCA2}} = 1.500$	$E_{t_{TQCA2}} = 1.731$	Out state = A state
In3 = A	$E_{t_{PQCA}} = 2.686$	$E_{t_{PQCA}} = 2.843$	$E_{t_{PQCA}} = 3.00$	$E_{t_{PQCA}} = 3.231$	$E_{t_{PQCA}} = 3.462$	
In1 = A	$E_{t_{TQCA1}} = 1.343$	$E_{t_{TQCA1}} = 1.343$	$E_{t_{TQCA1}} = 1.500$	$E_{t_{TQCA1}} = 1.731$	$E_{t_{TQCA1}} = 1.731$	$E_{t_{PQCA}} = 2.632$
In2 = A	$E_{t_{TQCA2}} = 1.289$	$E_{t_{TQCA2}} = 1.378$	$E_{t_{TQCA2}} = 1.378$	$E_{t_{TQCA2}} = 1.378$	$E_{t_{TQCA2}} = 2.229$	Out state = A state
In3 = B	$E_{t_{PQCA}} = 2.632$	$E_{t_{PQCA}} = 2.721$	$E_{t_{PQCA}} = 2.878$	$E_{t_{PQCA}} = 3.109$	$E_{t_{PQCA}} = 3.96$	
In1 = A	$E_{t_{TQCA1}} = 1.289$	$E_{t_{TQCA1}} = 1.289$	$E_{t_{TQCA1}} = 1.378$	$E_{t_{TQCA1}} = 2.229$	$E_{t_{TQCA1}} = 2.229$	$E_{t_{PQCA}} = 2.578$
In2 = A	$E_{t_{TQCA2}} = 1.289$	$E_{t_{TQCA2}} = 1.378$	$E_{t_{TQCA2}} = 1.378$	$E_{t_{TQCA2}} = 1.378$	$E_{t_{TQCA2}} = 2.229$	Out state = A state
In3 = C	$E_{t_{PQCA}} = 2.578$	$E_{t_{PQCA}} = 2.664$	$E_{t_{PQCA}} = 2.756$	$E_{t_{PQCA}} = 3.607$	$E_{t_{PQCA}} = 4.458$	
In1 = A	$E_{t_{TQCA1}} = 1.049$	$E_{t_{TQCA1}} = 1.049$	$E_{t_{TQCA1}} = 1.759$	$E_{t_{TQCA1}} = 2.534$	$E_{t_{TQCA1}} = 2.534$	$E_{t_{PQCA}} = 2.338$
In2 = A	$E_{t_{TQCA2}} = 1.289$	$E_{t_{TQCA2}} = 1.378$	$E_{t_{TQCA2}} = 1.378$	$E_{t_{TQCA2}} = 1.378$	$E_{t_{TQCA2}} = 2.229$	Out state = A state
In3 = D	$E_{t_{PQCA}} = 2.338$	$E_{t_{PQCA}} = 2.427$	$E_{t_{PQCA}} = 3.137$	$E_{t_{PQCA}} = 3.912$	$E_{t_{PQCA}} = 4.763$	
In1 = A	$E_{t_{TQCA1}} = 1.049$	$E_{t_{TQCA1}} = 1.049$	$E_{t_{TQCA1}} = 1.759$	$E_{t_{TQCA1}} = 2.534$	$E_{t_{TQCA1}} = 2.534$	$E_{t_{PQCA}} = 2.098$
In2 = A	$E_{t_{TQCA2}} = 1.049$	$E_{t_{TQCA2}} = 1.759$	$E_{t_{TQCA2}} = 1.759$	$E_{t_{TQCA2}} = 1.759$	$E_{t_{TQCA2}} = 2.534$	Out state = A state
In3 = E	$E_{t_{PQCA}} = 2.098$	$E_{t_{PQCA}} = 2.808$	$E_{t_{PQCA}} = 3.518$	$E_{t_{PQCA}} = 4.293$	$E_{t_{PQCA}} = 5.068$	
In1 = A	$E_{t_{TQCA1}} = 1.343$	$E_{t_{TQCA1}} = 1.343$	$E_{t_{TQCA1}} = 1.500$	$E_{t_{TQCA1}} = 1.731$	$E_{t_{TQCA1}} = 1.731$	$E_{t_{PQCA}} = 2.382$
In2 = B	$E_{t_{TQCA2}} = 1.039$	$E_{t_{TQCA2}} = 1.120$	$E_{t_{TQCA2}} = 1.120$	$E_{t_{TQCA2}} = 1.120$	$E_{t_{TQCA2}} = 1.971$	Out state = A state
In3 = A	$E_{t_{PQCA}} = 2.382$	$E_{t_{PQCA}} = 2.463$	$E_{t_{PQCA}} = 2.620$	$E_{t_{PQCA}} = 2.851$	$E_{t_{PQCA}} = 3.702$	
In1 = A	$E_{t_{TQCA1}} = 1.343$	$E_{t_{TQCA1}} = 1.343$	$E_{t_{TQCA1}} = 1.500$	$E_{t_{TQCA1}} = 2.378$	$E_{t_{TQCA1}} = 2.378$	$E_{t_{PQCA}} = 1.662$
In2 = B	$E_{t_{TQCA2}} = 2.257$	$E_{t_{TQCA2}} = 0.319$	$E_{t_{TQCA2}} = 0.319$	$E_{t_{TQCA2}} = 0.319$	$E_{t_{TQCA2}} = 4.082$	Out state = B state
In3 = B	$E_{t_{PQCA}} = 3.600$	$E_{t_{PQCA}} = 1.662$	$E_{t_{PQCA}} = 1.819$	$E_{t_{PQCA}} = 2.697$	$E_{t_{PQCA}} = 6.460$	
In1 = A	$E_{t_{TQCA1}} = 1.289$	$E_{t_{TQCA1}} = 1.289$	$E_{t_{TQCA1}} = 1.378$	$E_{t_{TQCA1}} = 2.658$	$E_{t_{TQCA1}} = 2.658$	$E_{t_{PQCA}} = 1.608$
In2 = B	$E_{t_{TQCA2}} = 2.257$	$E_{t_{TQCA2}} = 0.319$	$E_{t_{TQCA2}} = 0.319$	$E_{t_{TQCA2}} = 0.319$	$E_{t_{TQCA2}} = 4.082$	Out state = B state
In3 = C	$E_{t_{PQCA}} = 3.546$	$E_{t_{PQCA}} = 1.608$	$E_{t_{PQCA}} = 1.697$	$E_{t_{PQCA}} = 2.977$	$E_{t_{PQCA}} = 6.740$	

In1 = A	$E_{t_{TQCA1}} = 1.049$	$E_{t_{TQCA1}} = 1.049$	$E_{t_{TQCA1}} = 1.759$	$E_{t_{TQCA1}} = 2.534$	$E_{t_{TQCA1}} = 2.534$	$E_{t_{PQCA}} = 1.368$
In2 = B	$E_{t_{TQCA2}} = 2.257$	$E_{t_{TQCA2}} = 0.319$	$E_{t_{TQCA2}} = 0.319$	$E_{t_{TQCA2}} = 0.319$	$E_{t_{TQCA2}} = 2.386$	
In3 = D	$E_{t_{PQCA}} = 3.666$	$E_{t_{PQCA}} = 1.368$	$E_{t_{PQCA}} = 2.078$	$E_{t_{PQCA}} = 2.853$	$E_{t_{PQCA}} = 4.920$	Out state = B state
In1 = A	$E_{t_{TQCA1}} = 1.049$	$E_{t_{TQCA1}} = 1.049$	$E_{t_{TQCA1}} = 1.759$	$E_{t_{TQCA1}} = 2.534$	$E_{t_{TQCA1}} = 2.534$	$E_{t_{PQCA}} = 1.749$
In2 = B	$E_{t_{TQCA2}} = 2.018$	$E_{t_{TQCA2}} = 0.700$	$E_{t_{TQCA2}} = 0.700$	$E_{t_{TQCA2}} = 0.700$	$E_{t_{TQCA2}} = 2.691$	
In3 = E	$E_{t_{PQCA}} = 3.067$	$E_{t_{PQCA}} = 1.749$	$E_{t_{PQCA}} = 2.459$	$E_{t_{PQCA}} = 3.234$	$E_{t_{PQCA}} = 5.225$	Out state = B state
In1 = A	$E_{t_{TQCA1}} = 1.039$	$E_{t_{TQCA1}} = 1.039$	$E_{t_{TQCA1}} = 1.120$	$E_{t_{TQCA1}} = 1.971$	$E_{t_{TQCA1}} = 1.971$	$E_{t_{PQCA}} = 2.078$
In2 = C	$E_{t_{TQCA2}} = 1.039$	$E_{t_{TQCA2}} = 1.120$	$E_{t_{TQCA2}} = 1.120$	$E_{t_{TQCA2}} = 1.120$	$E_{t_{TQCA2}} = 1.971$	
In3 = A	$E_{t_{PQCA}} = 2.078$	$E_{t_{PQCA}} = 2.159$	$E_{t_{PQCA}} = 2.240$	$E_{t_{PQCA}} = 3.091$	$E_{t_{PQCA}} = 3.942$	Out state = A state
In1 = A	$E_{t_{TQCA1}} = 1.039$	$E_{t_{TQCA1}} = 1.039$	$E_{t_{TQCA1}} = 1.120$	$E_{t_{TQCA1}} = 1.971$	$E_{t_{TQCA1}} = 1.971$	$E_{t_{PQCA}} = 1.358$
In2 = C	$E_{t_{TQCA2}} = 2.257$	$E_{t_{TQCA2}} = 0.319$	$E_{t_{TQCA2}} = 0.319$	$E_{t_{TQCA2}} = 0.319$	$E_{t_{TQCA2}} = 2.386$	
In3 = B	$E_{t_{PQCA}} = 3.296$	$E_{t_{PQCA}} = 1.358$	$E_{t_{PQCA}} = 1.439$	$E_{t_{PQCA}} = 2.290$	$E_{t_{PQCA}} = 4.357$	Out state = B state
In1 = A	$E_{t_{TQCA1}} = 2.257$	$E_{t_{TQCA1}} = 2.257$	$E_{t_{TQCA1}} = 0.319$	$E_{t_{TQCA1}} = 2.386$	$E_{t_{TQCA1}} = 2.386$	$E_{t_{PQCA}} = 0.638$
In2 = C	$E_{t_{TQCA2}} = 2.257$	$E_{t_{TQCA2}} = 0.319$	$E_{t_{TQCA2}} = 0.319$	$E_{t_{TQCA2}} = 0.319$	$E_{t_{TQCA2}} = 2.386$	
In3 = C	$E_{t_{PQCA}} = 4.514$	$E_{t_{PQCA}} = 2.576$	$E_{t_{PQCA}} = 0.638$	$E_{t_{PQCA}} = 2.705$	$E_{t_{PQCA}} = 4.772$	Out state = C state
In1 = A	$E_{t_{TQCA1}} = 2.018$	$E_{t_{TQCA1}} = 2.018$	$E_{t_{TQCA1}} = 0.700$	$E_{t_{TQCA1}} = 2.691$	$E_{t_{TQCA1}} = 2.691$	$E_{t_{PQCA}} = 1.019$
In2 = C	$E_{t_{TQCA2}} = 2.257$	$E_{t_{TQCA2}} = 0.319$	$E_{t_{TQCA2}} = 0.319$	$E_{t_{TQCA2}} = 0.319$	$E_{t_{TQCA2}} = 2.386$	
In3 = D	$E_{t_{PQCA}} = 4.275$	$E_{t_{PQCA}} = 2.337$	$E_{t_{PQCA}} = 1.019$	$E_{t_{PQCA}} = 3.010$	$E_{t_{PQCA}} = 5.077$	Out state = C state
In1 = A	$E_{t_{TQCA1}} = 2.018$	$E_{t_{TQCA1}} = 2.018$	$E_{t_{TQCA1}} = 0.700$	$E_{t_{TQCA1}} = 2.691$	$E_{t_{TQCA1}} = 2.691$	$E_{t_{PQCA}} = 1.400$
In2 = C	$E_{t_{TQCA2}} = 2.018$	$E_{t_{TQCA2}} = 0.700$	$E_{t_{TQCA2}} = 0.700$	$E_{t_{TQCA2}} = 0.700$	$E_{t_{TQCA2}} = 2.691$	
In3 = E	$E_{t_{PQCA}} = 4.036$	$E_{t_{PQCA}} = 2.088$	$E_{t_{PQCA}} = 1.400$	$E_{t_{PQCA}} = 3.391$	$E_{t_{PQCA}} = 5.382$	Out state = C state
In1 = A	$E_{t_{TQCA1}} = 0.541$	$E_{t_{TQCA1}} = 0.541$	$E_{t_{TQCA1}} = 1.242$	$E_{t_{TQCA1}} = 2.025$	$E_{t_{TQCA1}} = 2.025$	$E_{t_{PQCA}} = 1.580$
In2 = D	$E_{t_{TQCA2}} = 1.039$	$E_{t_{TQCA2}} = 1.120$	$E_{t_{TQCA2}} = 1.120$	$E_{t_{TQCA2}} = 1.120$	$E_{t_{TQCA2}} = 1.971$	
In3 = A	$E_{t_{PQCA}} = 1.580$	$E_{t_{PQCA}} = 1.661$	$E_{t_{PQCA}} = 2.362$	$E_{t_{PQCA}} = 3.145$	$E_{t_{PQCA}} = 3.996$	Out state = A state
In1 = A	$E_{t_{TQCA1}} = 0.541$	$E_{t_{TQCA1}} = 0.541$	$E_{t_{TQCA1}} = 1.242$	$E_{t_{TQCA1}} = 2.025$	$E_{t_{TQCA1}} = 2.025$	$E_{t_{PQCA}} = 0.860$
In2 = D	$E_{t_{TQCA2}} = 2.257$	$E_{t_{TQCA2}} = 0.319$	$E_{t_{TQCA2}} = 0.319$	$E_{t_{TQCA2}} = 0.319$	$E_{t_{TQCA2}} = 2.386$	
In3 = B	$E_{t_{PQCA}} = 2.798$	$E_{t_{PQCA}} = 0.860$	$E_{t_{PQCA}} = 1.561$	$E_{t_{PQCA}} = 2.344$	$E_{t_{PQCA}} = 4.411$	Out state = B state

(Continued)

Table B.5 (Continued)

Out In	Output cell in A state	Output cell in B state	Output cell in C state	Output cell in D state	Output cell in E state	Select final result
In1 = A	$E_{t_{TQCA1}} = 1.759$	$E_{t_{TQCA1}} = 1.759$	$E_{t_{TQCA1}} = 0.441$	$E_{t_{TQCA1}} = 2.441$	$E_{t_{TQCA1}} = 2.441$	$E_{t_{PQCA}} = 0.760$
In2 = D	$E_{t_{TQCA2}} = 2.257$	$E_{t_{TQCA2}} = 0.319$	$E_{t_{TQCA2}} = 0.319$	$E_{t_{TQCA2}} = 0.319$	$E_{t_{TQCA2}} = 2.386$	Out state = C state
In3 = C	$E_{t_{PQCA}} = 4.016$	$E_{t_{PQCA}} = 2.078$	$E_{t_{PQCA}} = 0.760$	$E_{t_{PQCA}} = 2.760$	$E_{t_{PQCA}} = 4.827$	
In1 = A	$E_{t_{TQCA1}} = 1.602$	$E_{t_{TQCA1}} = 1.602$	$E_{t_{TQCA1}} = 1.500$	$E_{t_{TQCA1}} = 1.473$	$E_{t_{TQCA1}} = 1.473$	$E_{t_{PQCA}} = 1.792$
In2 = D	$E_{t_{TQCA2}} = 2.257$	$E_{t_{TQCA2}} = 0.319$	$E_{t_{TQCA2}} = 0.319$	$E_{t_{TQCA2}} = 0.319$	$E_{t_{TQCA2}} = 2.386$	Out state = D state
In3 = D	$E_{t_{PQCA}} = 3.859$	$E_{t_{PQCA}} = 1.921$	$E_{t_{PQCA}} = 1.819$	$E_{t_{PQCA}} = 1.792$	$E_{t_{PQCA}} = 3.859$	
In1 = A	$E_{t_{TQCA1}} = 1.602$	$E_{t_{TQCA1}} = 1.602$	$E_{t_{TQCA1}} = 1.500$	$E_{t_{TQCA1}} = 1.473$	$E_{t_{TQCA1}} = 1.473$	$E_{t_{PQCA}} = 2.173$
In2 = D	$E_{t_{TQCA2}} = 2.018$	$E_{t_{TQCA2}} = 0.700$	$E_{t_{TQCA2}} = 0.700$	$E_{t_{TQCA2}} = 0.700$	$E_{t_{TQCA2}} = 2.691$	Out state = D state
In3 = E	$E_{t_{PQCA}} = 3.620$	$E_{t_{PQCA}} = 2.302$	$E_{t_{PQCA}} = 2.200$	$E_{t_{PQCA}} = 2.173$	$E_{t_{PQCA}} = 4.164$	
In1 = A	$E_{t_{TQCA1}} = 0.541$	$E_{t_{TQCA1}} = 0.541$	$E_{t_{TQCA1}} = 1.242$	$E_{t_{TQCA1}} = 2.025$	$E_{t_{TQCA1}} = 2.025$	$E_{t_{PQCA}} = 1.082$
In2 = E	$E_{t_{TQCA2}} = 0.541$	$E_{t_{TQCA2}} = 1.242$	$E_{t_{TQCA2}} = 1.242$	$E_{t_{TQCA2}} = 1.242$	$E_{t_{TQCA2}} = 2.025$	Out state = A state
In3 = A	$E_{t_{PQCA}} = 1.082$	$E_{t_{PQCA}} = 1.783$	$E_{t_{PQCA}} = 2.484$	$E_{t_{PQCA}} = 3.267$	$E_{t_{PQCA}} = 4.050$	
In1 = A	$E_{t_{TQCA1}} = 0.541$	$E_{t_{TQCA1}} = 0.541$	$E_{t_{TQCA1}} = 1.242$	$E_{t_{TQCA1}} = 2.025$	$E_{t_{TQCA1}} = 2.025$	$E_{t_{PQCA}} = 0.982$
In2 = E	$E_{t_{TQCA2}} = 1.759$	$E_{t_{TQCA2}} = 0.441$	$E_{t_{TQCA2}} = 0.441$	$E_{t_{TQCA2}} = 0.441$	$E_{t_{TQCA2}} = 2.441$	Out state = B state
In3 = B	$E_{t_{PQCA}} = 2.300$	$E_{t_{PQCA}} = 0.982$	$E_{t_{PQCA}} = 1.683$	$E_{t_{PQCA}} = 2.0691$	$E_{t_{PQCA}} = 4.466$	
In1 = A	$E_{t_{TQCA1}} = 1.759$	$E_{t_{TQCA1}} = 1.759$	$E_{t_{TQCA1}} = 0.441$	$E_{t_{TQCA1}} = 2.441$	$E_{t_{TQCA1}} = 2.441$	$E_{t_{PQCA}} = 0.882$
In2 = E	$E_{t_{TQCA2}} = 1.759$	$E_{t_{TQCA2}} = 0.441$	$E_{t_{TQCA2}} = 0.441$	$E_{t_{TQCA2}} = 0.441$	$E_{t_{TQCA2}} = 2.441$	Out state = C state
In3 = C	$E_{t_{PQCA}} = 3.518$	$E_{t_{PQCA}} = 2.200$	$E_{t_{PQCA}} = 0.882$	$E_{t_{PQCA}} = 2.882$	$E_{t_{PQCA}} = 4.882$	
In1 = A	$E_{t_{TQCA1}} = 1.602$	$E_{t_{TQCA1}} = 1.602$	$E_{t_{TQCA1}} = 1.500$	$E_{t_{TQCA1}} = 1.473$	$E_{t_{TQCA1}} = 1.473$	$E_{t_{PQCA}} = 1.914$
In2 = E	$E_{t_{TQCA2}} = 1.759$	$E_{t_{TQCA2}} = 0.441$	$E_{t_{TQCA2}} = 0.441$	$E_{t_{TQCA2}} = 0.441$	$E_{t_{TQCA2}} = 2.441$	Out state = D state
In3 = D	$E_{t_{PQCA}} = 3.361$	$E_{t_{PQCA}} = 2.043$	$E_{t_{PQCA}} = 1.941$	$E_{t_{PQCA}} = 1.914$	$E_{t_{PQCA}} = 3.914$	
In1 = A	$E_{t_{TQCA1}} = 1.602$	$E_{t_{TQCA1}} = 1.602$	$E_{t_{TQCA1}} = 1.500$	$E_{t_{TQCA1}} = 1.473$	$E_{t_{TQCA1}} = 1.473$	$E_{t_{PQCA}} = 2.946$
In2 = E	$E_{t_{TQCA2}} = 1.602$	$E_{t_{TQCA2}} = 1.500$	$E_{t_{TQCA2}} = 1.500$	$E_{t_{TQCA2}} = 1.500$	$E_{t_{TQCA2}} = 1.473$	Out state = E state
In3 = E	$E_{t_{PQCA}} = 3.204$	$E_{t_{PQCA}} = 3.102$	$E_{t_{PQCA}} = 3.000$	$E_{t_{PQCA}} = 2.973$	$E_{t_{PQCA}} = 2.946$	

Table B.6 Output results of Coulomb energy values for the quinary OR gate in all states (× 1e-20 (J)).

Out In	Output cell in A state	Output cell in B state	Output cell in C state	Output cell in D state	Output cell in E state	Final result
In1 = E	$E_{t_{TQCA1}} = 1.473$	$E_{t_{TQCA1}} = 1.473$	$E_{t_{TQCA1}} = 1.500$	$E_{t_{TQCA1}} = 1.602$	$E_{t_{TQCA1}} = 1.602$	$E_{t_{PQCA}} = 2.946$
In2 = A	$E_{t_{TQCA2}} = 1.473$	$E_{t_{TQCA2}} = 1.500$	$E_{t_{TQCA2}} = 1.500$	$E_{t_{TQCA2}} = 1.500$	$E_{t_{TQCA2}} = 1.602$	
In3 = A	$E_{t_{PQCA}} = 2.946$	$E_{t_{PQCA}} = 2.973$	$E_{t_{PQCA}} = 3.000$	$E_{t_{PQCA}} = 3.102$	$E_{t_{PQCA}} = 3.204$	Out state = A state
In1 = E	$E_{t_{TQCA1}} = 1.473$	$E_{t_{TQCA1}} = 1.473$	$E_{t_{TQCA1}} = 1.500$	$E_{t_{TQCA1}} = 1.602$	$E_{t_{TQCA1}} = 1.602$	$E_{t_{PQCA}} = 2.173$
In2 = A	$E_{t_{TQCA2}} = 2.691$	$E_{t_{TQCA2}} = 0.700$	$E_{t_{TQCA2}} = 0.700$	$E_{t_{TQCA2}} = 0.700$	$E_{t_{TQCA2}} = 2.018$	
In3 = B	$E_{t_{PQCA}} = 4.146$	$E_{t_{PQCA}} = 2.173$	$E_{t_{PQCA}} = 2.200$	$E_{t_{PQCA}} = 2.302$	$E_{t_{PQCA}} = 3.620$	Out state = B state
In1 = E	$E_{t_{TQCA1}} = 2.691$	$E_{t_{TQCA1}} = 2.691$	$E_{t_{TQCA1}} = 0.700$	$E_{t_{TQCA1}} = 2.018$	$E_{t_{TQCA1}} = 2.018$	$E_{t_{PQCA}} = 1.400$
In2 = A	$E_{t_{TQCA2}} = 2.691$	$E_{t_{TQCA2}} = 0.700$	$E_{t_{TQCA2}} = 0.700$	$E_{t_{TQCA2}} = 0.700$	$E_{t_{TQCA2}} = 2.018$	
In3 = C	$E_{t_{PQCA}} = 5.382$	$E_{t_{PQCA}} = 3.391$	$E_{t_{PQCA}} = 1.400$	$E_{t_{PQCA}} = 2.718$	$E_{t_{PQCA}} = 4.036$	Out state = C state
In1 = E	$E_{t_{TQCA1}} = 2.534$	$E_{t_{TQCA1}} = 2.534$	$E_{t_{TQCA1}} = 1.759$	$E_{t_{TQCA1}} = 1.049$	$E_{t_{TQCA1}} = 1.049$	$E_{t_{PQCA}} = 1.749$
In2 = A	$E_{t_{TQCA2}} = 2.691$	$E_{t_{TQCA2}} = 0.700$	$E_{t_{TQCA2}} = 0.700$	$E_{t_{TQCA2}} = 0.700$	$E_{t_{TQCA2}} = 2.018$	
In3 = D	$E_{t_{PQCA}} = 5.225$	$E_{t_{PQCA}} = 3.234$	$E_{t_{PQCA}} = 2.459$	$E_{t_{PQCA}} = 1.749$	$E_{t_{PQCA}} = 3.067$	Out state = D state
In1 = E	$E_{t_{TQCA1}} = 2.534$	$E_{t_{TQCA1}} = 2.534$	$E_{t_{TQCA1}} = 1.759$	$E_{t_{TQCA1}} = 1.049$	$E_{t_{TQCA1}} = 1.049$	$E_{t_{PQCA}} = 2.098$
In2 = A	$E_{t_{TQCA2}} = 2.534$	$E_{t_{TQCA2}} = 1.759$	$E_{t_{TQCA2}} = 1.759$	$E_{t_{TQCA2}} = 1.759$	$E_{t_{TQCA2}} = 1.049$	
In3 = E	$E_{t_{PQCA}} = 5.068$	$E_{t_{PQCA}} = 4.293$	$E_{t_{PQCA}} = 3.518$	$E_{t_{PQCA}} = 2.808$	$E_{t_{PQCA}} = 2.098$	Out state = *E* state
In1 = E	$E_{t_{TQCA1}} = 1.473$	$E_{t_{TQCA1}} = 1.473$	$E_{t_{TQCA1}} = 1.500$	$E_{t_{TQCA1}} = 1.602$	$E_{t_{TQCA1}} = 1.602$	$E_{t_{PQCA}} = 1.914$
In2 = B	$E_{t_{TQCA2}} = 2.441$	$E_{t_{TQCA2}} = 0.441$	$E_{t_{TQCA2}} = 0.441$	$E_{t_{TQCA2}} = 0.441$	$E_{t_{TQCA2}} = 1.759$	
In3 = A	$E_{t_{PQCA}} = 3.914$	$E_{t_{PQCA}} = 1.914$	$E_{t_{PQCA}} = 1.941$	$E_{t_{PQCA}} = 2.043$	$E_{t_{PQCA}} = 3.361$	Out state = B state
In1 = E	$E_{t_{TQCA1}} = 1.473$	$E_{t_{TQCA1}} = 1.473$	$E_{t_{TQCA1}} = 1.500$	$E_{t_{TQCA1}} = 1.602$	$E_{t_{TQCA1}} = 1.602$	$E_{t_{PQCA}} = 1.792$
In2 = B	$E_{t_{TQCA2}} = 2.386$	$E_{t_{TQCA2}} = 0.319$	$E_{t_{TQCA2}} = 0.319$	$E_{t_{TQCA2}} = 0.319$	$E_{t_{TQCA2}} = 2.257$	
In3 = B	$E_{t_{PQCA}} = 3.859$	$E_{t_{PQCA}} = 1.792$	$E_{t_{PQCA}} = 1.819$	$E_{t_{PQCA}} = 1.921$	$E_{t_{PQCA}} = 3.859$	Out state = B state
In1 = E	$E_{t_{TQCA1}} = 2.691$	$E_{t_{TQCA1}} = 2.691$	$E_{t_{TQCA1}} = 0.700$	$E_{t_{TQCA1}} = 2.018$	$E_{t_{TQCA1}} = 2.018$	$E_{t_{PQCA}} = 1.019$
In2 = B	$E_{t_{TQCA2}} = 2.386$	$E_{t_{TQCA2}} = 0.319$	$E_{t_{TQCA2}} = 0.319$	$E_{t_{TQCA2}} = 0.319$	$E_{t_{TQCA2}} = 2.257$	
In3 = C	$E_{t_{PQCA}} = 5.077$	$E_{t_{PQCA}} = 3.010$	$E_{t_{PQCA}} = 1.019$	$E_{t_{PQCA}} = 2.337$	$E_{t_{PQCA}} = 4.275$	Out state = C state

(Continued)

Table B.6 (Continued)

Out In	Output cell in A state	Output cell in B state	Output cell in C state	Output cell in D state	Output cell in E state	Final result
In1 = E	$E_{t_{TQCA1}} = 2.534$	$E_{t_{TQCA1}} = 2.534$	$E_{t_{TQCA1}} = 1.759$	$E_{t_{TQCA1}} = 1.049$	$E_{t_{TQCA1}} = 1.049$	$E_{t_{PQCA}} = 1.368$
In2 = B	$E_{t_{TQCA2}} = 2.386$	$E_{t_{TQCA2}} = 0.319$	$E_{t_{TQCA2}} = 0.319$	$E_{t_{TQCA2}} = 0.319$	$E_{t_{TQCA2}} = 2.257$	Out state = D state
In3 = D	$E_{t_{PQCA}} = 4.920$	$E_{t_{PQCA}} = 2.853$	$E_{t_{PQCA}} = 2.078$	$E_{t_{PQCA}} = 1.368$	$E_{t_{PQCA}} = 3.306$	
In1 = E	$E_{t_{TQCA1}} = 2.534$	$E_{t_{TQCA1}} = 2.534$	$E_{t_{TQCA1}} = 1.759$	$E_{t_{TQCA1}} = 1.049$	$E_{t_{TQCA1}} = 1.049$	$E_{t_{PQCA}} = 2.338$
In2 = B	$E_{t_{TQCA2}} = 2.229$	$E_{t_{TQCA2}} = 1.378$	$E_{t_{TQCA2}} = 1.378$	$E_{t_{TQCA2}} = 1.378$	$E_{t_{TQCA2}} = 1.289$	Out state = E state
In3 = E	$E_{t_{PQCA}} = 4.763$	$E_{t_{PQCA}} = 3.912$	$E_{t_{PQCA}} = 3.137$	$E_{t_{PQCA}} = 2.427$	$E_{t_{PQCA}} = 2.338$	
In1 = E	$E_{t_{TQCA1}} = 2.441$	$E_{t_{TQCA1}} = 2.441$	$E_{t_{TQCA1}} = 0.441$	$E_{t_{TQCA1}} = 1.759$	$E_{t_{TQCA1}} = 1.759$	$E_{t_{PQCA}} = 0.882$
In2 = C	$E_{t_{TQCA2}} = 2.441$	$E_{t_{TQCA2}} = 0.441$	$E_{t_{TQCA2}} = 0.441$	$E_{t_{TQCA2}} = 0.441$	$E_{t_{TQCA2}} = 1.759$	Out state = C state
In3 = A	$E_{t_{PQCA}} = 4.882$	$E_{t_{PQCA}} = 2.882$	$E_{t_{PQCA}} = 0.882$	$E_{t_{PQCA}} = 2.200$	$E_{t_{PQCA}} = 3.518$	
In1 = E	$E_{t_{TQCA1}} = 2.441$	$E_{t_{TQCA1}} = 2.441$	$E_{t_{TQCA1}} = 0.441$	$E_{t_{TQCA1}} = 1.759$	$E_{t_{TQCA1}} = 1.759$	$E_{t_{PQCA}} = 0.760$
In2 = C	$E_{t_{TQCA2}} = 2.386$	$E_{t_{TQCA2}} = 0.319$	$E_{t_{TQCA2}} = 0.319$	$E_{t_{TQCA2}} = 0.319$	$E_{t_{TQCA2}} = 2.257$	Out state = C state
In3 = B	$E_{t_{PQCA}} = 4.827$	$E_{t_{PQCA}} = 2.760$	$E_{t_{PQCA}} = 0.760$	$E_{t_{PQCA}} = 2.078$	$E_{t_{PQCA}} = 4.016$	
In1 = *E*	$E_{t_{TQCA1}} = 2.386$	$E_{t_{TQCA1}} = 2.386$	$E_{t_{TQCA1}} = 0.319$	$E_{t_{TQCA1}} = 2.257$	$E_{t_{TQCA1}} = 2.257$	$E_{t_{PQCA}} = 0.638$
In2 = C	$E_{t_{TQCA2}} = 2.386$	$E_{t_{TQCA2}} = 0.319$	$E_{t_{TQCA2}} = 0.319$	$E_{t_{TQCA2}} = 0.319$	$E_{t_{TQCA2}} = 2.257$	Out state = C state
In3 = C	$E_{t_{PQCA}} = 4.772$	$E_{t_{PQCA}} = 2.705$	$E_{t_{PQCA}} = 0.638$	$E_{t_{PQCA}} = 2.576$	$E_{t_{PQCA}} = 1.257$	
In1 = E	$E_{t_{TQCA1}} = 2.229$	$E_{t_{TQCA1}} = 2.229$	$E_{t_{TQCA1}} = 1.378$	$E_{t_{TQCA1}} = 1.289$	$E_{t_{TQCA1}} = 1.289$	$E_{t_{PQCA}} = 1.608$
In2 = C	$E_{t_{TQCA2}} = 2.386$	$E_{t_{TQCA2}} = 0.319$	$E_{t_{TQCA2}} = 0.319$	$E_{t_{TQCA2}} = 0.319$	$E_{t_{TQCA2}} = 2.257$	Out state = D state
In3 = D	$E_{t_{PQCA}} = 4.615$	$E_{t_{PQCA}} = 2.548$	$E_{t_{PQCA}} = 1.697$	$E_{t_{PQCA}} = 1.608$	$E_{t_{PQCA}} = 3.546$	
In1 = E	$E_{t_{TQCA1}} = 2.229$	$E_{t_{TQCA1}} = 2.229$	$E_{t_{TQCA1}} = 1.378$	$E_{t_{TQCA1}} = 1.289$	$E_{t_{TQCA1}} = 1.289$	$E_{t_{PQCA}} = 2.578$
In2 = C	$E_{t_{TQCA2}} = 2.229$	$E_{t_{TQCA2}} = 1.378$	$E_{t_{TQCA2}} = 1.378$	$E_{t_{TQCA2}} = 1.378$	$E_{t_{TQCA2}} = 1.289$	Out state = E state
In3 = E	$E_{t_{PQCA}} = 4.458$	$E_{t_{PQCA}} = 3.607$	$E_{t_{PQCA}} = 2.756$	$E_{t_{PQCA}} = 2.667$	$E_{t_{PQCA}} = 2.578$	
In1 = E	$E_{t_{TQCA1}} = 2.025$	$E_{t_{TQCA1}} = 2.025$	$E_{t_{TQCA1}} = 1.242$	$E_{t_{TQCA1}} = 0.541$	$E_{t_{TQCA1}} = 0.541$	$E_{t_{PQCA}} = 0.982$
In2 = D	$E_{t_{TQCA2}} = 2.441$	$E_{t_{TQCA2}} = 0.441$	$E_{t_{TQCA2}} = 0.441$	$E_{t_{TQCA2}} = 0.441$	$E_{t_{TQCA2}} = 1.759$	Out state = D state
In3 = A	$E_{t_{PQCA}} = 4.466$	$E_{t_{PQCA}} = 2.0691$	$E_{t_{PQCA}} = 1.683$	$E_{t_{PQCA}} = 0.982$	$E_{t_{PQCA}} = 2.300$	

In1 = E	$E_{t_{TQCA1}} = 2.025$	$E_{t_{TQCA1}} = 2.025$	$E_{t_{TQCA1}} = 1.242$	$E_{t_{TQCA1}} = 0.541$	$E_{t_{TQCA1}} = 0.541$	$E_{t_{PQCA}} = 0.860$
In2 = D	$E_{t_{TQCA2}} = 2.386$	$E_{t_{TQCA2}} = 0.319$	$E_{t_{TQCA2}} = 0.319$	$E_{t_{TQCA2}} = 0.319$	$E_{t_{TQCA2}} = 2.257$	
In3 = B	$E_{t_{PQCA}} = 4.411$	$E_{t_{PQCA}} = 2.344$	$E_{t_{PQCA}} = 1.561$	$E_{t_{PQCA}} = 0.860$	$E_{t_{PQCA}} = 2.798$	Out state = D state
In1 = E	$E_{t_{TQCA1}} = 1.971$	$E_{t_{TQCA1}} = 1.971$	$E_{t_{TQCA1}} = 1.120$	$E_{t_{TQCA1}} = 1.039$	$E_{t_{TQCA1}} = 1.039$	$E_{t_{PQCA}} = 1.358$
In2 = D	$E_{t_{TQCA2}} = 2.386$	$E_{t_{TQCA2}} = 0.319$	$E_{t_{TQCA2}} = 0.319$	$E_{t_{TQCA2}} = 0.319$	$E_{t_{TQCA2}} = 2.257$	
In3 = C	$E_{t_{PQCA}} = 4.357$	$E_{t_{PQCA}} = 2.290$	$E_{t_{PQCA}} = 1.439$	$E_{t_{PQCA}} = 1.358$	$E_{t_{PQCA}} = 3.296$	Out state = D state
In1 = E	$E_{t_{TQCA1}} = 1.731$	$E_{t_{TQCA1}} = 1.731$	$E_{t_{TQCA1}} = 1.500$	$E_{t_{TQCA1}} = 1.343$	$E_{t_{TQCA1}} = 1.343$	$E_{t_{PQCA}} = 1.662$
In2 = D	$E_{t_{TQCA2}} = 2.386$	$E_{t_{TQCA2}} = 0.319$	$E_{t_{TQCA2}} = 0.319$	$E_{t_{TQCA2}} = 0.319$	$E_{t_{TQCA2}} = 2.257$	
In3 = D	$E_{t_{PQCA}} = 4.117$	$E_{t_{PQCA}} = 2.050$	$E_{t_{PQCA}} = 1.819$	$E_{t_{PQCA}} = 1.662$	$E_{t_{PQCA}} = 3.600$	Out state = D state
In1 = E	$E_{t_{TQCA1}} = 1.731$	$E_{t_{TQCA1}} = 1.731$	$E_{t_{TQCA1}} = 1.500$	$E_{t_{TQCA1}} = 1.343$	$E_{t_{TQCA1}} = 1.343$	$E_{t_{PQCA}} = 2.632$
In2 = D	$E_{t_{TQCA2}} = 2.229$	$E_{t_{TQCA2}} = 1.378$	$E_{t_{TQCA2}} = 1.378$	$E_{t_{TQCA2}} = 1.378$	$E_{t_{TQCA2}} = 1.289$	
In3 = E	$E_{t_{PQCA}} = 3.960$	$E_{t_{PQCA}} = 3.109$	$E_{t_{PQCA}} = 2.878$	$E_{t_{PQCA}} = 2.721$	$E_{t_{PQCA}} = 2.632$	Out state = E state
In1 = E	$E_{t_{TQCA1}} = 2.025$	$E_{t_{TQCA1}} = 2.025$	$E_{t_{TQCA1}} = 1.242$	$E_{t_{TQCA1}} = 0.541$	$E_{t_{TQCA1}} = 0.541$	$E_{t_{PQCA}} = 1.082$
In2 = E	$E_{t_{TQCA2}} = 2.025$	$E_{t_{TQCA2}} = 1.242$	$E_{t_{TQCA2}} = 1.242$	$E_{t_{TQCA2}} = 1.242$	$E_{t_{TQCA2}} = 0.541$	
In3 = A	$E_{t_{PQCA}} = 4.050$	$E_{t_{PQCA}} = 3.267$	$E_{t_{PQCA}} = 2.484$	$E_{t_{PQCA}} = 1.783$	$E_{t_{PQCA}} = 1.082$	Out state = E state
In1 = E	$E_{t_{TQCA1}} = 2.025$	$E_{t_{TQCA1}} = 2.025$	$E_{t_{TQCA1}} = 1.242$	$E_{t_{TQCA1}} = 0.541$	$E_{t_{TQCA1}} = 0.541$	$E_{t_{PQCA}} = 1.580$
In2 = E	$E_{t_{TQCA2}} = 1.971$	$E_{t_{TQCA2}} = 1.120$	$E_{t_{TQCA2}} = 1.120$	$E_{t_{TQCA2}} = 1.120$	$E_{t_{TQCA2}} = 1.039$	
In3 = B	$E_{t_{PQCA}} = 3.996$	$E_{t_{PQCA}} = 3.145$	$E_{t_{PQCA}} = 2.362$	$E_{t_{PQCA}} = 1.661$	$E_{t_{PQCA}} = 1.580$	Out state = E state
In1 = E	$E_{t_{TQCA1}} = 1.971$	$E_{t_{TQCA1}} = 1.971$	$E_{t_{TQCA1}} = 1.120$	$E_{t_{TQCA1}} = 1.039$	$E_{t_{TQCA1}} = 1.039$	$E_{t_{PQCA}} = 2.078$
In2 = E	$E_{t_{TQCA2}} = 1.971$	$E_{t_{TQCA2}} = 1.120$	$E_{t_{TQCA2}} = 1.120$	$E_{t_{TQCA2}} = 1.120$	$E_{t_{TQCA2}} = 1.039$	
In3 = C	$E_{t_{PQCA}} = 3.942$	$E_{t_{PQCA}} = 3.091$	$E_{t_{PQCA}} = 2.240$	$E_{t_{PQCA}} = 2.159$	$E_{t_{PQCA}} = 2.078$	Out state = E state
In1 = E	$E_{t_{TQCA1}} = 1.731$	$E_{t_{TQCA1}} = 1.731$	$E_{t_{TQCA1}} = 1.500$	$E_{t_{TQCA1}} = 1.343$	$E_{t_{TQCA1}} = 1.343$	$E_{t_{PQCA}} = 2.382$
In2 = E	$E_{t_{TQCA2}} = 1.971$	$E_{t_{TQCA2}} = 1.120$	$E_{t_{TQCA2}} = 1.120$	$E_{t_{TQCA2}} = 1.120$	$E_{t_{TQCA2}} = 1.039$	
In3 = D	$E_{t_{PQCA}} = 3.702$	$E_{t_{PQCA}} = 2.851$	$E_{t_{PQCA}} = 2.620$	$E_{t_{PQCA}} = 2.463$	$E_{t_{PQCA}} = 2.382$	Out state = E state
In1 = E	$E_{t_{TQCA1}} = 1.731$	$E_{t_{TQCA1}} = 1.731$	$E_{t_{TQCA1}} = 1.500$	$E_{t_{TQCA1}} = 1.343$	$E_{t_{TQCA1}} = 1.343$	$E_{t_{PQCA}} = 2.686$
In2 = E	$E_{t_{TQCA2}} = 1.731$	$E_{t_{TQCA2}} = 1.500$	$E_{t_{TQCA2}} = 1.500$	$E_{t_{TQCA2}} = 1.500$	$E_{t_{TQCA2}} = 1.343$	
In3 = E	$E_{t_{PQCA}} = 3.462$	$E_{t_{PQCA}} = 3.231$	$E_{t_{PQCA}} = 3.000$	$E_{t_{PQCA}} = 2.934$	$E_{t_{PQCA}} = 2.686$	Out state = E state

Index

Note: Page numbers in *italics* indicate figures and **Bold** page numbers indicate tables in the text.

a

alloyed quantum dots 6
architecture-dependent techniques 135
area–complexity cost function 142
area–delay cost function 142
Auger recombination 8

b

band gap 11
basic-extended hypothesis 157
binary QCA cell (bQCA)
  bistable parameters 141, **142**
  design parameters 196, **197**, 198, **198**
  extended circuits 199, **199**
  Hamiltonian matrix 35
  polarizations 24, *24*, 33–35
  processing in memory (PIM) cells (*see* processing in memory (PIM) cells, binary QCA)
  QCADesigner V2.0.3 software 141
  quantum matrix 33
  structure of 24
Bohr radius 2
bottom-up synthesis processes
  vapor-phase method
    chemical vapor deposition (CVD) 19
    heteroepitaxial growth 18
    molecular beam epitaxy (MBE) 18–19
    physical vapor deposition (PVD) 19
  wet chemical methods
    advantages and limitations 17
    hot-solution decomposition process 18
    microemulsion process 17–18
    nucleation 17
    precipitation techniques 17
    sol–gel process 17
bQCA. *See* binary QCA cell (bQCA)

c

cadmium sulfide quantum dots (CdSQDs) 5
carbon and graphene quantum dots (QDs) 4
carbon quantum dots (CQDs) 4
CdSe/ZnS QDs 8, 9
chemically reduced bovine serum albumin (BSA) 8
chemical vapor deposition (CVD) 19
clock 25, *25*
common memory designs 136
complementary metal–oxide–semiconductor (CMOS) technology 23
compound semiconductor-based nanostructures 1
core–shell quantum dots 5–6, *6*, 8, 9
core-type quantum dots (QDs) 4–5

e

electron acceptor mechanism 13
electron beam lithography 16, 119
electron donor mechanism 13
exciton Bohr radius (EBR) 11

f

fabrication of quantum dot cellular automata (QCA) cells 119

Computing with Multi-Value Logic in Quantum Dot Cellular Automata, First Edition.
Reza Sabbaghi-Nadooshan, Reza Akbari-Hasanjani, Leila Dehbozorgi, Majid Haghparast, and Hamid Reza Akbari-Hasanjani.

fault analysis
processing in memory (PIM) cells, binary QCA
of BPIM1 149, **150**, *150*, **151**
of BPIM2 149, *151*, **152**, 152–153
of BPIM3 153, *153*, **154**
of BPIM4 *154*, 154–156, **155**
extra-cell deposition technique 149
fault tolerance 147, 148, 196, **198**
processing in memory (PIM) cells, ternary QCA
fault tolerance 181–182, 196, *198*
first model of PIM cell (TPIM1) *182*, 182–184, **183**
second model of PIM cell (TPIM2) *184*, 184–185, **185**
third model of ternary QCA-based PIM cell (TPIM3) 185–187, **186**, **187**
focused ion beam (FIB) techniques 15–16
fuzzy logic design 83–85, *85*, *86*

g

germanium quantum dots (GeQDs) 5
graphene quantum dots (GQDs) 4

h

Hartree–Fock approximation 33, 35, 39, 44, 56, 101, 105
heavy metal-free quantum dots (QDs) 4
heteroepitaxial growth 18
high-angle annular dark field (HAADF) detector 9
hot-solution decomposition process 17, 18

i

inorganically passivated quantum dots 8

l

Lent method 119
line-based memories 136, 139
loop-based memories 136, 138–139

m

manufacturing defects 202
manufacturing technology of quantum dot cellular automata (QCA) 26, 119
memory structures
Akers logic array *137*, 138, 139
DFF structure 139
D flip-flop 136
fault-tolerant three-input majority gates 140, *140*
line-based memories 136, 139
loop-based memories 136, 138–139
NAND gate 139
NOR gate 139
parallel memory 139
PIM architecture 136, *137*, 138
PIM-based XOR gate 139
pipeline mechanism 139
RAM cell 136
serial memory architecture 139
SET/RESET control 139
ternary memory model 139–140
von Neumann architecture 136, *137*
microemulsion process 17–18
MIN–MAX functions 84
molecular beam epitaxy (MBE) 18–19
Moore's law 23

n

nucleation 17
n-value QCA cell 82–83, **83–84**, 201

o

optoelectronics 6, 19
organically capped quantum dots 7–8

p

Pauli exclusion principle 2
physical vapor deposition (PVD) 19
power consumption
quaternary QCA (QQCA) 96–97, *97*, *98*
quinary quantum dot cellular automata (QuQCA) system 97, *98*, 99, *99*
ternary quantum dot cellular automata (TQCA) 95–96, *96*
processing in memory (PIM) cells, binary QCA 202
first binary PIM model (BPIM1)
cost functions 143, *144*
fault analysis of 149, **150**, *150*, **151**
AND gate design 157–158, *158–160*
implementation in binary QCA 143, *144*
multiplexer structure 143
OR gate design 158, *161–163*, 162, 164
simulation results 143, *144*

second binary PIM model (BPIM2)
cost functions 143
fault analysis of 149, *151*, **152**, 152–153
flip-flop 143
AND gate design 157–158, *158–160*
implementation in binary QCA 143, *145*
simulation results 143, *145*
third binary PIM model (BPIM3)
block diagram 143, *146*
cost functions 146
fault analysis of 153, *153*, **154**
implementation in binary QCA 143, *146*
NAND gate design 164–165, *164–166*
NOR gate design 166, *167*, 168, *168*, 169
operating principle 145, **147**
fourth binary PIM model (BPIM4)
Akers array architecture 147
block diagram 147, *148*
cost functions 147
fault analysis of *154*, 154–156, **155**
implementation in binary QCA 147, *148*
NAND gate design 164–165, *164–166*
simulation results 147, *148*
XOR gate design 169, *170–172*, 171
processing in memory (PIM) cells, ternary QCA 202
basic processing design 173
data flip-flop (DFF) design 173–174, *174*, *175*
first model of PIM cell (TPIM1)
cost functions 175
design parameters 175
fault analysis *182*, 182–184, **183**
AND gate design 189, 190, *190*, *191*, 192
implementation in ternary QCA 174, *176*
OR gate design 192, *193*, *194*
simulation results 174, *176*
second model of PIM cell (TPIM2)
cost functions 178
design parameters 178
fault analysis *184*, 184–185, **185**
AND gate design 189, 190, *190*, *191*, 192
implementation in ternary QCA 176, *177*
OR gate design 192, *193*, *194*
simulation results 176, *177*
third model of ternary QCA-based PIM cell (TPIM3)
advantage 179
Akers structure 180, *180*
cost functions 178–179
design parameters 178, 179
fault analysis 185–187, **186**, **187**
implementation in ternary QCA 178, *179*
operating principle of switches, Akers cell 180, *180*
operating principle of switches, binary cell 180, *180*
operating principle of switches, ternary cell 180, *180*
simulation results 178, *179*
XOR gate design 192, *195*, 196

q

QQCA. *See* quaternary QCA (QQCA)
QQCASim software
cell modes 126, *128*
cells specifications 126, *128*
AND gate 129, *130*
majority gate structure 127, *129*
MATLAB 126
NOT gate 130–131, *132*
OR gate 130, *131*
quantum wire 131, *133*
software version 1.0.0.1 126, 128
symbolic model 126
quantum communication 88
quantum computation 87–88, *88*, 96, *97*, 105, 201
quantum computers 21
quantum dots (QDs)
advantages 1
alloyed quantum dots 6
applications 1
display technologies 19
optoelectronics 19
quantum computers 21
solar cell 20–21
white light-emitting diode 19–20
bottom-up synthesis processes (*see* bottom-up synthesis processes)
CdSe/ZnS QDs 8, 9
core–shell quantum dots 5–6, *6*, 8, 9
core-type QDs 4–5
emission mechanism 12, *12*

quantum dots (QDs) (*contd.*)
energy levels 2–3, *3*, 10
multishell structure 8–9
physical characteristics of 2–3
properties of
confinement effect 12
electronic properties 10–12, *12*
excitation and emission spectra 12
luminescence effect 13, *13*
optical properties 10–12, *12*
stability 13
quantum properties 22
semiconductor quantum dots 2–3, *3*, 10
shell structures characterization 9
surface passivation
inorganically passivated quantum dots 8
organically capped quantum dots 7–8
synthesis of 1
top-down synthesis processes (*see* top-down synthesis processes)
uncoated QD 4, *4*
x-ray photoelectron spectroscopy (XPS) 9
quantum dot-sensitized solar cells (QDSSCs) 21
quantum dots superlattice (QDSL)/ hyperlattice 11
quantum encryption. *See* quantum key distribution (QKD)
quantum entanglement 88
quantum gates 87, *88*, 126
binary majority gate 29–30, *29–31*, **30**, **31**
binary NOT gate 27, *28*, *29*, **30**
quantum wire 26–27, *26–28*
quantum information (QI) 87–89, *88*
quantum key distribution (QKD) 88–89
quantum measurement 24, 34, 37, 38, 40, 43, 48, 54, 72, 87, 111, 120
quantum state 19, 22, 46, 48, 87
quantum superposition 87
quantum well quantum dots (QWQDs) 8
quasi-Pauli matrices 38
quaternary QCA (QQCA) *63*
AND and OR functions
diagonal majority gate 108, *109*
external electrostatic energies 110
majority gate output 108, 109, **109**
MATLAB analysis 110
quaternary logic levels 108
standard majority gate 108, *109*
truth table 108, 110, **110**, **111**
with two inputs and one fixed input 109, *110*
validity of 108, 109, **110**
bQCA layer 63, *63*, 64
cell symbol 64, *64*
effect of two adjacent cells 92, *93*, 94
electron momentum matrices 42
external electrostatic energy 70–71, *71*, *72*
ground state energy 69–70
Hamiltonian matrix 44–46
Hartree–Fock approximation 105
input drive design
first structure 64–65, **66**, *66*
schematic diagram 66, *67*
second structure 67, **68**, *68*
truth table 66, **66**, 67, **68**
internal electrostatic energy 69
kinetic energy 70
NOT gate 106, **106**, *106*, **107**
output drive design
first structure 66, **67**, *67*
second structure 68, **69**
truth table 66, **67**, 68, **69**
polarization 64, *65*, 125
diagram 43, *44*
of three-particle system 43, **44**
power consumption 96–97, *97*, *98*
QQCASim software (*see* QQCASim software)
simulation of two adjacent cells 106, **107**, *107*, **108**
spin states of three-particle system 39, **40**
TQCA layer 63, *63*, 64
vectors of quaternary input states 43, **43**
qubits 22
quinary quantum dot cellular automata (QuQCA) system
AND and OR functions 112, 115, *115*, **116**, *117*, **118**
cell model 72, *72*
cell symbol 73, *73*
distinct states 72
effect of two adjacent cells *94*, 94–95
external electrostatic energy 80–81, *81*, *82*
extracellular energy 79
ground state energy 80

Hamiltonian matrix 56, 57
input drive design
first structure 74–75, *77*
second structure 78, *78*, **79**
truth table 75, **76**, 78, **79**
internal electrostatic energy 79–80
intracellular energy 79
kinetic energy 80
NOT gates 111–112, **112**, *112*, **113**
output drive design
first structure 75, *77*, **78**
input states 75, **78**
second structure 78, **79**, *79*
truth table 78, **79**
polarization 54, **56**, *56*, 72, 73, *74*
input states 54, **54–55**
lowering operator (S_-) 50–52
momentum matrices 52–54
quantum measurements 48
quantum states of four-electron system 48
raising operator (S_+) 50–52
spin states of four-electron system 46, **46**
S_z matrix 48, 49
power consumption 97, *98*, 99, *99*
simulation of two adjacent cells 112, **113**, *113*, **114**
ternary layer 72, *73*
TQCA cells 72, 73
QuQCA system. *See* quinary quantum dot cellular automata (QuQCA) system

s

semiconductor quantum dots 2–3, *3*, 10
silicon quantum dots (SiQDs) 4
single-walled carbon nanotubes (SWCNTs) 4
size quantization effect 11
solar cell 20–21
sol–gel process 17
spectral mixing 20
static RAM (SRAM) 135

t

ternary quantum dot cellular automata (TQCA)
AND and OR functions 103–104, **104**, *104*, **105**, *105*
calculated polarizations 59, *60*
cell proposed 59, *60*
design parameters 196, **197**, 198, **198**
distance between proposed cells 59, *61*
effect of two adjacent cells 91, *92*
energy calculations
external electrostatic energy 61–62
extracellular energy 59
ground state energy 60–61
internal electrostatic energy 59–60
kinetic energy 61
extended circuits 199, **199**
extreme ultra-violet (EUV) interference lithography 119
field emission scanning probe lithography (FE-SPL) method 119
Hamiltonian matrix 39, 96
Hartree–Fock approximation 101
Lent method 119
manufacturing technology 119
NOT gate 101, **102**, *102*
parameters 141, **142**
plasmonic lithography 119
polarization **35**, 35–39, *39*
positions of wells 59
power consumption 95–96, *96*
processing in memory (PIM) cells (*see* processing in memory (PIM) cells, ternary QCA)
quantum measurements 37
simulation of two adjacent cells 101–102, *102*, **103**
structure *24*, 24–25, *25*
TQCA simulator (*see* TQCA simulator)
top-down synthesis processes
electron beam lithography 16
etching processes 15
focused ion beam (FIB) techniques 15–16
schematic diagram 15, *16*
TQCA. *See* ternary quantum dot cellular automata (TQCA)
TQCA simulator 202
AND gate 122, *123*
cell options 120, *120*
majority gate 120, *121*, 122
OR gate 122, *124*
software version 1.0.11.1 120, *120*, 174
ternary wire 122, *126*, *127*

V

vapor-phase method
  chemical vapor deposition (CVD) 19
  heteroepitaxial growth 18
  molecular beam epitaxy (MBE) 18–19
  physical vapor deposition (PVD) 19

W

"Wannier–Mott" exciton 2
wet chemical methods 17
  advantages and limitations 17
  hot-solution decomposition process 18
  microemulsion process 17–18
  nucleation 17
  precipitation techniques 17
  sol–gel process 17
white light-emitting diode 19–20

Z

Z-contrast STEM 9
zero-dimensional structures. *see* quantum dots (QDs)

Printed and bound by CPI Group (UK) Ltd, Croydon, CR0 4YY

17/09/2026

14964335-0001